Electronic Systems Technician

Level Three

Trainee Guide
Third Edition

Prentice Hall

Boston   Columbus   Indianapolis   New York   San Francisco   Upper Saddle River
Amsterdam   Cape Town   Dubai   London   Madrid   Milan   Munich   Paris   Montreal   Toronto
Delhi   Mexico City   São Paulo   Sydney   Hong Kong   Seoul   Singapore   Taipei   Tokyo

National Center for Construction Education and Research
President: Don Whyte
Director of Product Development: Daniele Stacey
EST Project Manager: Matt Tischler
Production Manager: Tim Davis

Quality Assurance Coordinator: Debie Ness
Desktop Publishing Coordinator: James McKay
Editors: Chris Wilson, Matt Tischler
Production Specialist: Laura Wright

Writing and development services provided by Topaz Publications, Liverpool, NY
Project Manager: Thomas Burke
Desktop Publisher: Joanne Hart
Art Director: Megan Paye

Permissions Editors: Andrea LaBarge and Jaqueline Vidler
Writers: Thomas Burke, John Tianen, Casey Kournegay

Pearson Education, Inc.
Editorial Director: Vernon R. Anthony
Senior Product Manager: Lori Cowen
Operations Supervisor: Deidra M. Skahill
Art Director: Jayne Conte
Director of Marketing: David Gesell
Executive Marketing Manager: Derril Trakalo
Marketing Manager: Brian Hoehl
Senior Marketing Coordinator: Alicia Wozniak
Marketing Assistant: Crystal Gonzalez

Printer/Binder: Courier/Kendallville, Inc.
Cover Printer: Lehigh-Phoenix Color/Hagerstown
Text Font: Palatino and Univers
Cover Photo: © iStockphoto.com/Chris Fertnig

Credits and acknowledgments for content borrowed from other sources and reproduced, with permission, in this textbook appear in the Figure Credits section at the end of each module.

10 9 8 7 6 5 4 3

Prentice Hall
is an imprint of

www.pearsonhighered.com

ISBN 10:   0-13-257823-9
ISBN 13: 978-0-13-257823-3

Preface

To the Trainee

Electronic systems technicians (ESTs) represent a growing sector of highly-skilled labor that is in demand worldwide. The labor pool's need for ESTs is increasing faster than the availability of qualified applicants, making this field an attractive career choice for workers seeking high earnings and career advancement in an expanding profession.

ESTs are skilled in a variety of areas, and work in both residential and commercial settings. They are tasked with installing lighting, telecommunications equipment, and security systems. ESTs also install remote monitoring systems in commercial applications and can retrofit current systems with modernized remote monitoring technology. In multimillion dollar homes, ESTs install integrated systems that unify the control of lighting, climate, entertainment, and security applications. The skills and duties of electronic systems technicians are broad, varied, and in high demand.

New with *Electronic Systems Technician Level Three*

NCCER and Pearson are pleased to present the third edition of *Electronic Systems Technician Level Three*. This title is now in full-color and includes eight modules that have been updated to represent the most current technology and industry practices. The module order has been resequenced to provide fundamental training for electronic systems technicians and best prepare you for the various types of work that ESTs perform daily.

Electronics Systems Technician Level Three has updated the information available on rack systems and their installation in the "Rack Assembly" module. *Level Three* also adds a new module, "System Commissioning and User Training," which describes the various techniques used to activate systems and educate end-users on system functions. The revamped "Fundamentals of Crew Leadership" module has been included to provide trainees with the soft skills necessary to become a leader in the EST trade.

We invite you to visit the NCCER website at **www.nccer.org** for information on the latest product releases and training, as well as online versions of the *Cornerstone* newsletter and Pearson's Contren® product catalog.

Your feedback is welcome. You may email your comments to **curriculum@nccer.org** or send general comments and inquiries to **info@nccer.org**.

Contren® Learning Series

The National Center for Construction Education and Research (NCCER) is a not-for-profit 501(c)(3) education foundation established in 1995 by the world's largest and most progressive construction companies and national construction associations. It was founded to address the severe workforce shortage facing the industry and to develop a standardized training process and curricula. Today, NCCER is supported by hundreds of leading construction and maintenance companies, manufacturers, and national associations. The Contren® Learning Series was developed by NCCER in partnership with Pearson Education, Inc., the world's largest educational publisher.

Some features of NCCER's Contren® Learning Series are as follows:

- An industry-proven record of success
- Curricula developed by the industry for the industry
- National standardization providing portability of learned job skills and educational credits
- Compliance with Office of Apprenticeship requirements for related classroom training (CFR 29:29)
- Well-illustrated, up-to-date, and practical information

NCCER also maintains a National Registry that provides transcripts, certificates, and wallet cards to individuals who have successfully completed modules of NCCER's Contren® Learning Series. *Training programs must be delivered by an NCCER Accredited Training Sponsor in order to receive these credentials.*

Special Features

In an effort to provide a comprehensive user-friendly training resource, we have incorporated many different features for your use. Whether you are a visual or hands-on learner, this book will provide you with the proper tools to get started in the construction industry.

Introduction Page

This page is found at the beginning of each module and lists the Objectives, Performance Tasks, Trade Terms, and Required Trainee Materials for that module. The Objectives list the skills and knowledge you will need in order to complete the module successfully. The Performance Tasks give you the opportunity to apply your knowledge to the real-world duties that ESTs perform. The list of Trade Terms identifies important terms you will need to know by the end of the module. Required Trainee Materials list the materials and supplies needed for the module.

Notes, Cautions, and Warnings

Safety features are set off from the main text in highlighted boxes and organized into three categories based on the potential danger of the issue being addressed. Notes simply provide additional information on the topic area. Cautions alert you to a danger that does not present potential injury but may cause damage to equipment. Warnings stress a potentially dangerous situation that may cause injury to you or a co-worker.

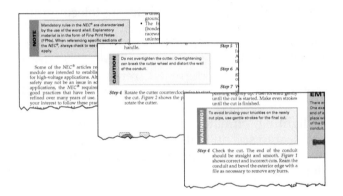

On Site

The On Site features offer technical hints and tips from the EST industry. These often include nice-to-know information that you will find helpful. On Sites also present real-life scenarios similar to those you might encounter as an EST.

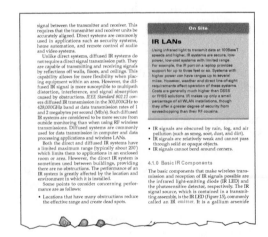

Going Green

Going Green looks at ways to preserve the environment, save energy, and make good choices regarding the health of the planet. Through the introduction of new construction practices and products, you will see how the "greening of America" has already taken root.

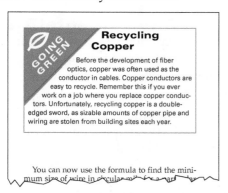

Think About It

The Think About It features introduce historical tidbits or modern information about the EST trade. Interesting and sometimes surprising facts about electronic systems are also presented.

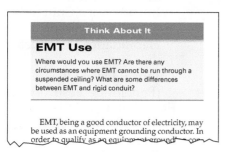

Review Questions

Review Questions are provided to reinforce the knowledge you have gained. This makes them a useful tool for measuring what you have learned.

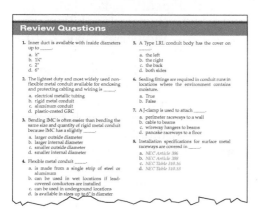

Color Illustrations and Photographs

Full-color illustrations and photographs are used throughout each module to provide vivid detail. These figures highlight important concepts from the text and provide clarity for complex instructions. Each figure is denoted in the text in *italic type* for easy reference.

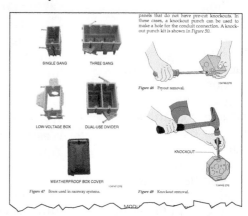

Trade Terms

Each module presents a list of Trade Terms that are discussed within the text and defined in the Glossary at the end of the module. These terms are denoted in the text with blue bold type upon their first occurrence. To make searches for key information easier, a comprehensive Glossary of Trade Terms from all modules is found at the back of this book.

Contren® Curricula

NCCER's training programs comprise more than 80 construction, maintenance, and pipeline areas and include skills assessments, safety training, and management education.

Boilermaking
Cabinetmaking
Carpentry
Concrete Finishing
Construction Craft Laborer
Construction Technology
Core Curriculum:
    Introductory Craft Skills
Drywall
Electrical
Electronic Systems Technician
Heating, Ventilating, and
    Air Conditioning
Heavy Equipment Operations
Highway/Heavy Construction
Hydroblasting
Industrial Coating and Lining
    Application Specialist
Industrial Maintenance
    Electrical and Instrumentation
    Technician
Industrial Maintenance
    Mechanic
Instrumentation
Insulating
Ironworking
Masonry
Millwright
Mobile Crane Operations
Painting
Painting, Industrial
Pipefitting
Pipelayer
Plumbing
Reinforcing Ironwork
Basic Rigger
Intermediate Rigger
Advanced Rigger
Scaffolding
Sheet Metal
Signal Person
Site Layout
Sprinkler Fitting
Tower Crane Operator
Welding

Green/Sustainable Construction
Building Auditor
Fundamentals of Weatherization
Introduction to Weatherization
Sustainable Construction
    Supervisor
Weatherization Crew Chief
Weatherization Technician
Your Role in the Green
    Environment

Energy
Introduction to the Power
    Industry
Introduction to Solar Photovol-
    taics
Introduction to Wind Energy
Power Industry Fundamentals
Power Generation Maintenance
    Electrician
Power Generation I&C
    Maintenance Technician
Power Generation Maintenance
    Mechanic
Power Line Worker
Solar Photovoltaic Systems
    Installer

Pipeline
Control Center Operations,
    Liquid
Corrosion Control
Electrical and Instrumentation
Field Operations, Liquid
Field Operations, Gas
Maintenance
Mechanical

Safety
Field Safety
Safety Orientation
Safety Technology

Management
Fundamentals of Crew Leadership
Project Management
Project Supervision

Supplemental Titles
Applied Construction Math
Careers in Construction
Tools for Success

Spanish Translations
Rigging Fundamentals
    (Principios Básicos de
    Maniobras)
Carpentry Fundamentals
    (Introducción a la
    Carpintería, Nivel Uno)
Carpentry Forms
    (Formas para Carpintería,
    Nivel Trés)
Concete Finishing, Level One
    (Acabado de Concreto,
    Nivel Uno)
Core Curriculum:
    Introductory Craft Skills
    (Currículo Básico:
    Habilidades Introductorias
    del Oficio)
Drywall, Level One
    (Paneles de Yeso, Nivel Uno)
Electrical, Level One
    (Electricidad, Nivel Uno)
Field Safety
    (Seguridad de Campo)
Insulating, Level One
    (Aislamiento, Nivel Uno)
Ironworking, Level One
    (Herrería, Nivel Uno)
Masonry, Level One
    (Albañilería, Nivel Uno)
Pipefitting, Level One
    (Instalación de Tubería
    Industrial, Nivel Uno)
Reinforcing Ironwork, Level
    One (Herreria de Refuerzo,
    Nivel Uno)
Safety Orientation
    (Orientación de Seguridad)
Scaffolding
    (Andamios)
Sprinkler Fitting, Level One
    (Instalación de Rociadores,
    Nivel Uno)

Acknowledgments

This curriculum was revised as a result of the farsightedness and leadership of the following sponsors:

Baltimore City Community College
DirecTV
Dish Network Service LLC
Independent Electrical Contractors of Oregon
Lincoln Educational Services
Lincoln College of Technology

M.C. Dean, Inc.
Northern New Mexico Independent Electrical Contractors (IEC)
Satellite Broadcasting and Communications Association (SBCA)
Tera-Byte Technologies, Inc.

This curriculum would not exist were it not for the dedication and unselfish energy of those volunteers who served on the Authoring Team. A sincere thanks is extended to the following:

Wayne Adair
George Bish
Stephen Clare
Raymond Edwards

Dave Gilson
David Lettkeman
Don Owens
Alton Smith

NCCER Partners

American Fire Sprinkler Association
Associated Builders and Contractors, Inc.
Associated General Contractors of America
Association for Career and Technical Education
Association for Skilled and Technical Sciences
Carolinas AGC
Carolinas Electrical Contractors Association
Center for the Improvement of Construction Management and Processes
Construction Industry Institute
Construction Users Roundtable
Construction Workforce Development Center
Design Build Institute of America
Merit Contractors Association of Canada
Metal Building Manufacturers Association
NACE International
National Association of Minority Contractors
National Association of Women in Construction
National Insulation Association
National Ready Mixed Concrete Association
National Technical Honor Society
National Utility Contractors Association
NAWIC Education Foundation
North American Technician Excellence

Painting & Decorating Contractors of America
Portland Cement Association
SkillsUSA
Steel Erectors Association of America
The Manufacturers Institute
U.S. Army Corps of Engineers
Women Construction Owners & Executives, USA
University of Florida, M.E. Rinker School of Building Construction

NCCER Business Partners

Contents

Note: *NFPA 70®, National Electrical Code®* and *NEC®* are registered trademarks of the National Fire Protection Association, Inc., Quincy, MA 02269. All *National Electrical Code®* and *NEC®* references in this textbook refer to the 2011 edition of the *National Electrical Code®*.

33307-11
Maintenance and Repair

Introduces the background information and the tasks involved in the maintenance and repair of low-voltage systems and equipment. Covers a systematic approach to system and component-level troubleshooting and the methods of identifying common types of repairs. (20 Hours)

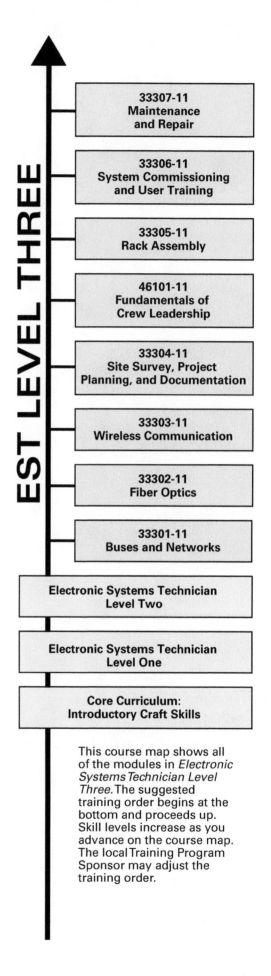

EST LEVEL THREE

- 33307-11
  Maintenance and Repair
- 33306-11
  System Commissioning and User Training
- 33305-11
  Rack Assembly
- 46101-11
  Fundamentals of Crew Leadership
- 33304-11
  Site Survey, Project Planning, and Documentation
- 33303-11
  Wireless Communication
- 33302-11
  Fiber Optics
- 33301-11
  Buses and Networks

Electronic Systems Technician Level Two

Electronic Systems Technician Level One

Core Curriculum: Introductory Craft Skills

This course map shows all of the modules in *Electronic Systems Technician Level Three.* The suggested training order begins at the bottom and proceeds up. Skill levels increase as you advance on the course map. The local Training Program Sponsor may adjust the training order.

Buses and Networks

33301-11

Trainees with successful module completions may be eligible for credentialing through NCCER's National Registry. To learn more, go to **www.nccer.org** or contact us at **1.888.622.3720**. Our website has information on the latest product releases and training, as well as online versions of our *Cornerstone* newsletter and Pearson's Contren® product catalog.

Your feedback is welcome. You may email your comments to **curriculum@nccer.org,** send general comments and inquiries to **info@nccer.org**, or use the User Update form at the back of this module.

V.1   4/11

Objectives

When you have completed this module, you will be able to do the following:

1. Describe the characteristics, connections, and uses for various types of data transmission media.
2. Explain the operating principles of network topologies.
3. Explain how information is transferred using different network topologies.
4. Explain device communication in an addressable network.
5. Describe the functions of routers.
6. Identify the protocols used with networks.
7. Explain power line carrier communications.

Performance Tasks

Under the supervision of the instructor, you should be able to do the following:

1. Identify IP addresses of the devices on a network.
2. Demonstrate various procedures for troubleshooting media access problems to a network.
3. Design a basic network.
4. Demonstrate PC configuration of IP and serial connections.

Trade Terms

Address
Addressable
Backbone
Backbone provider
Baseband
Basic input/output system (BIOS)
Baud rate
Binary
Bit
Bridge
Broadband
Broadcast
Bus
Byte
Carrier sense multiple access with collision detection (CSMA/CD)
Centralized control
Channel
Checksum

Circuit switching
Collision
Deterministic
Distributed control
Ethernet
File transfer protocol (FTP)
FireWire®
Gateway
Hub
Hybrid topology
Hypertext
Hypertext transfer protocol (HTTP)
Internet mail access protocol (IMAP)
Internet Relay Chat (IRC)
Media Access Control (MAC)
Multicast
Network basic input/output system (NetBIOS)

Network news transfer protocol (NNTP)
Network Operating System (NOS)
Network software
Network switch
Node
Open Systems Interconnection (OSI) Reference Model
Operating system (OS)
Packet
Packet switching
Parallel
Point of presence (POP)
Polling
Port
Post office protocol (POP)
Protocol
Register
Ring

Router
Ruggedized
Serial
Server
Simple mail transfer protocol (SMTP)
Stochastic
Stream
Telnet
Token
Token ring
Topology
Transfer medium
Transmission control protocol/Internet protocol (TCP/IP)
Universal serial bus (USB)
Usenet
Web browser
Wide area network (WAN)
Word

Contents ────────────────────────

Topics to be presented in this module include:

Figures and Tables

1.0.0 INTRODUCTION

The joining of two or more intelligent devices with a communication link forms a network. A network may also exist between the components inside a computer system or between equipment and controls in an industrial setting. A typical business office network would include a server and several client computers connected by a network switch. The switch allows the client devices to communicate with each other, share files, and share resources such as a workgroup printer (*Figure 1*). The router provides the means for the network clients to access the Internet. A router also typically serves as a firewall to prevent network intrusion.

Within a computer, multiple devices need to communicate with each other in an orderly fashion. Most computers possess a system or data bus, which provides a common communication pathway for all of the devices.

As computers continue to become a larger part of everyone's lives, the need for those computers to communicate with each other becomes greater.

In this module, you will learn more about connecting computers and components. You will see various methods for connecting computers in a network and learn how data is transferred between the nodes in a network.

At a very basic level, a network might consist of two PCs sharing a common printer. The two PCs could be connected by a cable or through a network switch, which is the device that enables networking. The printer would be physically connected to one of the PCs, which would act as the print server. When the other PC (the client) needed the printer, a user would simply go through the print commands as though the printer were connected to his/her PC.

In order to establish a network, the PC operating system (OS) must be configured to allow it. In later versions of Windows® such as Windows® 7, setup of a wired network is automatic when the devices are plugged into a data port on the wall or to a network switching device. Wireless networks, on the other hand, must be configured using the network "wizard" features built into the operating system. The same is true for older versions of Windows®, such as XP.

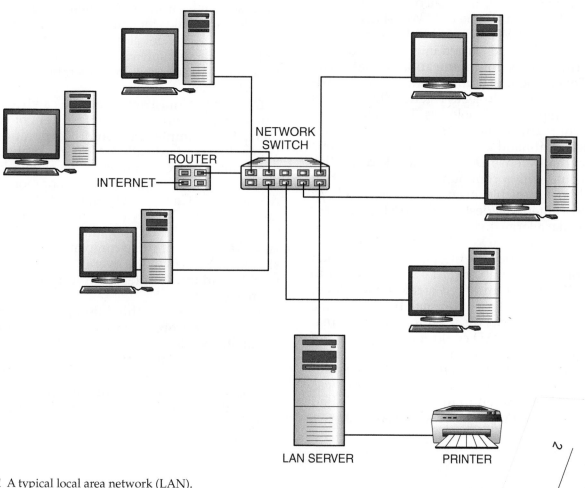

Figure 1 A typical local area network (LAN).

Buses and Networks

1.1.0 History of Networking

The first form of electrical digital communication was the telegraph. It used a series of dots and dashes to represent letters, numbers, and common punctuation marks. A complete network was established worldwide to pass information. Of course, there had to be points where one telegraph network would interface with another in order to extend the range or translate into another language.

Later, there was a typewriter-style device that was devised to pass five data bits between typewriter devices. To know when a letter was beginning, there was a start bit; a stop bit ended each letter. Teletype networks were established to pass information worldwide. In the 1950s and on into the 1980s, most commercial telegraph transmissions were actually done via teletype. The five-bit byte was the Baudot Code.

When large mainframe computers came on the scene, individual keyboards and display units were connected to the mainframe for input and output. There was a need for keyboard/display units called terminals to be connected at a distance of more than 50 feet from the mainframe. That is when modems originated.

After World War II, the Cold War started. Each side in the Cold War wanted to know what the other was doing and started intercepting or jamming the other's communication signals. The U.S. Government established the Defense Advanced Research Projects Agency (DARPA) to help develop new technologies. The Defense Department realized the importance of maintaining their command and control communications (C^3), so the Defense Department asked DARPA to come up with a communications scheme that would be survivable, even if a major switching or relay center was destroyed. DARPA's answer was to take a large quantity of data bits and convert them into a packet. A sender and receiver address was added to the packet. The packet assembler/disassembler (PAD) was designed to put it all together and then break it back down for the end computer or terminal. If a PAD in the particular route was disabled for whatever reason, the previous PAD could reroute the packet through a second path to reach its final destination. This technology was used within the military for about ten years. Its original name was ARPAnet, but the civilian version was called X.25. (By then, DARPA had become ARPA.)

ARPA identified a number of problems with the PADs and came up with another scheme they first called cell relay, or packet relay, and finally frame relay. This scheme did not require the PAD to break down each packet and reassemble it with new addresses in order to reach its final destination. Frame relay is still in wide use in both the government and private industry. However, another progression of ARPA involved the addressing of varying sized packets with permanent to and from addresses. The address never gets stripped or removed from the packet until the final destination, which removes it in order to use the information within the packet. This protocol was and still is called Transmission control protocol/Internet protocol (TCP/IP). The Internet, as we know it, is now the largest and most efficient worldwide communications network.

As computers began to be connected together to form networks, a problem became apparent. Each central processor had to use much of its valuable time in communication instead of its primary job, which was data processing. It was an inefficient use of the central processor. Front-end processors and cluster processors were developed to take that communication load off the central processor, which made it much more efficient. Simple traffic rules for connecting systems, called protocols, were developed and implemented in hardware and software. There was, however, no standardization for interoperability between manufacturers.

In the 1950s and 1960s, there was a virtual turf war going on between manufacturers of equipment. For example, Datapoint Corporation developed a standard called ARCnet. At about the same time, Xerox Corporation developed Ethernet, while Digital Equipment Corporation (DEC) developed DECnet. A bit later, International Business Machine (IBM) Corporation developed a standard called token ring and Apple developed AppleTalk. Each of these companies wanted their standard to become the overall standard for the world to adopt, so interoperability was not encouraged.

To help overcome this interoperability problem, organizations such as the American National Standards Institute (ANSI), the Electronics Industries Association (EIA), and the Institute of Electrical and Electronics Engineers (IEEE) supported the International Standards Organization (ISO) with the objective of worldwide interoperability.

In a relatively short time, networks progressed from university experiments to an essential component of modern business. Today's systems move data quickly between dissimilar computers, and information is processed without interfering with the computer's ability to do other work. The capabilities of today's information exchange systems have grown to the point that networks can easily cross national, commercial, and governmental boundaries. Few technologies are changing as rapidly as the computer industry, with new technologies being unveiled almost daily.

2.0.0 THE DATA HIGHWAY

Data is transferred between computers and between the components within a computer as a stream of data bits. A data bit is a binary value, which can be either one or zero. Electronically, a bit is represented as an electronic charge. This value can be transmitted through an electrical conductor from one device to another. Individually, a data bit means nothing. When data bits are combined into groups according to a standardized set of rules, they become meaningful.

The most common grouping for bits is the byte. A byte consists of eight bits. Using an eight-bit byte, a value ranging from 0 to 256 can be represented. This is sufficient for most communication purposes. Specific applications may use different numbers of bits to build a meaningful group. Such groups are commonly called words, regardless of their size. Typical word sizes in networking and computers are 16-bit, 32-bit, and 64-bit.

If all devices agree that a standard word format is to be used for communications, it simplifies the process of communicating information; each device knows how many bits to send at a time, and each device knows how many bits to expect.

Data may be communicated in one of two ways, each of which relies on using a standard word size. The two methods used for data communications are serial communication and parallel communication.

2.1.0 Serial Communication

Serial communication takes place one bit at a time. It is like a stream that is only wide enough for one bit (*Figure 2*). Bits may pass through quickly, one after another, but at any given time there can only be one bit entering the stream, and one bit exiting it. Because both the sending device and the receiving device know how many bits constitute a word, the grouping of the bits can be maintained. The receiving device has a set of memory loca-

tions, called registers, which act as a holding area for the incoming bits. When all of these registers have received data, the data is moved to another location to be processed. The registers are reset so they can receive the next incoming word. The sending device also has registers that are filled by another functional area of the sending device. When the registers are filled, the sending device starts sending the bits, one by one, through the communication stream.

2.1.1 RS-232, RS-422, and RS-485

A number of standards have been developed to handle data communication between components on a network. These standards include RS-232, RS-422, and RS-485. Of these standards, RS-232 (now EIA232) is the most common. Using an unbalanced line, RS-232 provides a standard data bus for serial communication between controllers, such as computers, and remote devices, such as printers. It also specifies signal voltages, signal timing, and signal functions, as well as an information exchange protocol. RS-232 has historically been used with data terminal equipment (DTE). It has been the primary connecting bus between computers and their modems, printers, and plotters.

RS-232 cables can be used for distances up to 50 feet. They were originally designed to connect modems, printers, and other peripheral devices to a computer. A 25-pin D-shell connector is attached to both ends. One end is male, the other female. To support the signaling requirements of modem communication, 22 pins were used. Eventually, the 22-pin requirement was reduced to nine, allowing smaller connectors to be used. These nine-wire cables require that some wires be used for multiple signals, but only one at a time. The emergence of the universal serial bus (USB) interface largely replaced RS-232 interfaces to PCs.

RS-232 is a single-driver, single-receiver standard. In order to accommodate multiple devices on a common bus, different standards were needed. This need led to the development of the RS-422 and RS-485 standards. RS-422 defines the electrical characteristics of the balanced voltage

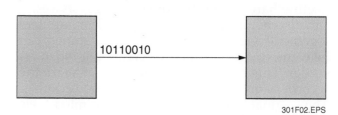

10110010

301F02.EPS

Figure 2 Serial communication.

digital interface circuit. This standard supports data transmission using balanced or differential signaling in a multi-drop system in which one driver can be connected to up to 10 receivers using twisted pair cable. It has a maximum cable length of 4,000 feet, as compared with 50 feet for RS-232.

Where RS-422 is a multi-drop standard (one driver, 10 receivers), RS-485 is a multi-point standard, providing for up to 32 drivers and 32 drops. Like RS-422, it has a 4000-foot range. Repeaters can be used to extend this distance. The full title of the RS-485 standard is *ANSI TIA/EIA-485 Electrical Characteristics of Generators and Receivers for Use in Balanced Digital Multipoint Systems*. Devices on an RS-485 network are addressable, so any node can be accessed independently of the others.

2.1.2 Universal Serial Bus (USB)

The USB is one of two high-bandwidth data communication standards that are now commonly used in computer and control systems. A single USB interface can support up to 127 devices, with a data transfer rate of 12Mbps, which is much faster than previous forms of serial communication. Multiple devices may be daisy-chained to the USB interface, or a hub (or multiple hubs) may be used to create a mini-network of devices. USB 2.0 increased the speed of the USB interface to 480Mbps. USB 3.0, with a 5.0 gigabit speed, became available in 2010. USB 3.0 devices are backward-compatible with USB 2.0. The length of USB 2.0 cables is limited to 5 meters (16.4 feet). The cable length for USB 3.0 is 3 meters (9.8 feet), although this length can be extended using a repeater extension cable or a USB bridge.

2.1.3 FireWire®

Another high-bandwidth serial communication standard is the *Institute of Electrical and Electronics Engineers, 1394*, commonly known as FireWire® or *i.Link®*. FireWire® allows the connection of a maximum of 63 devices, with a data transfer rate of 400Mbps (1394A) or 800Mbps (1394B). FireWire® was originally developed as a transfer medium for digital video and audio information. However, with its high bandwidth, it is popular for connecting hard drives and other storage devices. It can also be used to connect multiple computers together. For example, two or more computers can be connected to a FireWire®-compliant video recorder. This allows both computers to share the recorder and to communicate with one another. The original FireWire® specification limited cable length to 14 feet (4.5 meters). Version 2.0 of the standard extends cables to 45 feet (14 meters). Fiber optic cabling, used with in-wall wiring for transceivers now extends cable length to 225 feet (70 meters).

2.2.0 Parallel Communication

Parallel communication describes the process of transferring more than one bit of information at a time. The data is split into multiple streams, or channels (*Figure 3*). These channels can be thought of as running side-by-side, like the lanes on a highway. The data on the sending and receiving devices still passes through registers. However, instead of being fed through a single stream, each register has its own stream for sending data to the receiving device.

Due to the simultaneous nature of parallel communication, the data is transferred much more quickly. For example, if data is being communicated in 8-bit words, the information will be transferred in about one-eighth of the time it would take for a serial data transfer.

The disadvantage of parallel communication is the need for a separate wire for each data channel, or stream. This causes the cables to be too bulky and heavy for most uses. Therefore, parallel communication is typically only used for internal components in a computer system or for specific, short-distance applications.

2.3.0 Data Buses

Within a computer system, data must be moved quickly between different parts of the computer. Information must be transferred from the disk drive that stores programs and data into the random access memory (RAM) of the computer, where it can be held until needed by the central processing unit (CPU). It must be moved from the memory to the CPU. Graphical information must be moved from the CPU and memory to the video card that constructs the image you see on the monitor.

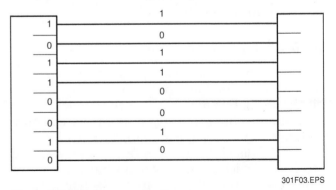

301F03.EPS

Figure 3 Parallel communication.

A data bus can be visualized as a multi-lane highway providing parallel data paths. It does not start at one component and stop at another. Several components are connected to the data bus, as shown in *Figure 4*. Each component has a specific identifier, or address. The bus carries all data messages. Each component is responsible for watching those messages and acting on the ones that carry its address.

This principle of watching a communication path for messages directed at a specific device is one of the core concepts of computer networks.

Multiple buses can exist in a single network, with one bus designated as the primary bus. Secondary buses extend from the primary bus to isolate controllers programmed for polling-type communication or to extend the physical length of the bus system. A communication module, sometimes called a bridge, is used to connect a secondary bus to a primary bus.

The bus is a simple three-wire cable with a shield that is physically wired to each controller in the network. Fiber-optic cable is also used. The controllers connected in a network can share information because they use a common language or set of rules known as a protocol. In a protocol, the computer words traveling along the bus must be configured in a specific way in order to be recognized. The protocol is designed to check the data for errors and make corrections; provide access for all devices on the network; and make sure that only one device transmits at a time. Each control system manufacturer has a unique protocol that governs communication along the bus.

The rate at which information moves on the network is known as the baud rate. The baud rate is stated in bits per second (BPS). The higher the baud rate, the faster the data transfers. In copper-wire networks, data can be transferred at baud rates upwards of 1000 million bits (gigabits) per second. Higher transfer rates are available with fiber optic cable.

3.0.0 TRANSFER MEDIUM

At its most basic level, a network is a connection between two devices. The most common method for connecting those devices is a cable, although wireless networks based on radio signals, microwave transmission, or infrared transmission are also used. The method used to carry the signal between the devices in the network is called the transfer medium. Proper selection of the transfer medium is the most critical decision to be made in planning a network. The wrong choice of medium can result in degraded performance, high maintenance costs, and compromised security.

The three predominant types of transfer media are fiber optic cable, twisted-pair wiring, and coaxial cable.

4.0.0 OSI REFERENCE MODEL

In 1977, the ISO created the Open Systems Interconnection (OSI) Reference Model. The OSI model is a functional guideline for communications tasks. It consist of seven layers, as shown in *Figure 5*.

Each layer of the OSI model performs a specific function. The layers and functions are chosen based on natural subtask divisions. Each layer communicates with the same layer in other computers through layers in its own computer. Upper layers use the services of the lower layers and provide services to the higher layers.

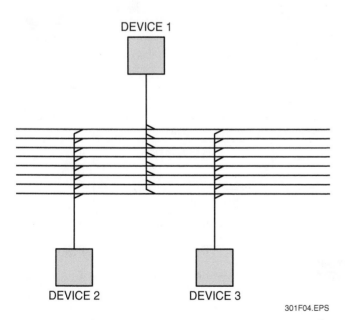

DEVICE 1

DEVICE 2   DEVICE 3

301F04.EPS

Figure 4 Data bus structure.

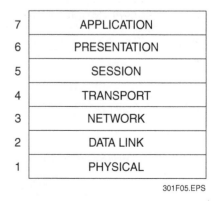

7	APPLICATION
6	PRESENTATION
5	SESSION
4	TRANSPORT
3	NETWORK
2	DATA LINK
1	PHYSICAL

301F05.EPS

Figure 5 OSI Reference Model.

In *Figure 6*, the transport layer (layer 4) of computer A wants to communicate with layer 4 of computer B. To do so, layer 4 of computer A requests a service provided by layer 3, the network layer. Layer 3, in turn, requests a data link service from layer 2 of computer A. From there, computer A, layer 2 requests a service of layer 1, the physical layer of computer A, and layer 4 of computer A is connected to computer B. The process sends the request over the network medium to computer B, where it ascends to layer 4 of computer B and gets processed.

The seven functional layers of the OSI modes are as follows:

1. *Physical layer* – The physical layer (layer 1) describes the electrical and physical connection between the communicating units, as shown in *Figure 7*. Layer 1 is the lowest level and all others depend on it to get the information onto a medium or to collect the information from a medium. Often the most visible layer, it is sometimes the most troublesome part of the system. It describes what a "1" or "0" is and how they are to be transmitted across a medium. It does not describe the medium itself, such as fiber optics, twisted pair, coaxial cable, infrared light through the air, or even wireless electrical signals through space. However, it does define what medium must be connected to make a physical connection between two or more devices on that medium. The medium itself is sometimes referred to as layer 0, because it depends on layer 1 specifications. One interesting piece of information is that the term *transmit data* is the information at the final output of a device, and that information is always correct. From the output of a device, everything else in any network must reconstruct the information to be exactly the same as it was when it left the output of the transmitter. If it doesn't, there is a problem with the reconstruction and it needs to be fixed.

2. *Data link layer* – The data link layer (layer 2) performs the accounting and traffic control functions that are necessary to transfer information onto (and incoming from) a communications link. It forms the information to be moved into bit strings (long lines of 1s and 0s) or blocks (packages of bit strings) of characters. The data link layer functions like a traffic signal to put every piece of information into the right place and release it to the next layer (stop, turn right, turn left, go straight ahead, slow down, speed up, yield to one or more packets, wait here a second, proceed).

3. *Network layer* – The network layer (layer 3) sets up a logical transmission path through a switched or dedicated network (*Figure 8*).

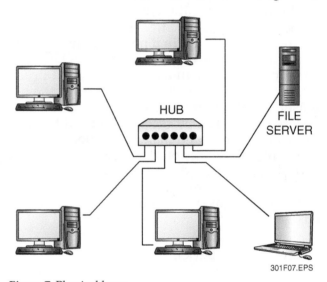

Figure 7 Physical layer.

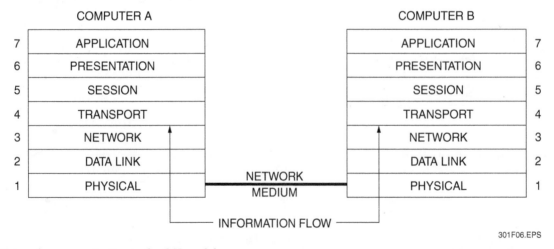

Figure 6 Network communication in the OSI model.

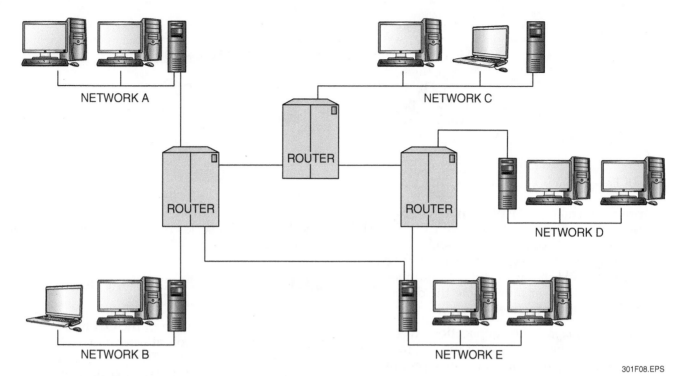

Figure 8 Network layer.

In local area networks (LANs), the path may only be theoretical, since the individual units are almost always electrically connected, and the paths are defined by the network topology. In large systems, however, several transmission paths and even alternative media, such as dialed telephone service versus dedicated service, may exist. The transmission path may provide continuous connection for two or more users of the network, or the connection may only be temporary, as in an Internet connection that is broken when the computer is logged off the internet. In a LAN, the network control function can exist in one place (star topology) or be distributed over the network (bus or ring topology).

4. *Transport layer* – The transport layer (layer 4) provides a common interface to the communications network. It translates whatever unique requirements the higher layers might have into something the network can understand. It detects and corrects errors in transmission and provides for the expedited delivery of priority messages. It also checks the data bytes, puts them into the proper order if necessary, and usually sends an acknowledgement back to the originating transport layer. It also attempts to reestablish contact in the event of a network failure. The transport layer checks each packet and character to make sure it is correct and in the correct order to be used at a higher level. This is where parity bits are added and sequence numbers are placed onto each packet. The receiving data link layer checks the individual packets for the proper parity and either gets them back into the proper sequence, or requests a retransmission of information using ACK (acknowledge) or NAC (non-acknowledge) characters for particular packets or characters. Several industry and governmental standards exist for a transport function in data communication devices. The most common OSI protocols are Transport Protocol Class 4 (TPC4) and TCP/IP. This is also where the quality of service (QOS), known as the H.323 standard, is invoked.

5. *Session layer* – The session layer (layer 5) is a coordinating function. It establishes the logical communication link between units and gradually feeds (buffers) the information to the device or program that performs the presentation function. It recognizes users and acknowledges both their arrival and departure. In some systems, the session layer can be a driving factor in system design; in others, it is a very small consideration. It is this layer where the DHCP or Static IP address assignments are performed and monitored.

6. *Presentation layer* – The presentation layer (layer 6) prepares the information for the application (*Figure 9*). An example of this function is the conversion of a file received from a computer using American Standard Code for Information Interchange (ASCII) into the

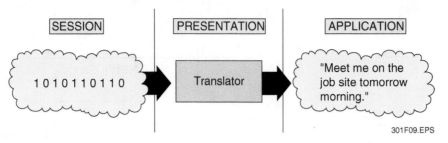

| SESSION | PRESENTATION | APPLICATION |

1010110110 → Translator → "Meet me on the job site tomorrow morning."

301F09.EPS

Figure 9 Presentation layer.

proper format for display on a system using rich text format (RTF), or even the Baudot Code used for older 5 bit-byte teletype communications. Each system uses different codes to represent a character, letter, or number. The presentation layer must know the differences and provide for them.

7. *Application layer* – The application layer (layer 7) is the upper most layer of the OSI model. It interacts with the user's software program. It is concerned with the information in the message and how well it serves the user. This is where application programs (those computer programs that do the work, such as Microsoft Word), call upon the communication services. If this is not handled properly, the entire system is useless, because the goal of the system is to serve the user when and how the user wants to be served. Typical protocols at this layer include File Transfer Access Management (FTAM) and Virtual Terminal (VT). FTAM is primarily within the computer such as the CPU. VT usually deals with input and output devices such as the keyboard, mouse, and monitor. The local printer connection usually drops down to layer 6.

4.1.0 Protocols

It is important to understand that the OSI model itself does not cause network communication to occur. Network communication requires a protocol calling for use of one or more layers of the OSI model. Protocols are like blueprints for building a house. NetWare®, ARCnet, Ethernet, and TCP/IP are examples of protocol blueprints. These protocols are discussed later in the module.

Some protocols specify use of several OSI layers, while others specify only one layer or even a portion of a single layer. The OSI model was designed to direct network development toward a standard. Many protocols were already in use when it was developed. Some of the existing protocols were back-fitted into the OSI model. To help everyone understand how protocols should conform to and comply with the OSI model, ISO has developed a series of protocol specifications (*Figure 10*).

5.0.0 RELATING NETWORK PROTOCOLS TO THE OSI MODEL

Within the data link layer of the OSI model, there are two control protocols at work. One is the LLC-MAC service description. This interface describes the services necessary to transfer network information to and from the physical network medium. Another is the logical link control (LLC), which interfaces with the network layer. The network-LLC service interface describes the services that the network system expects from the data link.

5.1.0 Network-LLC Service Interface

The service access point (SAP) is the actual protocol that handles communication between the network layer and the physical layer. The SAP works within the data link layer to take network information and translate it into the format needed for the physical layer. There is an SAP for every conversation that takes place between one device and another station or device on the network. All of the information that a station is passing to a specific destination must pass through the SAP associated with that destination.

The data unit is the actual block of information that is passed from one LLC to another LLC on another station or device. The complete technical term is *protocol data unit (PDU)* or frame. A PDU can be used to carry messages from one LLC to another and also to transmit commands. In its use for carrying messages from LLC to LLC, it is referred to as an LLC PDU. Under the *IEEE 802* standard, the PDU is the unit which is understood by the LLC on the sending and receiving node. The PDU is not the actual message that is sent, but contains the message. The format of the data message that is sent over the medium is controlled by the media access control (MAC). *Figure 11* shows the structure of an LLC PDU. When

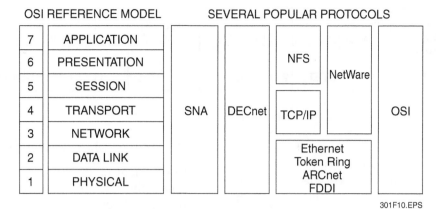

Figure 10 OSI model layers used by different protocols.

DESTINATION SAP — 1 OCTET | SOURCE SAP — 1 OCTET | CONTROL FIELD — 1 OR 2 OCTETS | INFORMATION FIELD — 0–n OCTETS

301F11.EPS

Figure 11 Logical link control protocol data unit.

discussing the PDU, the destination service access point (DSAP) is the SAP on the destination node, and the source service access point (SSAP) is the SAP on the sending node.

The DSAP may identify an individual destination node or a network group, or it may indicate that the message is to be broadcast to all nodes.

The SSAP uniquely identifies the sending station. This is necessary to allow the proper acknowledgement to be sent back to the sending station.

The control field can vary in size, depending on its usage. When it is being used for information transfer, supervisory commands, or responses, the control field is 2 octets, or 16 bits. When the control field is carrying a command word for the network, only 8 bits, or 1 octet, are used.

The information field, or the data field, varies in size from zero to a maximum size as defined by the physical layer. There are three PDU sizes allowed under the three common IEEE standards.

For a message to be sent over the network, the PDU must have a DSAP, SSAP, and a control field. Therefore, the minimum size for a valid LLC PDU is three octets.

5.2.0 LLC-MAC Service Interface

The MAC or physical layer performs two functions upon receiving data from the LLC or data link layer. The first function is data encapsulation and decapsulation. For data to pass from device

to device, the data must be placed into the correct message format, known as a transmission frame, for the physical medium. This differs from the PDU formats, and in fact, the transmission frame will carry the PDU as its data. The information added to the LLC frames is used to perform the following:

- Synchronize the receiving station
- Signal (delimit) the start and end of the frame
- Identify source and destination addresses
- Detect errors

The other function performed by the physical layer is media access management. These functions are commonly thought of as the carrier sense multiple access with collision detection method (CSMA/ CD) functions. These protocols describe how to get access to the medium, how to determine if a collision has occurred, and how to recover from a collision.

5.2.1 Transmission Frame

Data is encapsulated into a transmission frame to provide a standard format for information exchange over the network. *Figure 12* shows the structure of an example transmission frame. For this example, the transmission frame is the one defined by *IEEE 802.3*.

The preamble is used to indicate that a message is starting. It consists of 56 bits (7 octets) of alternating ones and zeros. It is followed by the start

PREAMBLE	START FRAME DELIMITER	DESTINATION ADDRESS	SOURCE ADDRESS	LENGTH COUNT	LLC/ INFORMATION	PAD	FRAME CHECK SEQUENCE
7 OCTETS	1 OCTET	2 OR 6 OCTETS	2 OR 6 OCTETS	2 OCTETS	0–n OCTETS	n–0 OCTETS	4 OCTETS

301F12.EPS

Figure 12 Transmission frame for an *IEEE 802.3* data frame.

frame delimiter, which is a single octet with a bit pattern of 10101011. The address fields, both destination and source, may be two-byte or six-byte fields, depending on the network. The destination address field can support multicast and broadcast messages.

Frames must meet a minimum frame size, defined by the type of network. For a 10Mbps baseband network, frames must be no smaller than 64 bytes. The control fields of the frame (preamble; start frame delimiter; destination and source address; length count; and frame check sequence) comprise 18 bytes, leaving 46 bytes to be filled. Data being sent in a frame is either a PDU or a command. A command is information being sent to the lower levels (the physical, data link, or network levels). A PDU sends information between the higher level layers. If the data being sent in this frame, either a PDU or a command, will not fill this minimum space, the pad field is used to make up the remaining space. If the pad field is used, the length count field contains the number of bytes used by the data.

The last field in the frame is the frame check sequence (FCS). This four-byte field is generated by the sending station after the other fields have been filled in. A cyclical redundancy check (CRC) is performed on the frame, resulting in a four-byte number. This is stored in the FCS field. When the message is received, the receiving station performs the same CRC check on the frame, and compares the result with the FCS field. If they match, the frame is accepted as valid. If they are different, a bad frame message is sent back to the source station and the frame is discarded.

5.3.0 Physical Medium Functions

The physical layer is responsible for three functions: data encoding, data decoding, and physical medium attachment.

Encoding and decoding describe the process of translating the data frames into the signals that are actually sent over the network. This will vary, depending on whether the network uses broadband or baseband communications, and the frequencies used for modulation. Access management and carrier sensing functions are performed in the physical layer of the OSI model.

5.3.1 Access Management

The physical layer performs two functions that manage the station's access to the network: carrier sensing and collision detection.

5.3.2 Carrier Sensing

The physical hardware of the network interface contains circuitry for detecting when the network medium is in use. The hardware will not put a message on the medium while it detects another message in transit.

5.3.3 Collision Detection

It is conceivable that two stations, each detecting that the medium is clear, could place a message on the network medium at the same time. When this occurs, it causes a collision. The physical hardware is responsible for detecting when these collisions occur, and for recovering from these collisions so that all messages reach their destination. Collision detection is based on timing, and the timing method depends on whether the network uses baseband or broadband transmission:

- *Baseband collision detection* – Because baseband transmission is based solely on the presence of a signal, the mechanics of collision detection are fairly simple. In effect, a collision is detected as an echo on the medium: when two messages collide, the result is a garbled signal that is returned to both of the originating stations (*Figure 13*). The time it takes for the collision to be detected by a given station is two times the propagation delay from the originating station to the point of collision.
- *Broadband collision detection* – Because it is based on modulating the frequency of a carrier signal, collision detection on a broadband network is more difficult. Collisions cannot be truly detected until the conflicting signals pass through the head end and return to the originating stations. Therefore, the time delay for detecting a collision is more than twice the propagation delay between stations. The message must be long enough that the source station is still transmitting when the collision is detected, otherwise the collision will go undetected.

NCCER – *Contren® Learning Series* 33301-11

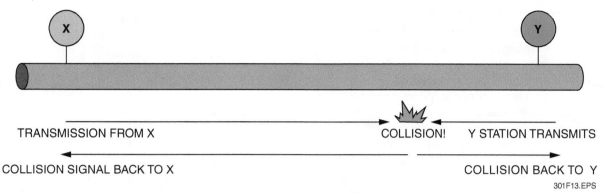

TRANSMISSION FROM X   COLLISION!   Y STATION TRANSMITS

COLLISION SIGNAL BACK TO X   COLLISION BACK TO Y

301F13.EPS

Figure 13 Baseband collision detection.

5.3.4 *Recovery from Collisions*

When a collision is detected, all transmitting stations stop and wait for a randomized period of time before trying again. It is this random interval after a collision that causes CSMA/CD networks to be stochastic in nature.

The amount of time that the station waits is determined randomly, and is called the backoff delay or slot time. The slot time is generally viewed as the period of time it takes for a signal to traverse the network and return. Because different stations may use shorter or longer delays, a technique called binary exponential backoff is used to balance out the delay times. Binary exponential backoff provides a range for the backoff delay based on increasing powers of two.

The more collisions a station detects during an attempted transmission, the longer it waits before trying again. This has the undesirable effect that another station may be able to steal the medium. After a network-defined maximum number of attempts have failed, an error condition is reported from the physical layer to the data link layer.

The degradation of performance in heavy traffic conditions is not a major factor for some networks. However, in a real-time network, where timely delivery of messages is crucial, such degradation is not acceptable. Therefore, CSMA/CD access is not commonly used in real-time networks.

6.0.0 NETWORK TOPOLOGIES

Network topology is the way in which the different elements of a network are wired together. Each topology has its own strengths and weaknesses with regard to network access control, speed of transmission, and security.

The simplest topology is a two-node, point-to-point topology, such as the connection of a computer and a printer. As shown in *Figure 14*, a single communication path connects the two nodes, or devices.

If a third node is added, only two additional communication paths, or links, are needed, as shown in *Figure 15*. However, as the number of nodes increases, so does the number of links required to allow each point to communicate with every other point.

6.1.0 Star Topology

The star topology is commonly used. In a star topology, one device acts as the switch of the system. All messages go through this switch, which relays the messages to the other devices. All other devices are connected to the switch, and to no other device (*Figure 16*).

NODE 1   NODE 2

301F14.EPS

Figure 14 Two-node, point-to-point network.

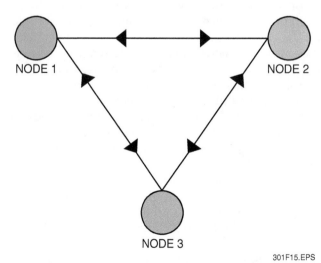

Figure 15 Three-node, point-to-point network.

The terms *hub*, *switch*, and *router* are often used interchangeably when discussing networks. In reality, they are very different devices. A hub is simply a means by which devices in a network can communicate with each other. It broadcasts any data packets it receives to all the devices on the network. This results in collisions, slowing down the network. Hubs operate at half-duplex, being unable to send and receive data at the same time, which limits their bandwidth. A network switch, on the other hand, operates at full duplex. It reads the destination address of a data packet and sends the packet only to the device(s) it is intended for. In doing so, it reduces the amount of unnecessary network traffic. Thus, it is able to operate in higher bandwidth networks. A router is designed to link networks, rather than devices on a network. A typical use is to connect a local area network (LAN) to the Internet.

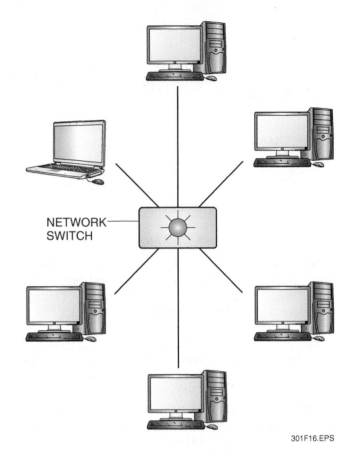

NETWORK SWITCH

301F16.EPS

Figure 16 Star network.

One problem with the star network is that if the switch fails, the entire network fails.

Another potential problem with a star network configuration is cost. If the nodes of the network are widely scattered, or if they must be located far from the switch, cabling costs will be high because each node must be connected directly to the switch.

6.2.0 Ring Topology

To prevent the potential bottlenecks of the star topology, some networks arrange their nodes in a ring topology. The ring topology (*Figure 17*) eliminates the hub or switch and arranges the nodes in a continuous chain, which is joined at the ends. Each node passes messages along to the next node on the ring. As each node receives messages from the preceding node, it checks the message's identifier to determine whether it should act on the message or merely pass it along. The messages are always passed in the same direction.

One potential drawback of the ring topology is its dependence on an unbroken chain. Each node is itself an integral part of the network. If any single node fails, it can impede all communication. Since messages travel in one direction only, a failed node prevents any messages originating in nodes before it from reaching the nodes that are beyond the failed one, as shown in *Figure 18*. The solution would be to reverse the direction of the message flow; however, it requires sophisticated network management to reverse the flow and ensure that messages are properly delivered to their destinations.

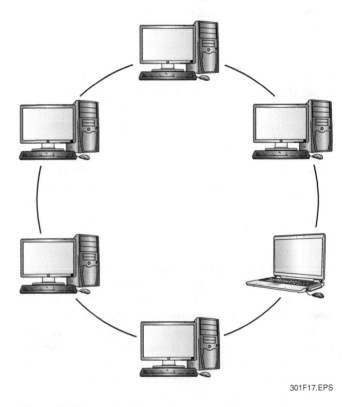

301F17.EPS

Figure 17 Ring network.

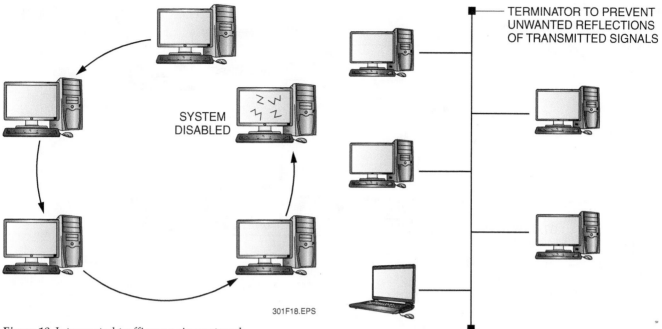

Figure 18 Interrupted traffic on a ring network.

Figure 19 Bus network.

6.3.0 Bus Topology

The bus topology works in the same manner as the system bus described earlier, using USB or FireWire®, for example. Messages pass in both directions on the bus (*Figure 19*). Devices are attached to the bus with a tap or drop. Each node monitors the traffic and acts on the messages that are destined for that node.

Devices can be easily connected to the bus without requiring changes to the other nodes. Because the node devices are not responsible for passing messages to other nodes, a bus network is not affected when a single node is disabled. The bus itself is open to access and has less control than other topologies, which creates security concerns.

6.4.0 Hybrid Topologies

Most systems require more flexibility than a single topology allows. Typically, elements of two or more topologies are combined to meet the needs of the specific system. This combination of topologies creates a hybrid topology.

A common hybrid topology is a ring-wired star network (*Figure 20*). A ring network acts as a backbone, connecting multiple network switches. Each of these switches has its own star-connected nodes. This eliminates the long wiring distances that may be required for widely distributed nodes. At the same time, the use of multiple switches eliminates the possibility that the failure of a single switch could bring down the entire network.

Another common hybrid configuration uses a bus to connect multiple branches of a network

that are spaced far apart. Each branch may be a self-contained network of any topology connected to the bus by a bridge. The bus provides a method of connecting smaller, existing networks into a much larger network.

7.0.0 NETWORK ACCESS CONTROL

In addition to the topology, a means of controlling access to a network is necessary. This media access control (MAC) only determines when the nodes can transmit information on the media. It does not control access to the data, nor does it protect that data from access or use by nodes that have no need for the data.

The intended use of the network is the primary factor in deciding what type of MAC is needed for the network. Four primary access control methods are used in real-time control networks.

Each access control technique falls into one of two categories: distributed control or centralized control. In a distributed control system, access to the medium is controlled at each station, or node. With centralized control, a single network switch controls all access to the medium.

Access control methods also fall into one of two categories based on how they handle data collisions on the network. A stochastic access method handles collisions by delaying for a random time period. A deterministic access method handles collisions by delaying for a period of time that is calculated using a specific, predictable set of rules.

 Buses and Networks

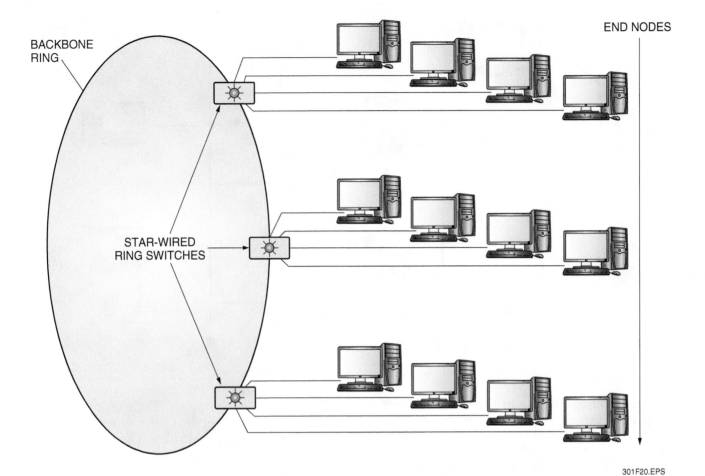

Figure 20 Ring-wired star network.

7.1.0 Random Access

There are several random access control techniques, although very few are now used because better techniques have been developed. One of the most common is the carrier sense multiple access with collision detection method (CSMA/CD). Carrier sensing is a check done by a node when it wants to send information. It checks the medium to see if there are any messages from other nodes being transmitted (*Figure 21*). If the data channel is clear, it sends the message.

During data transmission, the node monitors the network to ensure that it is the only station transmitting. If it detects garbled information, as shown in *Figure 22*, this is an indication that another station is transmitting, and the node immediately stops. In other words, a collision has been detected. The other station would also detect multiple transmissions and stop its transmission. The conflicting nodes wait for a random interval and then try again. By using a random interval, the odds are against the messages colliding again. Messages are transmitted so quickly that a complete message can be sent by one node while the other is waiting out its random interval.

Random access to the network medium can cause some problems. Heavy network traffic can cause a great number of collisions, which delays the delivery of messages. This is not acceptable in a control system that depends on timely communication between stations. For this reason, very few suppliers of control systems use CSMA/CD or any other stochastic access control method for real-time process control.

7.2.0 Polling

Polling is an access control technique in which a single device or station acts as the master for the network. This device polls (checks) each slave node in turn to see if it has any data to send. The master station acts as a gatekeeper to the network, allowing each station to send its message in its entirety before it continues polling the other stations. To prevent a single node from tying up the network for extended periods of time, maximum message sizes are typically enforced on polled networks. Polling can be used on any topology as long as all of the devices on the network are configured to recognize the master device.

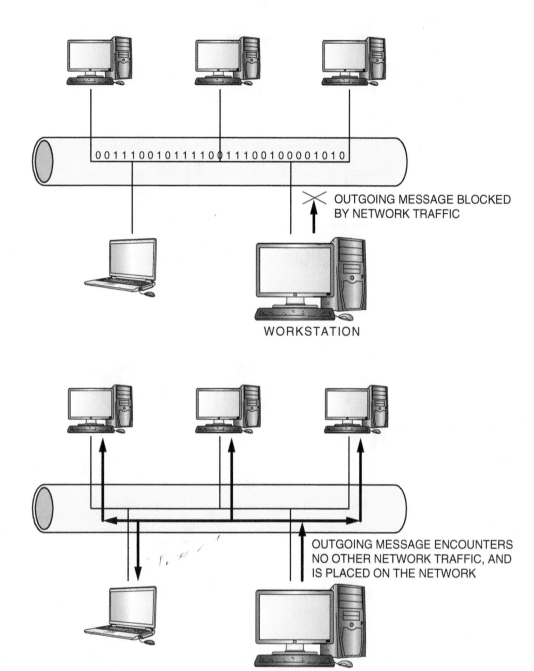

00111001011110011100100001010

OUTGOING MESSAGE BLOCKED
BY NETWORK TRAFFIC

WORKSTATION

OUTGOING MESSAGE ENCOUNTERS
NO OTHER NETWORK TRAFFIC, AND
IS PLACED ON THE NETWORK

WORKSTATION

301F21.EPS

Figure 21 Carrier sensing multiple access.

7.3.0 Dedicated Channel

A network could be established by creating a direct connection for each conceivable destination for a message. This creates, in effect, a point-to-point network. The cost of creating and maintaining a point-to-point network is very high, however, and makes this method of creating a dedicated channel very rare in actual implementation.

A **circuit-switched** network (*Figure 23*) establishes a direct, physical, peer-to-peer communication link between two nodes that need to communicate. This functions as a dedicated communication channel. Other nodes are physically locked out of the network while communication is taking place. The channel remains dedicated between two nodes only for as long as is necessary. For circuit switching to work, the network must be configured in a star topology so that all messages are required to pass through a single controlling device.

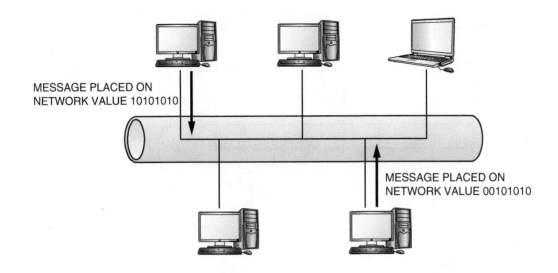

MESSAGE PLACED ON
NETWORK VALUE 10101010

MESSAGE PLACED ON
NETWORK VALUE 00101010

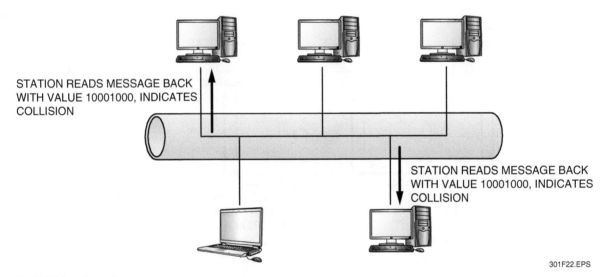

STATION READS MESSAGE BACK
WITH VALUE 10001000, INDICATES
COLLISION

STATION READS MESSAGE BACK
WITH VALUE 10001000, INDICATES
COLLISION

301F22.EPS

Figure 22 Collision detection.

7.4.0 Hierarchical Star Network

A hierarchical star configuration establishes direct, physical, peer-to-peer communication between two nodes that need to communicate between two or more switches. The public switched telephone network (PSTN) uses this hierarchical star network. Many data networks also use the hierarchical star configuration. Within this design, information is switched from one node to another, through one or more switches, to make the peer-to-peer connection.

Looking at *Figure 24*, node 1 of switch E is connected to node 5 of switch B, via switch A. At the same time, nodes 2 and 4 of switch E are connected together, as are nodes 2 to 4 and 3 to 6 of switch B. There are multiple connections operating si-

multaneously on each of the switches pictured. This is how the PSTN operates; each connection is dedicated during the time it is in use. When the conversation is terminated, the individual dedicated links are disconnected and are available for others to use.

The E to A link and the A to B link (or any of the other links) could be viewed as buses with multiple connections and multiple conversions to a number of other nodes throughout all the networks pictured. That is how the hierarchical star configuration works for data networks. Node 1 of switch E to node 5 of switch B connection could actually occur through the E to D link, the D to C link and the C to B link. It could route a number of ways to make the connections for peer-to-peer communication.

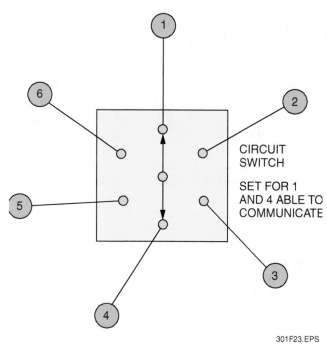

Figure 23 Circuit switching access control.

CIRCUIT SWITCH

SET FOR 1 AND 4 ABLE TO COMMUNICATE

301F23.EPS

7.5.0 Token Passing

Another form of media access control in use is token passing. Once the predominant network access control method for small systems, it has largely been replaced by the other methods discussed. However, the electronic systems technician can expect to encounter these networks, particularly in control systems and legacy installations.

A token is a data message that signifies that data transmission is not taking place on the network. Whenever the token is on the network medium, it signifies to all of the nodes that the medium is available for use. When a station sends a message, it holds the token as a way of preventing other stations from interfering with the transmission. For the token to be placed on the network again, the recipient of the message must send an acknowledgement back to the sender. When the sender receives the acknowledgement, it places the token back on the medium as a signal to the other stations that the medium is available once more. The use of a token to control access to the network creates a token ring network.

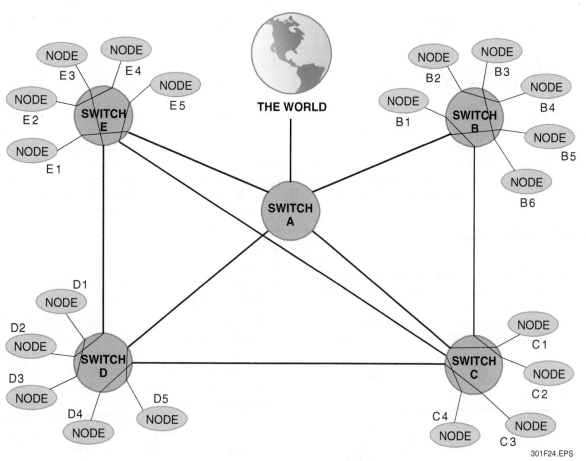

Figure 24 Hierarchical star network.

301F24.EPS

To work properly, the token must rotate through all of the stations in sequence. The sequence may be determined by physical connections, particularly in a ring topology network in which each station is only connected to two other stations. The sequence may also be set by the network software on the stations, so that each station checks for the token in a pre-arranged order, even though the token is not actually moving from station to station. *Figure 25* shows the path a token would take on a bus network.

Setting a maximum time limit for holding the token can prevent network logjams that might be caused by one station transmitting large amounts of data. When a station has reached the time limit, it stops transmitting. Once it receives the acknowledgement for the data it sent, it passes the token back to the network so the next station can send data. When the token comes back around, the station can send the next part of its data. On a network in which the maximum retention time, the number of stations, and the time needed to rotate between the stations are all known, it is possible to calculate the worst-case message delivery time under normal conditions. This enables a time limit to be set at each station to keep the network from slowing down too much.

To handle emergency conditions, most token ring networks allow a priority to be set. When a station has a priority message, it sets a flag on the token. Until that priority flag is cleared, the only stations that can grab the token and send messages are those with a higher priority. This effectively shuts out the other stations.

The combination of known transmission times and priority message routing makes the token ring network very desirable in control systems. The predictability of the token ring allows for the detailed planning of control system messages.

8.0.0 NETWORK SECURITY

The computer age brought its own forms of crime and mischief. The common term for a computer criminal is *hacker*. There are many forms of hackers. Some are simply computer hobbyists who try to enter a secure computer system just for the sake of doing it. They will generally leave a call-ing card of some kind. Others hack into computer networks in order to obtain or alter information for criminal purposes. In other cases, hackers access networks to leave a virus. An example of a virus used in this way is known as a Trojan horse. It is like a time bomb that is set to begin its damaging work at a specified day and time, or when a specific set of circumstances is met.

Security is an increasingly important consideration in the design and use of computer networks. The most basic form of network security is the password. Most networks require a user to enter a user name and password before obtaining access. However, these features alone are not enough to protect a network from a determined hacker. Biometric readers are devices that compare a person's physical characteristic to information in a data base. Examples are fingerprint readers, retina scanners, and facial recognition devices. Such devices have come into common use. Fingerprint recognition devices have even been incorporated into personal computers and laptops.

The dependence on computer-based communications, particularly electronic mail, introduces a risk to security. While in transit, normal email is vulnerable to anyone with some knowledge of electronic communications. To combat this, it is becoming more common to use data encryption for electronic mail messages. An encrypted message may only be read by a person who possesses the correct key to decipher the message.

Virus protection programs defend against most virus attacks by analyzing all of the files on the system for specific data patterns that match those of known virus programs.

For the small office or workgroup that is connecting to the Internet, a common method for providing security is to use a centralized gateway device. All Internet communication traffic is routed through this device, which provides several important services:

- Firewall protection
- Virus protection
- Remote access to the office network
- Data encryption

Security also includes the control of physical access to the computer and the files stored in it. Sensitive data may be revealed to others who happen to see an open document on the computer screen.

It is a common belief that deleting a file makes it impossible for anyone to see the contents of that file. This is a dangerously false assumption. The data from a deleted file remains on the computer's hard drive until it is overwritten with new information. Specialized programs are available

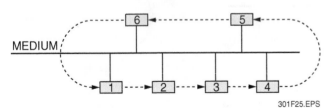

301F25.EPS

Figure 25 Physical bus with logical-ring token passing.

that act like a shredder, immediately destroying the data in a deleted file.

8.1.0 Firewalls

Firewalls are used to provide network access protection. A firewall is specialized software or hardware used to control network traffic. However, it is common to use both methods to enhance the level of protection. A common way to implement a hardware firewall is to use a router with a built-in firewall feature. For example, *Figure 26* shows a small office router. It connects up to four 802.3 connections with a single internet connection that is shared among the users. It includes a firewall and acts as a network address translator (NAT) for the nodes on a private network.

There are many software firewalls available. Many of them are free. Popular security suites, such as those marketed by McAfee® and Symantec™, include a firewall, along with anti-virus software and spyware. The Windows® operating system also includes a built-in firewall.

There are different ways in which a firewall can be implemented. In the packet filtering method, each packet of data that passes through the firewall is examined, generally to ensure that it complies with rules established by the user. The most common examination criteria are the packet source and destination address, its protocol, and its port address. A port address (port number) is used to identify the type of Internet traffic between two computers on the Internet. For example, the world wide web uses port 80, DNS uses port 53, and email uses port 25. These numbers are assigned by IANA (Internet Assigned Numbers Authority), which also controls and assigns domain names. A typical IP address for a 32-bit device might appear as 192.168.190.120:80, with the last two digits representing the port number.

301F26.EPS

Figure 26 Small office router, firewall, and NAT.

Network Access Control (NAC)

Sensitive networks, such as those used by large corporations and government agencies, employ suites of software designed to prevent network access by computers that do not meet certain standards. One of the criteria used to permit or deny access is the status of the client's antivirus software. If a client attempting access does not meet the network's established virus protection standards, it may be quarantined and then automatically upgraded to meet the network requirements. The NAC system can also be used to limit the level of network access for specific individuals or groups.

Packet filtering is the most common and least expensive method. It is built into most firewall routers. However, it is also the most vulnerable to hackers. It is considered to be "stateless" because it only looks at individual packets in isolation, without considering information that comes before or after the packet.

A more advanced version of packet filtering is stateful packet inspection (SPI). In SPI, packets are examined in groups. An SPI firewall is able to detect patterns that represent attempts at unauthorized access. SPI firewalls are used on small and medium sized networks.

Application layer firewalls, also called application gateways, go beyond packet filters. Application gateways examine packets in more detail and are able to prevent all unwanted traffic from reaching devices on a protected network.

A circuit-level gateway manages connections to a network based on TCP/IP addresses and port numbers. It differs from packet filtering in that it does not need to examine every packet once it establishes the connection.

A proxy server is an application layer firewall that is placed between the network servers and the outside word. A proxy server can be implemented either as a hardware or software application. A proxy server evaluates requests from a client device. It filters the request based on rules established for the network and may pass the request to the network server, reject the request, or fulfill the request without passing it on to the network server.

8.2.0 Antivirus Software

Viruses are self-replicating programs, sometimes called bugs, that can alter or destroy computer files and, in some cases, render computers useless. The various types of viruses, such as Trojan horses, worms, and spyware, are generically referred to as malware. Viruses can range from placing an unpleasant message on a computer screen to wiping out the content of the hard drive. Viruses are usually transmitted from system to system by innocent parties who may think they are merely transferring files to friends or co-workers. Once they are inside your computer, viruses attach themselves to executable files. They may lie dormant for a long time, then at some prearranged date and time, begin their destructive work on every computer that has received the innocent carrier file.

There is a certain amount of confusion when it comes to defining the terms *virus*, *Trojan horse*, and *worm*. The main thing to know is that any device connected to the Internet must have up-to-date anti-virus protection. The anti-virus protection will detect the intruder and tell you what you need to do to about it.

Virus protection programs are readily available at stores, through catalogs, and via Internet download. It is important to check every storage device you connect to your system and every file you download for the presence of viruses.

Please note that the virus detection software needs to be updated frequently in order to detect and counteract new viruses. There are some additional actions that can minimize the impact in case your computer is ever infected:

- Keep a set of emergency startup disks available.
- Back up your data files frequently to a storage device outside your computer.
- If your computer is ever infected, restore the data files (not the software programs) using the backup. Restore the programs from the program disks. If you did not receive a set of original CDs for the software applications on your computer when you bought it, then you should make your own backup set as soon as you activate your computer.

9.0.0 THE INTERNET

The TCP/IP protocols are the transport and network protocols of the Internet. The Internet is the global communications network that connects computer networks around the world.

9.1.0 Background

Unlike previous networking schemes, the Internet protocols were developed with the express purpose of connecting devices that were never originally meant to communicate with each other.

The Internet evolved out of a U.S. Department of Defense project called the Advanced Research Projects Agency Network (ARPAnet). Originally designed to connect several local area networks at research institutions involved in government-funded projects, ARPAnet soon encompassed several hundred defense contractors and universities. It used a protocol called network control protocol (NCP) to connect the systems and networks at the participating sites.

While the ARPAnet was still in development, the Department of Defense was also developing satellite and radio packet networks that they wanted to connect to the ARPAnet. These media had very specific network packet handling requirements that NCP could not support. The TCP/IP protocols were developed to support the transfer of information among the various networks in ARPAnet and the radio and satellite networks. The job of protocols is to handle the end-to-end message functions and routing functions of the resulting wide area network (WAN).

The growing use of TCP/IP as a network protocol laid the groundwork for a global internetwork, or Internet. Starting from the core of government agencies, contractors, and universities in the mid-1980s, the ARPAnet quickly grew as other businesses and agencies became part of the network. The current Internet is truly a network of networks. Each company that connects to the Internet typically does so by connecting its own internal, local area network(s) to the Internet via a gateway. Individuals connect to the Internet through an Internet service provider (ISP) that operates a network of clients who dial into a point of presence (POP) in their local area. These networks are in turn connected to wide area networks that may span cities or states. These wide area networks are then connected to backbone providers that provide the connectivity across the U.S. and form the core of the Internet.

The same communication forms used to build the Internet can be used in a closed environment to build an intranet, which may be physically or logically isolated from the Internet. Intranets are used as private networks in large companies.

9.2.0 Transmission Control Protocol/Internet Protocol (TCP/IP)

The TCP/IP protocols handle the packet management and the routing of Internet data messages. They assemble the data into packets for transmission to the data link layer of the network according to the OSI model.

9.2.1 Internet Protocol (IP)

The Internet protocol (IP) provides the data segmentation and the routing, or addressing, of the data. Data segmentation (packet switching) is particularly important if the maximum packet size varies between the source and destination network. Data segmentation breaks up the messages received from the higher layers into the size needed for transmission across the medium to the receiving network. It also reassembles received message packets into the size used by the higher layers. *Figure 27* shows the structure of an IP data packet. What the IP does not provide is any form of security or verification of message delivery.

IP is a connectionless, packet-based protocol. When IP receives a request, it starts transmitting data without establishing a connection. The data is broken down into packets and sent out with an address targeting the client. Packets do not necessarily follow the same transmission path. Depending on the amount of bandwidth available, packets may be routed across different networks.

9.2.2 IP Addressing

> **NOTE**
>
> The text and figures in this section primarily reflect the 32-bit, Version 4 IP, which at this time is the version in common use. Because of increasing demand for IP addresses, and the growth of 64-bit computer technology, Internet engineers developed a new IP address scheme known as Version 6. This version uses a 128-bit address. Its use is growing as 64-bit devices become more common.

Version 4 IP packets use a four-byte addressing scheme. Both the network address and the address of the station within that network are contained in the 32-bit IP address. IP addresses are divided into classes A through E. Classes A, B, and C are used primarily in standard application environments. Class D is used for multicasting, which is the sending of messages to multiple recipients simultaneously. Class E is reserved for experimental uses.

VERSION 4 BITS
IHL 4 BITS
TYPE OF SERVICE 1 OCTET
TOTAL LENGTH 2 OCTET
ID 2 OCTET
FLAGS 8 BITS
FRAGMENT OFFSET 13 OCTET
TIME TO LIVE 1 OCTET
PROTOCOL 1 OCTET
HEADER CHECKSUM 2 OCTETS
SOURCE ADDRESS 4 OCTETS
DESTINATION ADDRESS 4 OCTETS
OPTIONS 3 OCTETS
PADDING
DATA
VERSION 4 BITS
TRAFFIC CLASS 8 BITS
FLOW LABEL 20 BITS
PAYLOAD LENGTH 16 BITS
NEXT HEADER 8 BITS
HOP LIMIT 8 BITS
SOURCE ADDRESS 128 BITS
DESTINATION ADDRESS 128 BITS

301F27.EPS

Figure 27 IP data packets.

There is also a classless addressing and routing scheme in use. This is the result of Classless Inter-Domain Routing (CIDR). With CIDR, additional information is carried in the address header to allow more efficient use of subnetting. Subnetting allows private networks to break their networks down into smaller networks. Through the use of a subnet mask, the network can receive packets based on the higher order bits in an IP address, and then forward them on to nodes on a subnet known only to the private network router.

The class of an address determines how many bits are used for the network address and how many are used for the station address. *Figure 28* shows how the 32 bits of the Version 4 IP address are divided according to the first three high-order bits.

A Version 4 IP address is broken down into four parts, which can clearly be seen in the decimal representation of an IP address such as 192.168.1.1.

Depending on the class of network, one or more of these decimal values are used for Internet routing. Each decimal value is separated by a dot. The scheme is collectively referred to as dot notation.

In a class A network, only the first decimal value is used to identify the network. The remaining three values are used to identify hosts and subnets within the local network. Class A addresses can support up to 16 million different hosts.

The class B network uses the first two decimal values for network identification. The remaining two values are used for host identification and subnetting. The class B network can support up to 65,500 hosts.

The class C network uses the first three decimal values to identify the network. The remaining decimal portion allows up to 256 hosts to be supported.

Because early users of the various address schemes acquired network addresses without always considering growth, additional class A and

FIRST THREE BITS			NETWORK ADDRESS NUMBER OF BITS	HOST ADDRESS NUMBER OF BITS
0	0	0	7	24
0	1	0	14	16
1	1	0	21	8
0	1	1	EXTENDED ADDRESSING	

301F28.EPS

Figure 28 IP address format.

B addresses were often consumed. Subnetting has been used extensively to compensate for growth limitations. A subnet is part of an overall network, but it uses its own routers to direct traffic. These router tables are not stored in the DNS tables of Internet routers. Rather, the host network receives packets based on the network address, and the local router distributes them to the various hosts. This allows a company to have a single network address, and to better manage the rearrangement of internal network configurations that evolve over time.

A subnet mask is applied to the network address and is carried as part of the address header in the IP packet. The subnet mask also has a four-decimal value using the dot notation format. The subnet mask is in the form of 255.255.255.0, or some variant, as determined by the class of network being addressed. By applying the mask, the network portion of the IP address is converted into a string of 1-bits. By using subnetting, there are eight networks on a class C network and up to 32 hosts (only 30 are actually available for host assignment) on each subnet. A pure class C network can have 256 hosts but only one network.

Routing of IP packets is a shared function. The source station first determines whether the destination is a part of the local network. If it is, the message is routed directly to that destination station. If the destination is outside of the local network, the IP layer uses a stored table of routing information and destination addresses to determine which gateway device should receive the packet. The message is sent to the gateway, and the gateway is then responsible for relaying the message on to its destination.

9.2.3 Transmission Control Protocol (TCP)

The TCP layer of the network is responsible for providing the security and reliability that the IP layer does not provide. TCP loads the message data into packets that include information for proper handling of the message data, as shown in *Figure 29*.

TCP, or transmission control protocol, is a single, two-way communication protocol. It acts as a direct wire, or pipe. The TCP layer handles the sequencing of data, ensuring that the messages sent over the network are arranged in the correct order when they arrive. This is especially important in control systems that run over TCP/IP networks.

Another service provided by the TCP layer is data integrity, which includes both data verification in the form of a checksum, as well as message receipt acknowledgement.

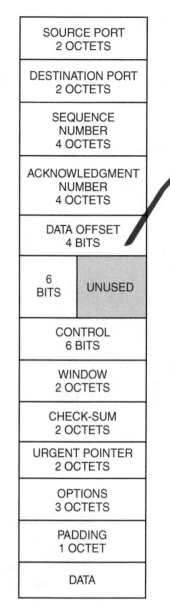

301F29.EPS

Figure 29 TCP data packet.

One other piece of information found in the TCP packet is the port. In Internet terms, a port is a number that designates the service to which the data is related. There are both source ports and destination ports. These services are defined in higher-level layers of the network, and indicate that the data is used for such services as the File transfer protocol (FTP), Internet Relay Chat (IRC), Telnet terminal services, post office protocol (POP), simple mail transfer protocol (SMTP), or the hypertext transfer protocol (HTTP). The use of port facilitates routing of time-sensitive information over the Internet. It is also used to interface services on one machine with those on another, such as accessing data from an email server application through a hypertext web page.

9.3.0 Internet Application Protocols

The growth of the Internet has led to the development of specific application-level protocols for communicating data among programs on the Internet. These protocols are the foundation of many of the common services available on the Internet.

9.3.1 Hypertext Transfer Protocol

Users of the world wide web on the Internet are familiar with the hypertext transfer protocol (HTTP). It is the technology that makes the Internet possible. As an application protocol, HTTP defines a method for sending a file request to another system running the HTTP protocol, which responds by sending that file back as a stream of data (*Figure 30*). The received data stream is directed to a specialized program, commonly known as a web browser. The browser interprets the data stream, building an on-screen window containing text, images, sound, and other types of information, formatted according to directions included within the requested file. The HTTP can also be used to execute a program or script on a remote computer and send the output of that program to the requesting system.

9.3.2 Simple Mail Transfer Protocol

SMTP is used to send email to a server. It is used both for the transfer of messages between mail servers, and for sending email from users' stations to their email servers. Note that SMTP works in one direction only—from the sender of email to the recipient. The recipient cannot request that mail be sent, but relies upon the sender to pass along messages. As a result, SMTP cannot be used by a network station to request or retrieve email destined for that station.

9.3.3 Post Office Protocol

The post office protocol (POP) is used to overcome the lack of a request or retrieval service in the SMTP. POP allows an email program running on a network node to communicate with a POP server and request that its messages be sent. For a client to have two-way email access, the user's email server must communicate with both a POP server to retrieve mail and an SMTP server to send messages (*Figure 31*).

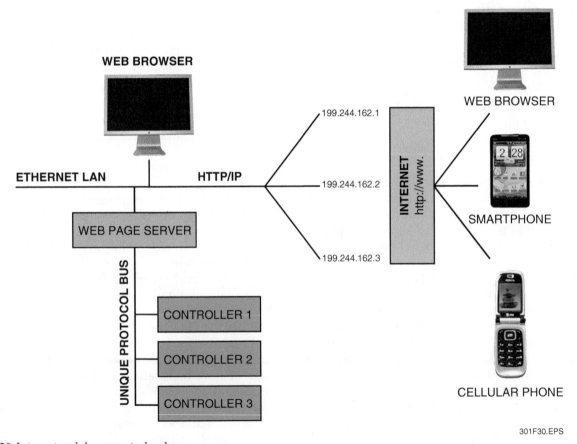

Figure 30 Internet web browser technology.

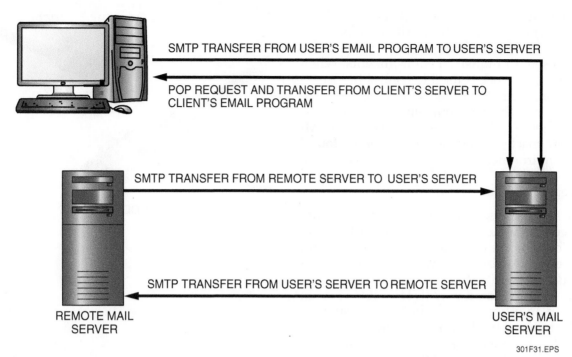

SMTP TRANSFER FROM USER'S EMAIL PROGRAM TO USER'S SERVER

POP REQUEST AND TRANSFER FROM CLIENT'S SERVER TO CLIENT'S EMAIL PROGRAM

SMTP TRANSFER FROM REMOTE SERVER TO USER'S SERVER

SMTP TRANSFER FROM USER'S SERVER TO REMOTE SERVER

REMOTE MAIL SERVER

USER'S MAIL SERVER

301F31.EPS

Figure 31 Two-way email communication.

9.3.4 Internet Mail Access Protocol

A proposed replacement for POP, the Internet mail access protocol (IMAP) adds features such as the ability to manipulate and search messages that are still on the server. IMAP is gaining more acceptance, though it is still not as universally used as POP.

9.3.5 Network News Transfer Protocol

A service similar to email is available through the Network news transfer protocol (NNTP). Usenet is a network of servers using primarily the NNTP protocol to transfer news messages across the Internet. Usenet news functions as a global bulletin board system in which any user can post and view messages that are viewable by anyone on the Internet. IMAP can also be used for viewing Usenet messages, but NNTP is still the most common method.

9.3.6 File Transfer Protocol

For those cases in which a user wants to move files to or from a location on the Internet, there is FTP. Through the use of an FTP-compliant application, users can copy files across the network as easily as copying files from one folder to another on their own systems.

10.0.0 ETHERNET

Ethernet is one of the most successful networking protocols ever implemented. While Ethernet deals only with the physical and data link functionality in the context of the ISO-OSI model, it is the basis for numerous minicomputer, personal computer, and real-time control networks. Ethernet led the way to the acceptance and growth of LANs.

Because of the presence of such a large amount of Ethernet equipment and its use as a protocol for several control systems, the design of real-time control networks must take Ethernet into consideration. Ethernet and *IEEE 802.3* devices can share the same cable and coexist, but they cannot communicate with each other. Given the similarities of the protocols, a number of network equipment manufacturers have developed interface cards that can operate over either protocol, but not simultaneously. It is important to know whether the control network has the correct interface for the type of network installed.

Another consideration is the life expectancy of the equipment. Any new device purchased should support *IEEE 802.3* protocols or both Ethernet and *IEEE 802.3*. Because a number of early control systems used Ethernet, it is probably best to specify compatibility with both to ensure that a useful life can be maintained.

Noise immunity is another critical issue. Baseband (office) Ethernet has been criticized for being overly sensitive to the electromagnetic interference (EMI) present in industrial environments. When an Ethernet network is used to control industrial automation, the network must be protected from the harsh operating environments found in industrial facilities. Dust, corrosives, vibration, and EMI created by machines are factors commonly found in such environments that can affect network integrity. Industrial Ethernet differs from baseband Ethernet in the following ways:

- Switches, routers, and other network hardware devices manufactured for industrial Ethernet applications are ruggedized so that they will not be affected by corrosive or damp environments, dust, and EMI.
- Unshielded twisted-pair (UTP) cable is often used in an office network, but it is sensitive to EMI produced by the types of machines found in industrial facilities. Category 5e, 6, or 6A shielded twisted-pair (STP) cable is preferred for industrial networks. Network cables should be protected by conduit when used in industrial environments.
- The RJ45 connectors used to terminate cable in office networks will deteriorate over time in many industrial settings. For that reason, a special M12 connector is used (*Figure 32*). A special RJ45 connector with a protective boot may also be used.
- To prevent problems in the office network from affecting the industrial Ethernet, the two networks are generally isolated from each other. While they may be able to communicate, they do not share the same bus.

Because most networks follow the *IEEE 802* standard for computer networks, the *IEEE 802.3* naming standard has become common for identifying networks. This naming standard combines elements that identify various options of the network. The shorthand used is SSMMMMDD.

- *S* – Speed (megabits per second)
- *M* – Media class (baseband or broadband)
- *D* – Maximum distance allowed (in hundreds of meters or some other, industry-specified designator)

Ethernet networks are divided into three groups—10/100Mbps, 100Mbps, and 1,000Mbps (gigabit)—based on their data transfer rates and distances:

- *10Base5* – A 10Mbps network on baseband media, at a maximum distance of 500 meters
- *10Base2* – A 10Mbps network on baseband media at a maximum distance of 200 meters

M12 CONNECTORS

RJ 45 CONNECTOR

301F32.EPS

Figure 32 M12 and RJ45 connectors.

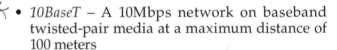

- *10BaseT* – A 10Mbps network on baseband twisted-pair media at a maximum distance of 100 meters
- *100BaseT* – A 100Mbps network on baseband twisted-pair media at a maximum distance of 100 meters
- *1000BaseT* – A 1,000Mbps (1 gigabit) network on baseband twisted-pair media at a maximum distance of 100 meters
- *10GBaseT* – A 10 gigabit network on twisted-pair copper cable with four twisted pairs to a maximum distance of 100 meters
- *100BaseF* – A 100Mbps network on baseband fiber optic cable at varying distances, depending on the type of cable

- *1000BaseF* – A 1,000Mbps network on fiber optic cable to a maximum distance of 500 meters
- *10GBaseF* – A 10 gigabit network of fiber optic cable to a maximum distance of 100 kilometers
- *10Broad36* – A 10Mbps network on broadband media at a maximum distance of 3.6 kilometers

11.0.0 MICROCOMPUTER-BASED LANs

The two main types of LANs are office LANs, which connect individual PCs to centralized file and print servers, and real-time control networks, often based on commercially available microcomputers or PCs.

In a control network (*Figure 33*), the individual PCs can act as control devices for equipment connected directly to the computer, or for devices that are connected to the network through a gateway. The devices connected through a gateway can be controlled from any PC on the network, depending on how the control system is configured.

Because of the nature of microcomputer architecture, networking services must be closely intertwined with the functions of the operating system (OS). The operating system software is directly responsible for managing the hardware resources of the computer, including the network and communications hardware. The OS is also responsible for coordinating the execution of all software running on the computer. In the past, it was necessary to install specialized network software to support network communications. As a result, even though network software was an add-on for early computers, it was referred to as the network operating system (NOS). Now it is often included as part of the operating system itself. The term *NOS* is still used when referring to the software that controls functions such as the following: execution of computer programs; hardware input/output operations; device and event scheduling; system diagnostics; data management; compilations; and other related services on any microcomputer-based LAN hardware device.

11.1.0 Basic Input/Output System (BIOS)

One of the core pieces of any operating system is its basic input/output system (BIOS), which is a set of instructions that clearly defines how the computer handles all input and output operations. Hardware and software developers create their products to be compatible with a given BIOS specification to ensure that their products are compatible with existing computers and software.

11.2.0 Operating Systems

As stated before, most operating systems have basic networking support as an integral feature. Communication between computers has become a necessity in business. The most common operating system in use on personal computers is

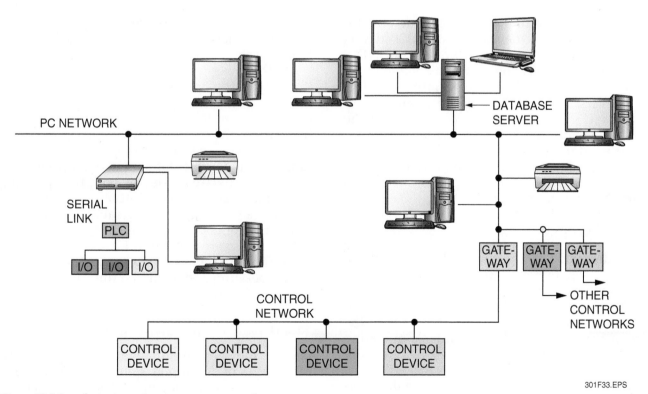

Figure 33 Manufacturing microcomputer network.

301F33.EPS

Microsoft Windows®, which is available in several versions that can be selected to match the intended use of the computer. However, Windows® is not the only available operating system. The following sections provide a basic overview of Windows® and various other operating systems, including UNIX®, LINUX®, and Mac OS®.

11.2.1 Microsoft Windows®

Microsoft's Windows® is the operating system found on most PCs used in personal and business applications. A new Windows® operating system is introduced every few years. Upgraded versions of existing operating systems are released in the interim. There are basic versions of the Windows® operating system designed for home users, as well as more robust versions designed for business use. The latter range from small business versions to enterprise versions designed for use in large networks. Special versions are designed specifically for use in network servers. All of these operating systems include built-in network client functionality so that they can easily connect to a server or share resources among themselves. They are commonly used as Internet servers, storing and distributing content for the Internet, managing users who dial in over phone lines, and managing corporate email systems.

11.2.2 UNIX®

UNIX® is the operating system on which the Internet was built. The development of TCP/IP was very closely linked to the development of UNIX®. Until the early 1990s, Internet users had to know UNIX®. UNIX® is still in use on larger systems and servers, and it has been used in some control systems. UNIX® suffers from confusion over the various versions that have been produced by different vendors. While they are all based on the same core programs, or kernel, these different versions are usually not compatible with each other.

11.2.3 LINUX®

Often described as a version of UNIX®, LINUX® is truly a separate operating system. Originally developed as a project by a university student, LINUX® is now maintained by a world-wide consortium of independent programmers with the aim of creating a reliable, extendable operating system for use on computers ranging from mainframes down to embedded devices that will allow users to control equipment from across the Internet.

11.2.4 Mac OS®

The Apple Macintosh® line of computers, commonly referred to as Mac® computers, uses a proprietary operating system called Mac OS®. The original Mac OS® predates Microsoft Windows®, having first appeared in computers around 1983. It provided the original graphical user interface and was the first to allow the user to navigate menus and program windows with a mouse. It is still in widespread use, particularly for graphic design, multimedia, and video production applications.

11.3.0 Networking Software/Network Operating Systems

Because of the explosive growth of networking in general and the Internet in particular, all of the commercially available operating systems currently in use support networking without the need for additional networking software. Some operating systems may not directly support network connections to other operating systems, but these are becoming increasingly rare, as TCP/IP is now the default protocol used for networking. Networks that contain devices with incompatible operating systems (older versions of Microsoft Windows® and Mac OS®, for example) may require additional networking software to enable network communications.

Older systems still in use may require a separate NOS. The network operating system works on two levels. At a high level, it may provide applications that are used specifically for network functions, such as file transfer programs or system-to-system communication programs.

On the lower level, the NOS extends the functionality of the operating system by installing a set of instructions called the network basic input/output system (NetBIOS). The NetBIOS adds to the BIOS of the computer by providing the internal instructions needed to support network communications and the standardized interface to the network hardware needed by both the operating system and applications software. These functions are a part of modern operating systems.

The network operating systems that are likely to still be in use are NetWare®, LAN Manager®, and AppleTalk®. Windows NT® was classified as a network operating system at one time, but it is now considered an operating system in its own right.

11.4.0 Real-Time Performance Issues

Because they were originally meant to extend the capabilities of desktop PCs, the performance of a network is usually judged against that of a comparable desktop function. For example, the time required to print a document over the network is compared to the time required to print that same document on a printer connected directly to the PC. The makers of network hardware and software have labored to bring their performance level up to meet these expectations. The result is a networking technology that is fast, reliable, and robust.

Manufacturing-oriented software vendors have taken note of these improvements and are producing networked control systems based on PC LANs. These systems provide cost-effective solutions in factory and distribution center applications. Real-time networks based on PC LANs are an excellent solution, providing the following guidelines are followed in implementing the control system:

- Keep the installation simple. Only those stations and devices that require real-time control should be on the network.
- Make certain that any traffic between the real-time control network and other networks or devices passes through a bridge to minimize the impact of noncritical messages on the performance of the control system.
- Obtain performance guarantees from the network hardware and software vendors.

12.0.0 ROUTERS, BRIDGES, AND GATEWAYS

There is often a need to connect real-time control networks to other real-time networks, general-use LANS, or even WANs. Most low-voltage systems today are connected to the Internet to allow for monitoring and control of building networks over the Internet.

Because of the limitations imposed on most LANs by hardware or protocols meant for short distances or confined locations, there is often a need to extend either the LAN distance or the functionality. This is usually accomplished by connecting two individual LANs. Another reason that LANs are often interconnected is to share data between devices on separate systems. If the differences in the protocols are minimal, a bridge will often suffice. If there is a need to support multiple paths through a series of interconnected networks, a router is required. Finally, if the requirement is to facilitate communication between two different protocol stacks, a gateway is generally used.

12.1.0 Routers

The primary use of a router is to route data packets between two networks. One of the most common uses of a router is to provide a link between a LAN and the Internet. The router examines each packet and interprets the packet content to determine the destination. If the packet is intended for the home network, it will direct the packet to the correct Ethernet address. If the packet is intended for another network, the router will repackage it for transmission to another router. *Figure 34* shows how a router is positioned in reference to the ISO-OSI model. By using the address information contained in the network layer, a scheme can be established in which packets are sent using only a single path. A router is different than a network switch. Switches usually work at OSI level 2 (data) using MAC addresses, while routers work

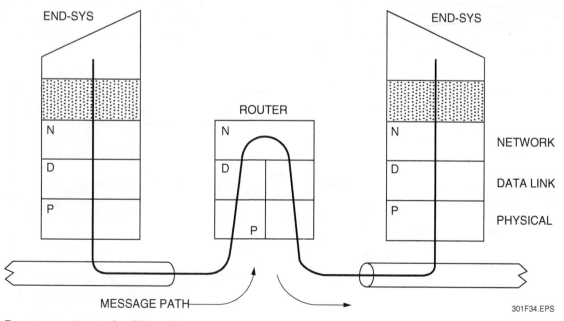

Figure 34 Router operation in the OSI model.

at level 3 (network) using network addresses such as IP addresses.

A router must be able to support a variety of networking protocols in order to communicate between networks using different protocols. A network using *IEEE 802.3/.2* protocols, for example, can equally support an OSI upper-layer stack, a TCP/IP stack, and a proprietary stack. Each of these protocols is different and cannot communicate with the others. Accordingly, a router for this system (if all three traffic streams need to be transported across the entire network) would have to support each protocol. Thus, the router is a very intelligent device because it must first recognize the network protocol and implement the appropriate routing scheme, which may be different for each network type. Firewalls are often incorporated into network routers in order to provide a layer of protection beyond that provided by firewalls built into software.

12.1.1 Brouters

A brouter is able to perform the functions of both a bridge and a router. Brouters have the multiple-protocol passing capabilities of routers and the ability to pass broadcast traffic very efficiently, something a router cannot do.

12.2.0 Bridges

The simplest form of LAN-to-LAN connection is through a bridge. *Figure 35* shows the function of a bridge in the context of the ISO-OSI model. Note that the bridge merely performs data-link layer data transfer. Thus, a bridge is very useful for performing physical layer transformations, such as coupling an *IEEE 802.4* network to an *IEEE 802.3* network. Bridges can also convert from one form of modulation to another or from one speed

to another. For example, a bridge could be used to move data from a seven-layer broadband network running at 10Mbps to a seven-layer carrier band network running at 5Mbps. In this case, both speed and modulation changes are accommodated. Bridges are also used to move data between 16Mbps and 4Mbps segments of *IEEE 802.5* token ring networks.

12.3.0 Gateways

Because standardized networks using typical protocols are not always available, a common way to provide LAN-to-LAN connectivity between real-time systems is through the use of gateways. A gateway is really just a computer that has been programmed to provide a translation between the protocol stacks. *Figure 36* shows how an OSI system can communicate with a proprietary system using a gateway.

While gateways are certainly common, they are not very desirable. In fact, gateways are the least desirable form of interfacing for real-time control networks. If a gateway is used, two protocols have to be learned, and multiple types of hardware and software must be maintained. Gateways are most typically used when interfacing networks that use different protocols.

12.4.0 Repeaters

Repeaters are used primarily in *IEEE 802.3* networks. Repeaters are used to make *IEEE 802.3* (or Ethernet) media transitions or to extend *IEEE 802.3* (or Ethernet) networks when distance limitations are reached. A repeater will resynchronize and regenerate the digital signals and send them on their way again. More commonly today, a fiber link is installed to extend the connections.

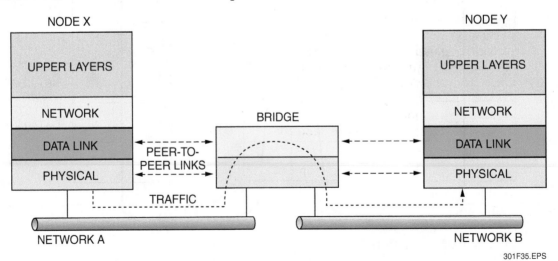

Figure 35 Bridge operation in the OSI model.

NCCER – *Contren® Learning Series* 33301-11

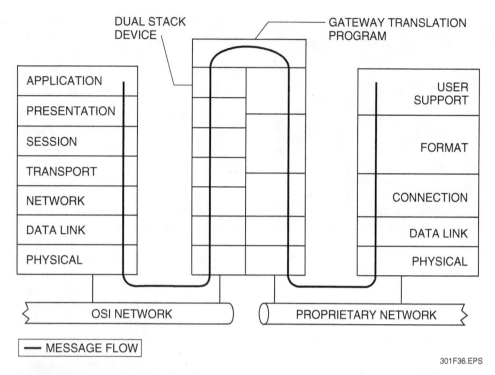

APPLICATION	DUAL STACK DEVICE	GATEWAY TRANSLATION PROGRAM	
PRESENTATION			USER SUPPORT
SESSION			FORMAT
TRANSPORT			
NETWORK			CONNECTION
DATA LINK			DATA LINK
PHYSICAL			PHYSICAL

OSI NETWORK   PROPRIETARY NETWORK

— MESSAGE FLOW

301F36.EPS

Figure 36 Gateway operation in the OSI model.

13.0.0 ADDRESSABLE SYSTEMS

Addressable systems use advanced technology and detection equipment for discrete identification of alarm signals at the detector level. An addressable system can pinpoint an alarm to the precise physical location of the initiating detector. The basic idea of an addressable system is to provide identification or control of individual initiation, control, or notification devices on a common circuit. Each component on the signaling line circuit (SLC) has an identification number or address. The addresses are usually assigned using DIP switches or other similar means. *Figure 37* shows the locations of DIP switches on addressable devices. The address is typically programmed using a computer at the programmer on the fire alarm panel (*Figure 38*).

The fire alarm control panel (FACP) constantly polls each device using a signaling line circuit (SLC). The response from the device being polled verifies that the wiring path is intact and that the device is in place and operational. Most addressable systems use at least three states to describe the status of the device: normal, trouble, and alarm. Smoke detection devices make the decision internally regarding their alarm state just like conventional smoke detectors. Output devices, like relays, are also checked for their presence and in some cases for their output status. Notification output modules also supervise the wiring to horns, strobes, and other devices, as well

DIP SWITCHES

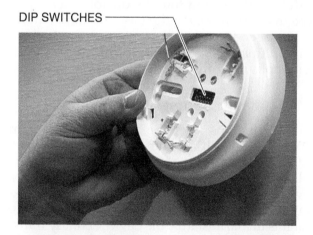

301F37.EPS

Figure 37 DIP switches on addressable devices.

Buses and Networks

KEYPAD

301F38.EPS

Figure 38 Fire alarm programmer.

as the availability of the power needed to run the devices in case of an alarm. When the FACP polls each device, it also compares the information from the device to the system program. For example, if the program indicates device 12 should be a contact transmitter but the device reports that it is a relay, a problem exists that must be corrected. Addressable fire alarm systems have been made with two, three, and four conductors. Generally, systems with more conductors can handle more addressable devices. Some systems may also contain multiple SLCs. These are comparable to multiple zones in a conventional hardwired system.

13.1.0 Analog Addressable Systems

Analog systems take the addressable system capabilities much further and change the way the information is processed. When a device is polled, it returns much more information than a device in a standard addressable system. For example, instead of a smoke detector transmitting that it is in alarm status, the device actually transmits the level of smoke or contamination present to the fire alarm control panel. The control panel then compares the information to the levels detected in previous polls. A slow change in levels (over days, weeks, or months) indicates that a device is dirty or malfunctioning. A rapid change, however, indicates a fire condition. Most systems have the capability to compensate for dirt buildup in the detectors. The system will adjust the detector sensitivity to the desired range. Once the dirt buildup exceeds the compensation range, the system reports a trouble condition. The system can also administer self-checks on the detectors to test their ability to respond to smoke. If the airflow around a device is too great to allow proper

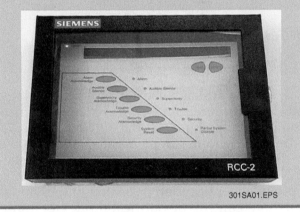
detection, some systems will generate a trouble report.

The information in some systems is transmitted and received in a totally digital format. Others transmit the polling information digitally but receive the responses in an analog current-level format.

The fire alarm control panel, not the device, performs the actual determination of the alarm state. In many systems, the LED on the detector is turned on by the panel and not by the detector. This ability to make decisions at the panel also allows the detector sensitivity to be adjusted at the panel. For instance, an increase in the ambient temperature can cause a smoke detector to become more sensitive, and the alarm level sensitivity at the panel can be adjusted to compensate. Sensitivities can even be adjusted based on the time of day or day of the week. Other detection devices can also be programmed to adjust their own sensitivity.

The ability of an analog addressable system to process more information than the three elementary alarm states found in simpler systems allows the analog addressable system to provide pre-alarm signals and other information. In many devices, five or more different signals can be received.

Most analog addressable systems operate on a two-conductor circuit. Most systems limit the number of devices to about one hundred. Because of the high data rates on these signaling line circuits, capacitance also limits the conductor lengths. Always follow the manufacturer's installation instructions to ensure proper operation of the system.

14.0.0 POWER LINE CARRIER (PLC) SYSTEMS

A building's electrical wiring can be used to create a network. In power line carrier (PLC) systems, data signals are impressed onto AC power wiring and used to control appliances, lighting, security cameras, and other devices. Each device to be controlled is plugged into a special receiver module. *Figure 39* shows an example of such a module used to control lighting or appliances. Controller modules send function commands, such as turn on or turn off, to designated receiver modules that are used to control a device such as a lamp or appliance. The broadcast goes out over the building's electrical wiring. Each receiver module has a unique identification and reacts only to commands it is programmed to act on.

PLC systems can be used for network lighting, security, appliances, and entertainment systems. Each of these systems requires a specialized PLC module. System programming and control can be done from infrared remote controls, wall-mounted keypads, or plug-in control modules like the one shown in *Figure 40*. A computer can also be incorporated into these networks using a special interface module (*Figure 41*). The addition of the computer provides for more sophisticated control and permits the devices on the network to be controlled via the Internet or a cell phone. Specialized software is available for programming computer-controlled PLC systems.

PLC systems are available for use in residential, light commercial, and institutional facilities. These systems are especially useful in structures that do not have a network of ethernet cable installed because they require only an AC power outlet. PLC systems can be designed into new construction as well. They are easily expandable by adding equipment up to the limit of the system.

301F39.EPS

Figure 39 PLC controller module.

301F40.EPS

Figure 40 PLC control panel.

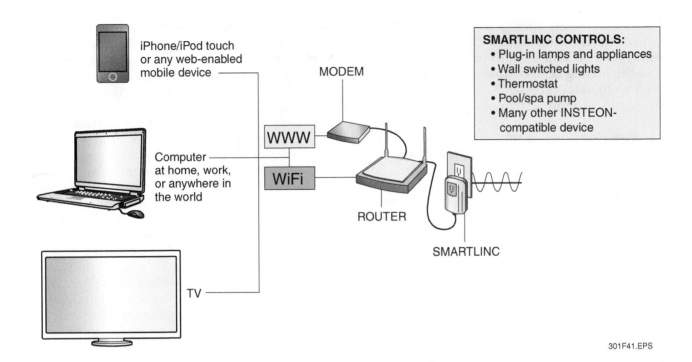

SMARTLINC CONTROLS:
• Plug-in lamps and appliances
• Wall switched lights
• Thermostat
• Pool/spa pump
• Many other INSTEON-
  compatible device

iPhone/iPod touch
or any web-enabled
mobile device

MODEM

WWW

WiFi

Computer
at home, work,
or anywhere in
the world

ROUTER

SMARTLINC

TV

301F41.EPS

Figure 41 PLC network interface module.

15.0.0 POWER OVER ETHERNET (POE) SYSTEMS

Power line carrier systems are a good solution for situations where it is not convenient to run data cable to all the locations where it might be desired. Power over Ethernet (POE) systems are designed to solve the opposite problem—situations where AC power is not available to all desired network nodes. POE is ideal for surveillance cameras, Wi-Fi access points, and voice over IP phones that must be placed at various locations around a property where there is no power source.

POE devices can obtain power either directly from a POE network switch (*Figure 42*) or from a special POE injector (*Figure 43*). The switch method would most likely be used on a new installation where network switches would need to be purchased. The injector would be used when adding POE to an existing installation in order to avoid the cost of replacing network switches. The system operates by imposing a nominal –48 volts DC onto the data cable. In 10/100 Base systems, it would use the unused pair. In gigabit applications, all four pairs are required for data, so the voltage travels on all four pairs.

On Site

Coax to Ethernet Adapter

This adapter is used to connect the incoming TV coax to the customer's router. The device allows a customer to view video programming from one DVR on any compatible DVR or HD receiver. Through this connection, TV programming can be viewed on a network computer and video or audio from the computer can be directed to the TV network.

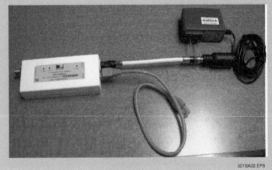

301SA02.EPS

Figure 42 48-port POE network switch.

301F42.EPS

301F43.EPS

Figure 43 POE injector.

SUMMARY

The proprietary networking systems of years past have given way to shared networks based on common standards that establish compatibility between systems made by different vendors. This module examined the underlying standards for most common control networks and explored some of the criteria to consider when setting up a network, such as network topology, access control, and network security.

The Internet must be considered in any discussion of networking because of its growing role in network communication. The Internet and its standards for communication are rapidly becoming the predominant force in monitoring network status and controlling network activity in homes and businesses.

Network interconnection devices, such as switches, routers, bridges, and gateways are critical elements of a network. These devices allow communication among the devices on a network and enable networks to communicate with each other.

Fire alarm, security, and entertainment systems have, in recent years, become parts of larger networks that manage the facilities of entire buildings from a central location within the building. Linking these networks to the Internet allows such systems to be controlled by computer or phone from anywhere in the world.

Specialized systems such as power line carrier and power over ethernet systems have been developed to compensate for situations in which it is not feasible to run data cable or where remote devices do not have access to power.

1. A byte consists of _____.
   a. 2 bits of data
   b. 4 bits of data
   c. 6 bits of data
   d. 8 bits of data

2. RS-485 is a form of _____.
   a. parallel communication
   b. serial communication
   c. Internet protocol
   d. addressable system

3. RS-422 is a _____ standard.
   a. multi-port
   b. multi-drop
   c. multi-point
   d. single-driver

4. RS-232, RS-422, and RS-485 are types of _____.
   a. cables
   b. cable connectors
   c. serial data buses
   d. data transfer protocols

5. Using fiber-optic cable, FireWire® can be used to communicate at distances up to _____.
   a. 14 feet
   b. 45 feet
   c. 75 feet
   d. 225 feet

6. Which of these statements accurately describes a bus?
   a. It is a form of serial communication.
   b. It is used to translate from one protocol to another.
   c. It carries data messages to devices on parallel data paths.
   d. It is used to extend a network when it reaches its maximum length.

7. The method used to carry a signal between the devices in a network is called the _____.
   a. web server
   b. data link
   c. transfer medium
   d. Ethernet

8. The layer of the OSI model that performs traffic control functions is the _____.
   a. physical layer
   b. network layer
   c. transport layer
   d. data link layer

9. In the OSI reference model, any layer can communicate _____.
   a. with the same layer on another device
   b. with any layer on another device
   c. only with layers above it
   d. only with layers below it

10. The topology in which all devices are connected to a network switch and no other device is called a _____.
    a. ring topology
    b. bus topology
    c. star topology
    d. hybrid topology

11. An advantage of using a bus topology is _____.
    a. it provides centralized control of information
    b. it provides direct connection to other nodes
    c. new devices can be easily connected
    d. it provides a high level of security

12. A data message that signifies that data transmission is not taking place on the network is called a _____.
    a. bit
    b. byte
    c. token
    d. blocker

13. Packet filtering is one method of implementing a _____.
    a. transfer protocol
    b. firewall
    c. network protocol
    d. gateway

14. The process in which data messages are broken up into smaller message units is known as data segmentation, or _____.

    a. address format
    b. packet switching
    c. IP addressing
    d. domain routing

15. Data integrity on the Internet is provided by the _____.

    a. Internet protocol (IP)
    b. transmission control protocol (TCP)
    c. hypertext transfer protocol (HTTP)
    d. file transfer protocol (FTP)

16. A network that supports data transmission rates of 1 gigabit is known as a _____.

    a. 10BaseT network
    b. 10Base2 network
    c. 100BaseT network
    d. 1000BaseT network

17. The device that can provide translation between protocol stacks is the _____.

    a. gateway
    b. bridge
    c. regenerator
    d. router

18. A fire alarm system can report the exact location of a fire within a building if its devices are _____.

    a. unbalanced
    b. addressable
    c. matched
    d. balanced

19. A device that gets its power over a data cable is a(n) _____.

    a. power line carrier module
    b. instant phone jack
    c. power over Ethernet device
    d. addressable device

20. In a power over Ethernet system, power can be provided by the network switch.

    a. True
    b. False

Address: A unique identifier for a node or device on a network. On the Internet, the address is composed of four numbers, which define the location of the node in a hierarchical numbering system.

Addressable: A device with a discrete identification that identifies the type of device and its location.

Backbone: A high-capacity network link between two major sections of a network.

Backbone provider: One of a very few select network service providers that own and operate the major Internet backbone networks.

Baseband: A transmission technique in which all of the available bandwidth is dedicated to a single communications channel. Only a single message transfer can occur at a given time.

Basic input/output system (BIOS): A core set of instructions defining how a computer system performs read-and-write operations on its components.

Baud rate: The rate at which information moves on the network.

Binary: Having one of only two possible values, such as on or off, or 1 or 0.

Bit: A single binary value within a computer. Computer equipment, communications paths, and software are described according to the number of bits they handle.

Bridge: A device for transferring information from one network to another.

Broadband: A transmission technique in which the bandwidth is divided into multiple channels. This allows multiple message transfers to take place simultaneously.

Broadcast: Transmission from a single source to all locations on a network.

Bus: A communication channel made up of multiple parallel paths.

Byte: A logical grouping of bits within a computer, representing a value.

Carrier sense multiple access with collision detection (CSMA/CD): A common method used for controlling access to the network medium.

Centralized control: Network access control provided by a single node or entity on the network.

Channel: A path for information transfer.

Checksum: A value calculated from the content of a message. It is used by the receiving device to verify that the data has not been altered during its transfer from source to receiver.

Circuit switching: A method for routing information on the network by establishing a virtual direct connection between two nodes.

Collision: A condition where two or more messages attempt to use the network at the same time.

Deterministic: Based upon a predictable set of rules.

Distributed control: Network access control split up among the individual nodes on the network, in which each node is responsible for determining when it may place a message on the medium.

Ethernet: A common networking protocol that led to the growth and acceptance of local area networks. It uses CSMA/CD access control and was originally based on bus topology.

File transfer protocol (FTP): A message format for packaging information and transferring files over the Internet.

FireWire®: Trade name for the *IEEE 1394* standard for high-speed serial communications.

Gateway: A hardware device that can pass information packets from one network environment to another (such as from an Ethernet network to a token ring).

Hub: A device for networking two or more similar devices.

Hybrid topology: A network that uses elements of two or more topologies together to configure the nodes.

Hypertext: Text containing active links to other documents.

Hypertext transfer protocol (HTTP): A method of requesting and sending text files that include page layout information and hypertext links to other documents.

Internet mail access protocol (IMAP): A protocol for handling email messages while they remain on a remote server.

Internet Relay Chat (IRC): An Internet-based system for communicating with others in real time through text-based chat screens.

Media Access Control (MAC): The means of controlling access to the actual medium or wire of the network.

Multicast: Transmission from a single source to multiple, finite destinations on a network.

Network basic input/output system (NetBIOS): A core set of instructions required by an operating system for communicating with the network hardware.

Network news transfer protocol (NNTP): A protocol for sending, retrieving, and transferring bulletin-board messages across the Internet.

Network operating system (NOS): A program consisting of the commands and instructions that allow computers to function as a network.

Network software: Software that runs on a computer to provide additional network services that are not provided by either the operating system or a network operating system.

Network switch: A device used to manage traffic between devices on a network.

Node: An element on the network that functions as an autonomous part of the network.

Open Systems Interconnection (OSI) Reference Model: A seven-layer model developed by the International Standards Organization to describe how to connect any combination of devices for the purposes of communication.

Operating system (OS): The program that runs constantly on a computer to provide the basic services and control needed to execute other programs.

Packet: A data unit created at the network layer of the OSI model. It contains the data and control information necessary to transfer a message from one network to another.

Packet switching: A method for sharing a data transmission medium by breaking each message into smaller pieces.

Parallel: Two or more paths running alongside each other without meeting.

Point of presence (POP): A computer system equipped with multiple modems that allows users to dial in and connect to the Internet.

Polling: A method for controlling access to the network by individually checking each node to see if it needs to send a message.

Port: Either a physical connection to a system, or, in the case of an Internet server, a logical address where data for a specific application or protocol is routed.

Post office protocol (POP): A protocol for storing email messages on a remote server and retrieving them with a client program.

Protocol: A common language or set of rules allowing controllers connected in a network to share information.

Register: A special, high-speed area of memory that is typically used as a staging area for either processing by the CPU or communication with another device.

Ring: A network topology in which all of the nodes are wired in a continuous sequence.

Router: A hardware device that provides a communication path from one node of a network to another, with both nodes using the same communication protocol.

Ruggedized: Refers to equipment built using materials and methods that allow it to be used in harsh environments.

Serial: A communication path with a single channel or stream, where all information must flow sequentially, one bit at a time.

Server: A computer used to manage a network. It generally contains high-reliability components and larger, faster storage devices.

Simple mail transfer protocol (SMTP): An Internet protocol for passing messages to an email server.

Stochastic: Based upon random time values.

Stream: A continuous flow of data through a channel or path.

Telnet: A program that allows a user to log in to a remote computer as if they were physically present at a terminal on the computer.

Token: A data packet passed from node to node in a specific sequence. The node that has the token is allowed to transmit data on the network.

Token ring: A network topology in which a token must be passed to a terminal or workstation by the network controller before it can transmit.

Topology: The physical layout of a network, especially how the nodes are connected to one another.

Transfer medium: The physical medium, usually some type of wire or cable, that connects the nodes on a network.

Transmission Control Protocol/Internet Protocol (TCP/IP): The set of protocols that forms the foundation for the Internet and defines how messages are passed between nodes on the Internet.

Universal Serial Bus (USB): A communication standard for connecting peripherals to a computer.

Usenet: A worldwide bulletin board system that can be accessed through the Internet.

Web browser: A program used for viewing documents on the Internet using the hypertext transfer protocol.

Wide area network (WAN): A network whose elements are separated by distances great enough to require the use of telephone lines.

Word: In computer terminology, a grouping of bits to form a value. A word may consist of one or more bytes.

Appendix

MANUFACTURING AUTOMATION PROTOCOL (MAP)

In the early 1980s, several companies began looking into large-scale factory and office automation initiatives. Two companies carried their work on automation far enough to form standard protocols that were accepted throughout their respective industries.

General Motors developed the General Motors Manufacturing Automation Protocol (GM-MAP) for communication and control of manufacturing systems. GM made the protocol public in March 1984 to increase their viability with vendors and to ensure that the standard was applicable throughout the manufacturing industry. This public standard, renamed the Manufacturing Automation Protocol (MAP), was refined and disseminated. In January 1985, a MAP user's group was formed to maintain the standard.

Simultaneously, Boeing Computer Services was working on a standard networking protocol for office automation services. Their Technical and Office Protocols (TOP), unveiled in 1984, used many of the same network protocols as the MAP. The two groups merged in 1986 to form the Manufacturing Automation Protocol/Technical and Office Protocols (MAP/TOP) User's Group of the Society of Manufacturing Engineers (SME). This, in turn, has become a part of the Corporation for Open Systems, a nonprofit organization dedicated to the promotion of open (as opposed to proprietary) systems.

Both MAP and TOP are based on the OSI model and follow the layered structure of the OSI model very closely (*Figure A1*).

To ensure maximum compatibility with commercial hardware and software, each layer of the OSI model is implemented via one or more ISO standards. The sole exception to this is the Manufacturing Message Format Specification (MMFS) that was used in the 2.0 release of MAP. In the current implementation, MMFS has been replaced by the ISO 9506 MMS. This has eliminated the only notable incompatibility in the MAP/TOP specification.

There is an alternative form of MAP, known as Mini-MAP, or enhanced performance architecture (EPA), that is designed for time-critical applications. Support of the Mini-MAP requires the Mini-MAP nodes to be on a specific control segment of the network (*Figure A2*).

LAYER		SPECIFICATION

LAYER 7
APPLICATION
    ASCE
    FTAM
    MMS
    MANAGEMENT
		ISO 9506 MMS
		ISO 8571 FTAM
		ISO 8824&5 ASN.1
		CMIP MGT

LAYER 6 PRESENTATION		ISO KERNEL 8822, 8823
LAYER 5 SESSION		ISO/IS/8326, 1984/1986 ISO/IS/8327, 1984/1986
LAYER 4 TRANSPORT		ISO/IS/8072, 1985 ISO/IS/8073, 1985 TP4
LAYER 3 NETWORK		(1987/8 DOCUMENTS) ISO CLNS NETWORK ISO ES-IS ROUTING
LAYER 2 DATA LINK	LLC MAC	ISO/DLS 8802/2 CLASS 1 ISO/DIS 8802/4, 1985 ISO/DIS 8802/1 SECT 5 REV H REV H, JUNE 1985
LAYER 1 PHYSICAL	PHYSICAL	ISO/DIS 8802/4 10Mbps DUO-BINARY BROADBAND 5Mbps CARRIER BAND

301A01.EPS

FigureA1

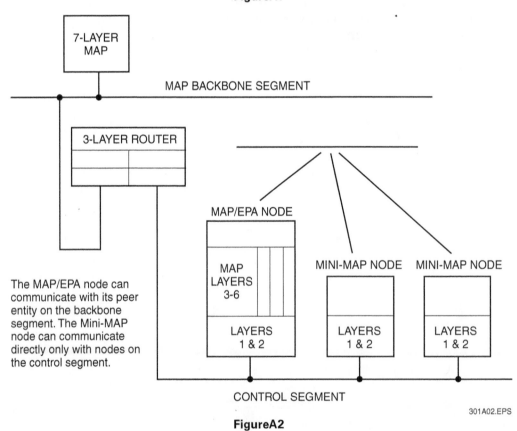

The MAP/EPA node can communicate with its peer entity on the backbone segment. The Mini-MAP node can communicate directly only with nodes on the control segment.

301A02.EPS

FigureA2

NCCER – *Contren® Learning Series* 33301-11

Additional Resources

This module presents thorough resources for task training. The following resource material is suggested for further study.

Computer Networks. Upper Saddle River, NJ: Pearson Education.

Figure Credits

Topaz Publications, Inc., Module opener, Figures 30, 32 (bottom photo), SA01, Figures 37 and 38

Linksys® A Division of Cisco Systems, Inc., Figure 26

Tyco Electronics, Figure 32 (top photo)

Smarthome, Figures 39 and 40

Ray Edwards, SA02

NETGEAR, Figure 42

Black Box Networking Services, Figure 43

CONTREN® LEARNING SERIES — USER UPDATE

NCCER makes every effort to keep its textbooks up-to-date and free of technical errors. We appreciate your help in this process. If you find an error, a typographical mistake, or an inaccuracy in NCCER's Contren® materials, please fill out this form (or a photocopy), or complete the online form at www.nccer.org/olf. Be sure to include the exact module number, page number, a detailed description, and your recommended correction. Your input will be brought to the attention of the Authoring Team. Thank you for your assistance.

Instructors – If you have an idea for improving this textbook, or have found that additional materials were necessary to teach this module effectively, please let us know so that we may present your suggestions to the Authoring Team.

NCCER Product Development and Revision
3600 NW 43rd Street, Building G, Gainesville, FL 32606

Fax: 352-334-0932
Email: curriculum@nccer.org
Online: www.nccer.org/olf

❑ Trainee Guide   ❑ AIG   ❑ Exam   ❑ PowerPoints   Other _____

Craft / Level: _____   Copyright Date: _____

Module Number / Title: _____

Section Number(s): _____

Description: _____

_____

_____

_____

_____

Recommended Correction: _____

_____

_____

_____

_____

Your Name: _____

Address: _____

_____

Email: _____   Phone: _____

Fiber Optics

33302-11

V.1 4/11

Objectives

When you have completed this module, you will be able to do the following:

1. Explain the basic principles of fiber optic systems.
2. Identify the uses of various types of fiber optic cables and devices.
3. Explain the features of fiber optic connectors and splices.
4. Describe the design, operation, and performance of a fiber optic system.
5. Explain the requirements for installation of fiber optic cabling and support equipment.
6. Perform a fiber optic termination.
7. Test a fiber optic link.

Performance Tasks

Under the supervision of the instructor, you should be able to do the following:

1. Perform a fiber optic termination.
2. Test a fiber optic link.

Trade Terms

Attenuation	Insertion loss	Multi-mode	Receiver sensitivity
Cladding	Laser	Multiplexing	Return loss
Duplex	Mode	Numerical aperture	Simplex
		Propagation	Splitter

Contents ─────────────────────────

Topics to be presented in this module include:

Figures and Tables

1.0.0 INTRODUCTION

Fiber optics is the science of transmitting and receiving light through glass or plastic fibers. One of the simpler applications of fiber optics is the popular novelty lamp in which a tight bundle of optical fibers spreads out to form an illuminated mushroom pattern. Fiber optic systems used for transmission of data are more complex, but the basic working concept is the same. *Figure 1* shows a short fiber optic cable with connectors and dust caps.

Modern fiber optic science traces its roots back to the late nineteenth century. In 1870 the British physicist John Tyndall used sunlight and a stream of water to demonstrate how light can be channeled within a medium to follow a designated path. In the 1880s, Henry Wheeler experimented with the concept of piping light from a central source into individual rooms for interior illumination. Around that same time, Alexander Graham Bell developed what he called a photophone – a device that used light to transmit voice waves.

Though none of these early experiments yielded much commercial success, they served to lay the groundwork for fiber optic science as we know it today. The breakthrough that revolutionized fiber optic technology was the development of fiber cladding. This outer layer effectively contains the light, reflecting it back into the optical fiber's core. *Figure 2* shows the behavior of light in both unclad and clad fibers. An unclad fiber loses light when it contacts another material. A clad fiber can contact other materials or objects with no loss of light.

1.1.0 Benefits

A fiber optic system transmits information from one point to another in the form of light. Fiber optic systems are useful for transmitting large amounts of data between many fixed points, such as in a telephone landline network. They are also used to connect a limited number of fixed points, such as a computer and a remote terminal.

There are several types of fiber optic cable used to make these connections. The distance of transmission and the amount of information to be transmitted are two factors to consider when selecting a fiber optic cable. A system's effective transmission distance is influenced by the transmitter's power, by receiver sensitivity, and by signal losses within the cable and connections

The quantity of information that can be carried by a fiber optic strand depends on the number of signals or waves that can be sent, the speed at which they are sent, and the distance they travel.

Fiber optic cables are capable of handling high-speed signals reliably over very long distances.

Along with high transmission speeds and low signal losses over distance, other benefits to using fiber optic systems include the following:

- *Compact size/light weight* – Fiber optic cables have a smaller profile and lower weight per foot than traditional metal wires or cables.

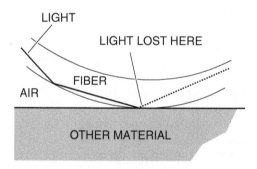

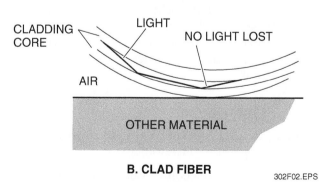

302F02.EPS

Figure 2 Clad versus unclad fiber.

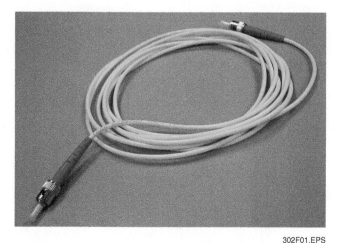

302F01.EPS

Figure 1 Fiber optic patch cable with SC connectors.

- *Free from electromagnetic interference (EMI) and crosstalk* – Fiber optic transmissions are not susceptible to interruption or distortion by electrical noise or transmissions from adjacent wires or fibers.
- *Very wide bandwidth capabilities* – Fiber optic systems can carry extremely large amounts of data. The higher the carrier frequency used, the greater its potential signal bandwidth (the range of usable frequencies in a system). Fiber optic systems operate at 10^{13} Hz to 10^{14} Hz, compared to radio frequencies of 10^8 Hz to 10^9 Hz.
- *Economical* – The cost of fiber optics is continuing to decline while the price of copper-based systems is steadily increasing.
- *Sustainability* – Supplies from copper mines are dwindling worldwide, while sand, the chief component of fiber optic cabling, is inexpensive and plentiful.
- *Safety* – Fiber optic lines are free from short-circuits and spark hazards. This is a definite advantage in environments where combustible chemicals may be present.
- *Corrosion* – The glass in fiber optic cables is impervious to the corrosive elements that damage metal-based systems.
- *Security* – A fiber optic communication system cannot be tapped without physically disrupting the service. This is a priority consideration for government and military installations, and other applications in which system security is important.

1.2.0 Applications

Fiber optics technology provides solutions in many important applications:

- Long-distance telecommunications systems
- High-definition cable television signals
- High-speed internet communications
- High-speed data transfer for computer local area networks (LANs)
- Military battlefield communications and weapons guidance
- Non-invasive surgery

2.0.0 FIBER OPTICS THEORY

In order to become proficient at installing or maintaining fiber optic systems, it is important to understand how these systems work. A basic fiber optic communication system is made up of the following parts:

- *Fiber optic cable* – A relatively small, flexible cable used to carry a signal in the form of a modulated light beam. Cables can range from a few feet up to several miles in length. The cable may contain a single fiber optic strand, or many strands bundled together as one unit.
- *Fiber optic strand* - A single fiber optic light guide used to transport a modulated light signal from the light source transmitter to a light-detecting receiver.
- *Transmitter* – A source of visible or invisible (infrared) radiation that produces the light signal. Usually this is a light-emitting diode (LED), laser device, or vertical cavity surface-emitting laser (VCSEL) that can convert an electrical signal into a modulated light signal. VCSEL is commonly pronounced vixel.
- *Receiver* – This is a photosensitive detector used to receive the light signal and convert it back into an electrical form.
- *Connectors* – These components connect the fibers to other fibers and to the transmitter and receiver.

Figure 3 shows a basic fiber optic communication system.

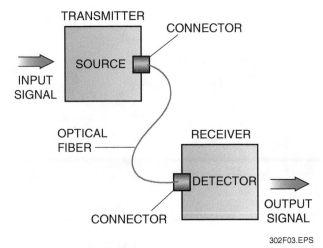

Figure 3 Basic fiber optic system.

302F03.EPS

A fiber optic link is a single strand of optical fiber with a connector on each end. It is the basic building block of a fiber optics communication system.

2.1.0 Light Generation and Coupling

Light coupling is the process of directing light into the optical fiber. The inventions of LEDs and lasers have made light coupling an easier and more reliable process. These innovations allow modern optical systems to use narrow, highly focused beams of light to transmit signals through optical fibers.

The process of coupling light into fibers can be simplified when the transmission light source is comparable in size to the fiber core. Very small core fibers work best with laser diodes, which can supply a very narrow, intense beam of light. Larger core fibers are more suited to LED light sources. LEDs have a longer life span and are less expensive than lasers, but operate at slower transmission speeds. VCSELs are low-cost lasers that provide a good compromise between more powerful laser diodes and weaker and less focused LEDs.

2.2.0 Light Transmission

Two key concepts in the study of light transmission are the following:

- Signal propagation – The characteristic movement of a light signal through a medium.
- Signal attenuation – The degradation of a light signal as it travels through a medium.

2.2.1 Signal Propagation

The refraction (bending) of light has been noted and studied for many centuries. However, it was not until 1850 that Jean B. L. Foucault proved that refraction of light was caused by a change in the propagation velocity (the speed at which light travels through a medium). This propagation velocity is known as the refractive index. The refractive index (N) of a material is the ratio of the speed of light in a vacuum compared to the speed of light in a certain material:

$$N = C_{vac}/C_{mat}$$

The speed of light in a material is always slower than in a vacuum, so the refractive index is always greater than 1.0.

Light tends to travel in a straight line until it passes through a substance with a different refractive index. When light passes through such a substance, bending occurs. The amount of bending that occurs depends not only on the refractive indexes of the two substances, but also the angle at which the light hits the surface between them. The angle at which the light enters the material is known as the angle of incidence and the angle at which it exits the material is known as the angle of refraction. These angles are measured from a line perpendicular to the surface, as shown in *Figure 4*.

Figure 4 shows what happens to light as it moves from air into glass and out again. The bending of light occurs whether the surface of the glass is flat or curved. However, if both the entry and exit surfaces of the glass are flat, as in *Figure 4A*, the net refraction effect is zero. In other words, the light will exit at the same angle at which it entered, although it may be displaced. If one or both surfaces are curved, as in *Figure 4B*, the net effect is that of a lens. In other words, the light rays exit from the lens at a different angle than they entered.

Refraction is not possible if the light's angle of incidence is large, exceeding what is called the critical angle. When this happens, light cannot escape the glass. It exhibits what is called total internal reflection, bouncing back into the glass. Internal reflection is the property that confines light within an optical fiber.

Optical fibers have two main sections. The inner core is the central portion of the fiber and is the area through which the light actually travels. The cladding is the outer portion surrounding the core. The cladding has a smaller refractive index than the core. As a result, when light meets with the cladding, it is reflected back into the core by total internal reflection. *Figure 5* shows an example of light channeled within an optical fiber.

Light rays entering a fiber at an angle above a certain value will be accepted and guided within the core of the fiber. This angle is known as the acceptance angle. As shown in *Figure 6*, only light rays inside the acceptance angle will enter and be guided along the fiber. This angle is not actually given in specification literature. It is expressed as the numerical aperture (NA). The NA is the range of angles over which the system can accept or emit light.

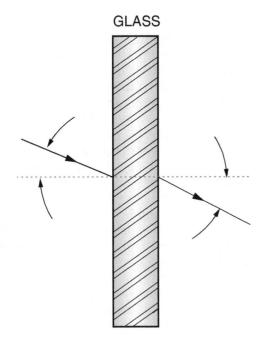

A. GLASS PLATE

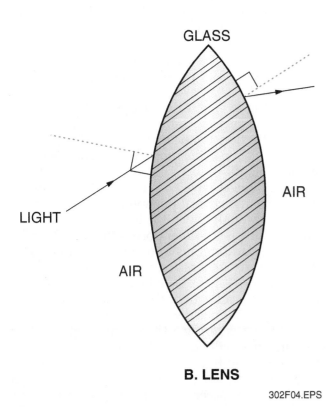

B. LENS

302F04.EPS

Figure 4 Light refraction through glass plate and lens.

2.2.2 Signal Attenuation

Fiber optic systems are very efficient, but not 100 percent efficient. Some signal attenuation (loss) occurs during transmission. Some materials in the fiber may absorb a small amount of light. Some light scatters out of the inner core, and some will escape the core as a result of the operating environment. The single major factor in signal attenuation is the wavelength of the transmitted light, as shown in *Figure 7*.

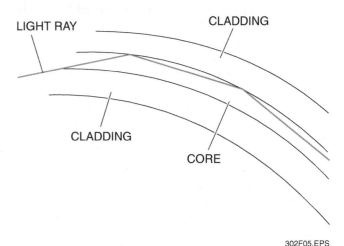

302F05.EPS

Figure 5 Light guided within an optical cable.

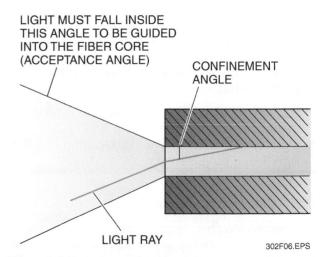

302F06.EPS

Figure 6 Acceptance angle.

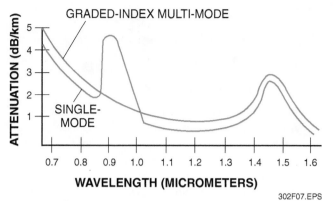

Figure 7 Signal loss over a range of wavelengths.

The ratio of output signal strength to input signal strength defines the attenuation in a system. Each optical fiber has a characteristic attenuation that is measured in decibels per unit of length (normally decibels per kilometer). The total attenuation in the fiber equals the characteristic attenuation times the length.

As shown in *Figure 7*, the attenuation of a light signal in a fiber depends on its wavelength. The curves shown are fairly typical of the two major types of telecommunication fibers, which are discussed in detail later in this module. The absorption peak at 900 nanometer (nm) is caused by the peculiarities of single-mode fiber; the peaks at 1,450nm are caused by traces of water remaining in the fiber as an impurity. Otherwise, the curve is fairly smooth in the range that is shown.

Fiber signal loss is a major determining factor in choosing the operating wavelength. Attenuation is very low at 1,350nm, and even lower at 1,600nm. An attenuation of -0.5dB/km is typical in the 1,300nm region; 1 percent of the light entering the fiber remains after 40km for a –20dB loss. At the –0.2dB/km attenuation available at 1,600nm, 1 percent of the input light remains after 100km. This allows long-distance transmission without amplifiers or repeaters, an important consideration for telecommunication applications.

There are some important factors that limit the transmission speed of fibers. A ray of light can enter the fiber at any of a number of different angles, as seen in *Figure 8*. A ray traveling a straight path through a fiber will do so faster than a ray reflected side-to-side, due to the shorter distance traveled. For example, in a 1km fiber with a 100µm core, the difference in travel distance between a straight-through ray of light and a ray entering the fiber at a 5-degree angle is 3.8m.

The resulting distances of varying paths of light are significant. For the current example, the first ray of light would exit the fiber 12.7 nanoseconds (ns) before the last. Thus, what was an

instantaneous pulse at the point of transmission is a 12.7ns pulse at reception. The spreading out of the signal is known as pulse dispersion.

Pulse dispersion is linearly proportional to the distance traveled. The dispersion is measured in nanoseconds per kilometer. For commercial applications, the dispersion is indicated as an analog bandwidth limit (for example, a 100µm core fiber has a 20MHz/km bandwidth) or a digital data transmission rate maximum. As fiber length increases, both bandwidth and data rate decrease due to pulse dispersion.

Single-mode fibers operate at significantly higher data rates. The low-dispersion characteristics of single-mode fibers will be discussed in detail later in this module.

2.3.0 Operational Considerations

There are several operational considerations involved in evaluating a fiber optic system. Major factors include speed, capacity, and proper alignment.

2.3.1 Speed and Capacity

The low attenuation of a fiber optic system is similar to that of an electrical system. In fact, the appeal of optical fibers to the telecommunication industry is not the low attenuation, but the large bandwidth and high-speed transmission capabilities. These properties give fiber optic systems the ability to handle more information more quickly than a conventional electrical system.

In telecommunication applications, the amount of information the system can handle is extremely important. In a digital system, information movement is measured in bits. A bit is simply a unit of information. Assuming the information is traveling between the same two points, a fiber optic system may provide the ability for a 100 megabyte per second (MBps) rate, while a conventional sys-

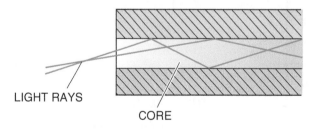

RAYS ENTERING AT DIFFERENT ANGLES TRAVEL DIFFERENT DISTANCES THROUGH THE SAME LENGTH OF FIBER

302F08.EPS

Figure 8 Different light paths in a fiber core.

Fiber Optics

tem might provide for 1MBps. The end result is a reduction in cost, since it is cheaper to build one 100MBps system than to build a hundred 1MBps systems. The same savings can be seen in an analog communications environment. The amount of information an analog system can carry is measured as frequency. The large bandwidth capability of fiber optic systems gives them the advantage over electrical wire systems.

In a typical analog system using coaxial cable, there is significant signal loss with higher signal frequencies, as shown in *Figure 9*. However, for a fiber optic system operating in its normal frequency range, signal losses are independent of frequency. The losses shown in *Figure 9* are measured in decibels per kilometer of cable. Signal losses associated with fiber optic couplings and connections are not taken into account on this scale.

Pulse dispersion has a major impact on speed and capacity because of its impact on the transmission time of the information. In the 1km fiber example, the instantaneous input pulse would actually have a 12.7ns output pulse. Since the receiver can only receive one output pulse at a time, the transmission rate would be limited to the delayed rate of 80MBps. In reality, the system would operate even slower because the input pulse is not instantaneous.

2.3.2 Alignment

There are many considerations when connecting fibers. Very tight tolerances must be maintained and proper alignment is crucial. In most copper wire systems, the losses resulting from connections are minimal. In fiber optics, signal losses at connectors, splitters, light source connections, and even splices can be significant. These losses are a major consideration for system designers.

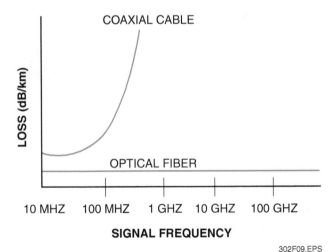

Figure 9 Loss versus frequency.

302F09.EPS

Hybrid-Fiber Coaxial (HFC) Networks

Many communications companies, such as cable TV, are using hybrid-fiber coaxial cable (HFC). This allows them to carry video, voice, and data in the same system. HFC networks use optical fiber cable and coaxial cable in different portions of a network. Fiber optic cable can be run from a CATV head-end to nodes located near customers. Coaxial cable is then run to homes and businesses. Several reasons that are often cited for HFC use include the following:

- Fiber optic cable can carry more data than coaxial cable alone.
- Higher bandwidth provides a reverse path for interactive data.
- Fiber optic cable is more reliable than coaxial cable.

Figure 10 shows some common connection errors and the resulting losses. In *Figure 10A*, the fibers are misaligned laterally (axially). Transmitted power is lost when light escapes from the offset portion of the fiber. The axial displacement reduces the effective area of the receiving fiber, reducing optical power.

The end separation shown in *Figure 10B* permits the escape of light. The amount of optical power lost depends on the separation distance.

Figure 10C shows angular misalignment of the fibers. This creates effects similar to the combined effects of lateral misalignment and end separation, but compounds the losses due to the angular offset.

In *Figure 10D*, the fiber end finishes are irregular rather than smooth. Small but cumulative losses are produced as a result of minute effects similar to *Figures 10A, 10B,* and *10C*.

Figure 10E shows losses due to fiber distortion or fiber size differences. The incompatible core sizes result in significant signal losses.

A sixth coupling loss not shown in *Figure 10* is known as Fresnel (pronounced fraynel) reflection. This occurs when the transmitted light changes refractive indexes between coupling media.

Reflections as well as refractions are caused by such faulty connections, adding to the minute, cumulative losses in optical power.

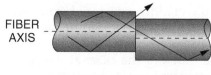

A. AXIAL DISPLACEMENT

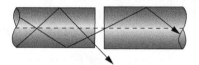

B. END SEPARATION

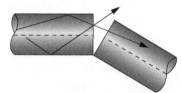

C. ANGULAR MISALIGNMENT

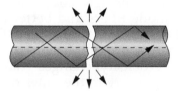

D. IRREGULAR END FINISH

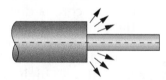

E. FIBER DISTORTION OR SIZE DIFFERENCE

302F10.EPS

Figure 10 Fiber optic connector problems.

3.0.0 FIBER OPTIC COMPONENTS

A conventional electrical system requires a power source, conductor, electrical loads, and some means of connecting these devices. A fiber optic system is similar in this respect. A fiber optic system requires a source (the transmitter), a conductor (the fiber), loads (the receiver), and a means of connection (connectors, splitters, and splices). We will discuss the function, construction, and operation of each of these important components.

3.1.0 Optical Fiber

As discussed earlier, the reflection or refraction of light depends on the indexes of refraction of the two media, and on the angle at which light strikes the surface between them. An optical fiber works

on these principles. Once light begins to reflect down the fiber, it continues to do so under normal circumstances.

Keep in mind the difference between optical fiber and fiber optic cable. The optical fiber is the signal-carrying member, like the metallic conductor in a wire. Fiber optic cable is the optical fiber encased in coverings that protect it from environmental and mechanical damage.

3.1.1 Fiber Construction

The two basic parts of an optical fiber are the core and the cladding. The core, at the center of the fiber, carries the beam of light. The surrounding cladding provides the change in refractive index that causes total internal reflection of light through the core. The refractive index of the cladding is usually less than 1 percent lower than that of the core. Using a 100- to 200-power microscope, you can see both the core and the cladding. However, these two sections of the fiber are inseparably joined during the manufacturing process.

Most fibers have an additional coating around the cladding called a buffer. This coating usually consists of one or more layers of polymer. It protects the fiber from shocks that might affect its optical or physical properties. This protective coating does not influence the propagation of light within the fiber. This buffer coating is removed at the ends of the fiber where it is to be terminated into a connector or a splice is to be performed.

Fibers have extremely small diameters. For comparison, a human hair has a diameter of approximately 100μm. Fiber sizes are usually expressed by first giving the core size, followed by the cladding size. For example, 50/125 means that the fiber has a core diameter of 50μm and a cladding diameter of 125μm. The symbol μm is an abbreviation for micro-meter but is usually referred to as micron, since 1 micron is a length equal to one millionth of a meter.

3.1.2 Modes

A mode is one path a light ray can travel through a material. In the case of fiber optics, the material is the glass core of the fiber.

The number of modes supported by a fiber can range from one to more than 100,000. If there are several paths through which the light may travel, the fiber is designated as multi-mode.

Low-order modes are shorter paths that enter and track more closely along the fiber's center axis. High-order modes are associated with light ranging in the outer portion of the fiber core, near the cladding boundary. High-order modes are longer paths—the light travels farther than when in low-order modes. Light in high-order modes is more apt to be lost due to imperfections like connector separations, misalignments, or sharp bends in the fiber.

3.1.3 Fiber Classification

Optical fibers are classified in three ways: by material makeup, by refractive index, and by mode properties.

The three most common fiber materials are glass, plastic-clad silica (PCS), and plastic.

Glass fibers have a glass core and glass cladding. This is the most common fiber material. The glass used in fibers is ultra-pure, ultra-transparent silicon dioxide or fused quartz. Impurities are purposely added to the pure glass used in the core and cladding to achieve the desired refraction index. The refraction index is typically higher in the core than in the cladding.

Plastic-clad silica fibers have a glass core and plastic cladding. Their performance is good, but not as good as all-glass fibers. Unlike glass fibers, fibers containing plastic do not have a buffer coating surrounding the cladding.

Plastic fibers have both a plastic core and plastic cladding. Compared with other fibers, plastic fibers are limited in loss and bandwidth. However, their very low cost and ease of use make them attractive in applications where high bandwidth and low loss are not a concern.

The second way to classify fibers is by the refractive index of the core and the modes that the fiber propagates. Common classifications of this type are:

- Multi-mode step-index
- Single-mode step-index
- Multi-mode graded-index

Multi-mode step-index fibers are the simplest fibers and were the first to find practical uses. The fiber core has a refractive index that is slightly higher than the cladding material, confining the light by total internal reflection in the core. The term *step-index* comes from the abrupt change in the refractive index of the fiber at the core-cladding boundary, the interface that confines light within the core. The degree of refractive index difference depends on the fiber design and material, but it is typically small. Less than 1 percent difference usually gives adequate light guiding in glass fibers.

Single-mode step-index fibers are designed to carry only one mode, eliminating the problem of mode dispersion. Limiting a fiber to one mode is usually achieved by greatly reducing the fiber's core diameter. The small core size puts tight requirements on light coupling into the fiber. For handling reasons, single-mode fiber cladding is usually at least 125µm in diameter (a dozen or more times the core diameter). The core diameter is about 8 microns.

Multi-mode graded-index fibers were developed to offer easier light coupling than small-core single-mode fibers and better bandwidth than multi-mode step-index fibers. Graded-index fibers get their name from the way the refractive index (density) changes gradually from the center of the core to the outer ranges of the core, where it meets the cladding. There is still an abrupt change in density at the core/cladding interface to produce total internal reflection. The center of the core is denser than the outer portion of the core. As a matter of fact, the density and refractive index goes lower as it approaches the core cladding interface. This causes the light rays, especially the high order modes, to bend rather than bounce. Those modes that are traveling at the center of the core travel at a slower speed than those traveling on the outer ring of the core. By providing a lower density and refractive index at the outer ring of the core, the modes traveling in the outer ring of the core travel faster than those in the center of the core. This tends to even out the time versus distance of travel between the modes going straight and those that are bending throughout the total length of the fiber optic strand. It allows all rays traveling in all modes to arrive at nearly the same time, when exiting the end of the fiber optic strand. The result is a greatly reduced modal dispersion for the graded-index multi-mode fibers, over step-index multi-mode fibers.

3.2.0 Cabling

Typically an optical fiber is packaged before it is usable. Packaging involves cabling the fiber or group of fibers in protective outer layers. These layers are much like the insulation and protective coating surrounding a copper wire. Cabling protects the fibers from damage or degradation, and also allows for ease of handling.

Many types of fiber optic cables are available. Design considerations include tensile strength, ruggedness, durability, flexibility, environmental resistance, and temperature extremes. For example, an outdoor telephone cable must withstand extremes of heat and cold, sunlight and rain, ice deposits that weigh it down, high winds that stress it, and rodents that chew on it underground. It must be more rugged than cable connecting equipment within the controlled environment of a telephone switching station. Similarly, a cable running under an office carpet, where people walk or stand on it and chairs roll over it, has different requirements than a cable running inside a tray or conduit.

Figure 11 shows the main parts of a simple single-fiber cable. Although cables come in many varieties, most have these components in common:

- Optical fiber
- Buffer
- Strength members
- Jacket

3.2.1 Buffer

Even with the polymer coating, fibers require additional protection against stress, moisture, and chemicals. This can be done by encasing the fiber within a buffer tube.

The simplest form of buffer is an acrolate coating applied over the cladding. This buffer may be applied by the fiber manufacturer. An additional buffer is added by the cable manufacturer. The cable buffer will be one of two types: loose tube or tight buffer. *Figure 12* shows the two buffer types, and *Table 1* summarizes the properties associated with each type.

The loose-tube type uses a hard plastic tube with an inside diameter several times larger than that of the fiber. One or more fibers lie within the buffer tube. The tube isolates the fiber from the rest of the cable and the mechanical forces acting on it. The buffer becomes the load-bearing member. As the cable expands and shrinks with changes in temperature, it does not affect the fiber as much. Loose-buffered cable is used in outdoor applications.

A tight buffer has a plastic coating applied directly over the fiber. This construction provides better crush and impact resistance, but it does not protect the fiber from the stresses of temperature variations. Because the plastic expands and

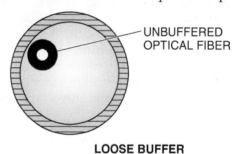

LOOSE BUFFER

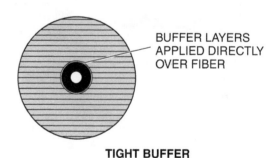

TIGHT BUFFER

302F12.EPS

Figure 12 Loose and tight buffers.

Table 1 Cable Buffer Properties

Cable Parameter	Cable Structure	
	Loose Tube	Tight Buffer
Bend radius	Larger	Smaller
Diameter	Larger	Smaller
Tensile strength installation	Higher	Lower
Impact resistance	Lower	Higher
Crush resistance	Lower	Higher
Attenuation change at low temperatures	Lower	Higher

302T01.EPS

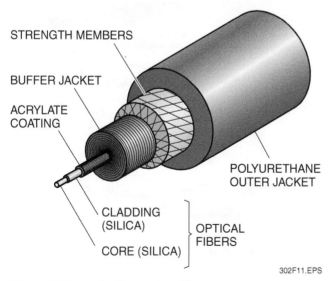

302F11.EPS

Figure 11 Parts of a fiber optic cable.

contracts at a different rate than the fiber, variations in temperature can result in loss-producing microbends. Tight-buffered cable is normally used in indoor applications.

Another advantage of a tight buffer is that it is more flexible and permits a tighter turn radius. This advantage can make tight-tube buffers useful for indoor use where temperatures are relatively stable and the ability to make tight turns inside walls and around corners is desired.

3.2.2 Strength Member

Strength members lend mechanical strength to the fiber. During and after installation, the strength members bear the tensile stresses applied to the cable so the fiber is not broken. The most common strength members are Kevlar® (aramid yarn), steel, and fiberglass epoxy rods. Kevlar® is often used when individual fibers are placed within their own jackets. Steel and fiberglass members are used in multi-fiber cables. Steel offers higher strength than fiberglass, but is not used in all-dielectric applications. Steel is subject to induced voltages from high-voltage lines, whereas fiberglass is not.

3.2.3 Jacket

Like conventional wire insulation, the jacket provides protection from the effects of abrasion, oil, ozone, acids, alkalis, solvents, and other hazards. The choice of jacket material depends on the cost and the degree of resistance required. Some popular jacket materials are PVC, polyethylene, polypropylene, polyurethane, nylon, and Teflon®.

When a cable contains several layers of jacketing and protective material, the inner layers are called jacketing and the outer layer is called the sheath. This terminology is especially common in the telephone industry.

3.3.0 Types of Cables

Though optical fibers can be used in a variety of environments, the cable covering that protects them is usually designed for very specific conditions. Many different types of cables have evolved for different applications (*Figure 13*). The types of cables can be broken down into the two broad categories: indoor cables and outdoor cables.

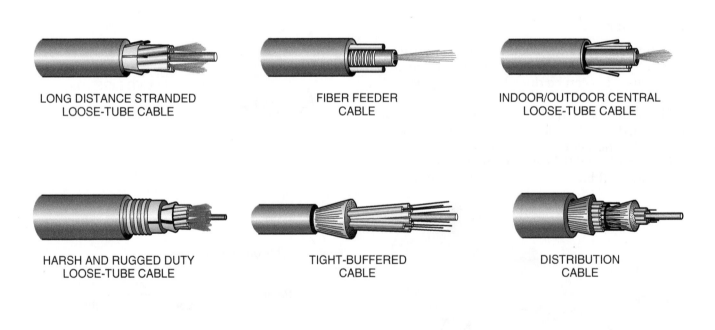

LONG DISTANCE STRANDED
LOOSE-TUBE CABLE

FIBER FEEDER
CABLE

INDOOR/OUTDOOR CENTRAL
LOOSE-TUBE CABLE

HARSH AND RUGGED DUTY
LOOSE-TUBE CABLE

TIGHT-BUFFERED
CABLE

DISTRIBUTION
CABLE

BREAKOUT
CABLE

SIMPLEX/DUPLEX
CORDAGE

302F13.EPS

Figure 13 Typical loose- and tight-buffered cables.

3.3.1 Indoor Cables

Indoor cables are not required to withstand the harsh environments of outdoor cables. However, indoor applications are numerous and varied enough to require several different types of cables. Cables for indoor applications include the following:

- Simplex cables
- Duplex cables
- Multi-fiber cables
- Heavy-duty, light-duty, and plenum-duty cables
- Undercarpet cables

Simplex cables contain a single fiber. Simplex is a term used in electronics to indicate one-way transmission. Since a fiber carries signals in only one direction, from transmitter to receiver, a simplex cable allows only one-way communication. *Figure 14A* shows a cross section of a typical simplex cable as well as a simplified block diagram showing a one-way communication application.

Duplex cables contain two optical fibers. *Figure 14B* shows cross sections of two typical duplex cables, as well as a simplified block diagram of a two-way communication application.

Duplex refers to two-way communication. One fiber carries signals in one direction; the other fiber carries signals in the opposite direction. This could also be done using two separate simplex cables. Duplex cable is used instead of two simplex cables for appearance and convenience. Ripcord constructions are popular because they allow the two cables to be easily separated.

Multi-fiber cables contain more than two fibers. They allow signals to be carried throughout a building. The fibers are usually in pairs or groups of six, with each pair carrying signals in opposite directions. A 12-fiber cable therefore permits six duplex circuits. *Figure 15* shows typical multi-fiber cables.

Heavy and light refer to the duty rating of the cable. Heavy-duty cables can withstand rougher handling than light-duty cables, which is of particular concern during installation. Heavy-duty cables usually have a thicker jacket than light-duty cables.

A plenum is air space between walls, under a structural floor, or above a drop ceiling in which environmental air is flowing. According to the *National Electric Code®* (*NEC®*), it is a compartment or chamber to which one or more air ducts are

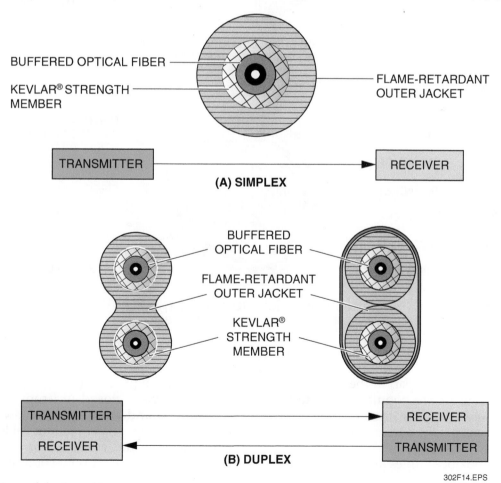

302F14.EPS

Figure 14 Simplex and duplex cables.

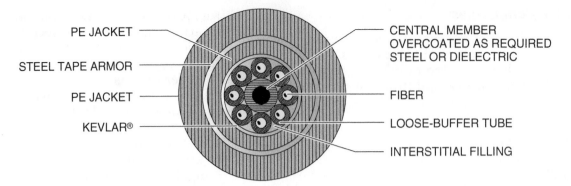

PE JACKET

STEEL TAPE ARMOR

PE JACKET

KEVLAR®

CENTRAL MEMBER
OVERCOATED AS REQUIRED
STEEL OR DIELECTRIC

FIBER

LOOSE-BUFFER TUBE

INTERSTITIAL FILLING

**SIECOR STANDARD (LOOSE TUBE) CABLE
8 FIBER – DOUBLE JACKET – STEEL TAPE ARMOR**

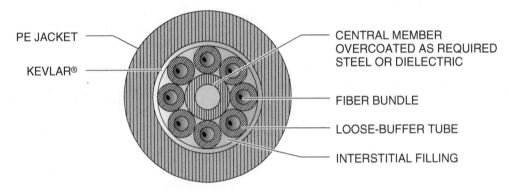

PE JACKET

KEVLAR®

CENTRAL MEMBER
OVERCOATED AS REQUIRED
STEEL OR DIELECTRIC

FIBER BUNDLE

LOOSE-BUFFER TUBE

INTERSTITIAL FILLING

**SIECOR MINI-BUNDLE (LOOSE TUBE) CABLE
43–48 FIBERS – SINGLE JACKET**

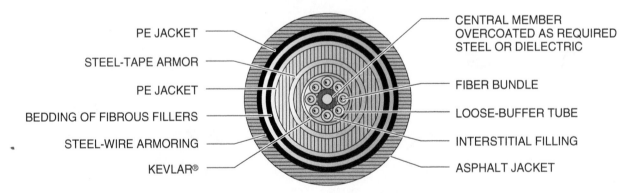

PE JACKET

STEEL-TAPE ARMOR

PE JACKET

BEDDING OF FIBROUS FILLERS

STEEL-WIRE ARMORING

KEVLAR®

CENTRAL MEMBER
OVERCOATED AS REQUIRED
STEEL OR DIELECTRIC

FIBER BUNDLE

LOOSE-BUFFER TUBE

INTERSTITIAL FILLING

ASPHALT JACKET

**SIECOR MINI-BUNDLE (LOOSE TUBE) CABLE
48 FIBERS – UNDERWATER CABLE**

302F15.EPS

Figure 15 Multi-fiber cables.

connected and that forms part of the air distribution system. Plenums are popular places to run signal, power, and telephone lines. Unfortunately, plenums are also places where fires can easily spread throughout a building. Certain jacket materials give off noxious fumes and/or toxic gases when burned.

The *NEC*® requires that cables run in plenums must either be enclosed in fireproof conduit or be insulated and jacketed with low-smoke and fire-retardant materials. Plenum cables are cables with materials and construction that meets the specification for use without conduit.

Undercarpet cable is run across a floor under a carpet. A popular use is in an open office where the work area is separated only by partitions or arrangements of desks and equipment. An important factor is the ability to rearrange or reconfigure the cable runs as the office needs change.

The cross section of an undercarpet cable illustrated in *Figure 16* shows two optical fibers and three fiber-reinforced plastic strength members enclosed in a polyvinyl chloride jacket. Each fiber is jacketed with polyester. This construction gives a very low profile, only about 0.075" high, which prevents unsightly bulges in the carpet.

3.3.2 Outdoor Cables

Outdoor cables must withstand harsher conditions than most indoor cables. Outdoor cables are used in applications such as the following:

- *Aerial runs* – Cables strung from utility poles
- *Direct burial* – Cables placed directly in a trench and covered with earth
- *Indirect burial* – Similar to direct burial, but the cable is inside a duct or conduit
- *Submarine* – The cable is underwater, including transoceanic applications

Most outdoor cables have additional protective sheaths. For example, a cable designed for direct burial may have a layer of steel armor to protect against rodents that might chew through plastic jackets and into the fiber. Most outdoor designs use water-blocking gels and water-blocking grease to fill the cable and eliminate air pockets. This prevents water from seeping into the cable, where it could freeze, expand, and damage the cable. The fibers float in a gel that will not freeze and damage the fiber.

Most outdoor cables contain many fibers. Many multi-fiber cables divide the fibers among several buffer tubes. The strength member is typically a steel or fiberglass rod in the center, although woven steel strands in the outer sheath are used in some designs.

Another kind of multi-fiber cable is ribbon cable. In this design, parallel fibers are sandwiched between double-sided adhesive polyester tape.

Each ribbon can be stacked with others to make a rectangular array. For example, a stack of 12 ribbons with 12 fibers each creates an array of 144 fibers (12 × 12). This array is placed in a loose tube, which in turn is covered by two layers of polyethylene. Each polyethylene layer contains steel wires serving as strength members. Depending on the application, additional layers, such as steel armor, cover the polyethylene. *Figure 17* shows such a ribbon cable.

In cables containing many fibers, some fibers are kept as installed spares. These are used to replace fibers that may fail in the future. Others are saved for future system expansions. Having extra fibers in place saves future installation costs of additional fibers and cables.

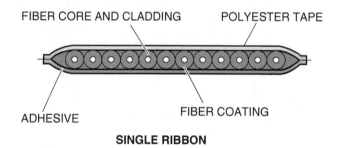

SINGLE RIBBON

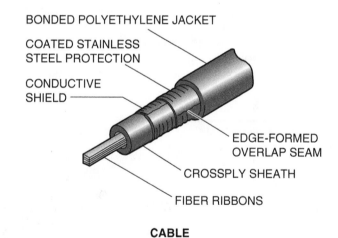

CABLE

302F17.EPS

Figure 17 Ribbon cable.

Figure 16 Undercarpet cable.

302F16.EPS

3.4.0 Cable Characteristics

The most important cable characteristics to the electronic systems technician are the available lengths, identification of fibers and cables, and the maximum tensile load that the cable can withstand. Most modern fiber cables and the associated tubes are divided into six, or multiples of six, individual strands. Six fiber optic strands allow for the primary equipment, the redundant equipment, and a redundant spare fiber for replacement in case of failure.

3.4.1 Lengths

Cables come on reels in various lengths. Typical lengths for outdoor cables are 1km and 2km, although lengths of 5km and 6km are available for single-mode fibers. Long cables are desirable for long-distance applications, since each splice creates additional signal losses in the system. Long cables mean fewer splices and less overall signal loss.

Indoor cables are available in shorter standard lengths. Some indoor cables can be custom-ordered for job-specific lengths and configurations.

3.4.2 Color-Coding

Fiber coatings and buffer tubes are normally color-coded to make identification of each fiber easier. The color codes are defined by *ANSI/TIA/EIA-598*. In a long-distance link, one must ensure that fiber A in the first cable is spliced to fiber A in the second cable, B to B, C to C, and so forth. Color coding simplifies fiber identification (see *Table 2*).

3.4.3 Tensile Loads

Most cable manufacturers specify the maximum tensile loads that can be applied to the cable. Two loads are usually specified. The installation load is the short-term load that the cable can withstand during the process of installation. This load includes the temporary tensile loads exerted by pulling the fiber through duct or conduit, around corners, and so forth. The maximum specified installation load limits the length of cable that can be installed at one time for a particular application. Different applications offer different installation load conditions. The installation must be carefully planned to avoid overstressing the cable.

The second type of load is the long-term or operating load. During its installed life, a cable cannot withstand loads as heavy as the temporary stresses experienced during installation. The specified operating load, also called the static load, is therefore much less than the installation load.

Table 2 Color Codes

Fiber or Tube #	Color	Fiber or Tube #	Color
1	Blue	13	Blue w/ black tracer
2	Orange	14	Orange w/ black tracer
3	Green	15	Green w/ black tracer
4	Brown	16	Brown w/ black tracer
5	Slate	17	Slate w/ black tracer
6	White	18	White w/ black tracer
7	Red	19	Red w/ black tracer
8	Black	20	Black w/ yellow tracer
9	Yellow	21	Yellow w/ black tracer
10	Violet	22	Violet w/ black tracer
11	Rose	23	Rose w/ black tracer
12	Aqua	24	Aqua w/ black tracer

4.0.0 UNDERSTANDING LIGHT TRANSMISSION

Fiber optic cable is used to carry a modulated light signal. The light signal is inserted into the fiber using an emitter. There are three types of emitters commonly used: edge-emitting laser diodes, VCSELs, and light-emitting diodes (LED). These small devices are part of a larger electronic device known as a transmitter. Transmitters convert an electrical current into light. To modulate the light signal and thereby imprint information onto it, the light must be turned off and on very quickly. In a fiber optic system, the signal must also have a high level of radiance. Transmitter functionality is characterized by signal types, speed, and operating wavelength.

4.1.0 Signal Types

Fiber optic transmitters are designed to generate either an analog or a digital signal. To differentiate between the two signal types, it may be helpful to think of a light bulb with both dimmer and on/off controls. Analog signaling requires a linear reproduction of the input signal. In the bulb analogy, the data would be conveyed as an analog signal by gradually altering (modulating) the brightness of the light over time. Digital signaling, on the other hand, relies on the presence or absence of a light signal to convey the data. In the bulb example, a digital signal would simply involve turning the bulb off or on in a precise pattern.

Analog signals in a fiber optic system are subject to the effects of distortion. Digital signals are not.

4.2.0 Speed

For a digital signal, the speed of transmission is measured as the data rate. The data rate is the maximum number of bits that can be transmitted per second without exceeding a certain error rate, normally one per billion.

In the analog system, transmission speed is measured as bandwidth. For analog modulation, bandwidth is normally defined as the point at which signal amplitude drops 3dB below the normal level, which is equivalent to a 50 percent reduction in power.

In light transmission, the limiting characteristic of a light source is its rise time. The rise time is the amount of time it takes the light output to rise to 90 percent of its steady-state level from 10 percent. The following formula shows the relationship of bandwidth to rise time:

$$BW = 0.35 \div \text{rise time}$$

The rise time required for a specific application is an important consideration when choosing the light source. Lasers have a considerably faster rise time than LEDs, but they are also more expensive. VCSELs are often a good compromise between cost and fast rise time.

4.3.0 Light Sources

As previously mentioned, the three major types of fiber optic light sources are LEDs, VCSELs, and edge-emitting laser diodes. All are very small semiconductor chips made of gallium arsenide (GaAs) or other semiconductor material.

4.3.1 LEDs

An LED (*Figure 18*) spontaneously emits light when current is passed through it. By introducing materials such as indium, aluminum, or phosphorus, the wavelength of the emitted light can be controlled.

The cone of radiation from LED devices is considerably larger than the cone of acceptance of optical fibers. Therefore, efficient coupling of light into small-core fibers with low numerical apertures is difficult unless special designs are employed.

LED light sources have the following characteristics:

- The output power is relatively insensitive to device temperature, making LEDs less complex and requiring little or no temperature compensation.

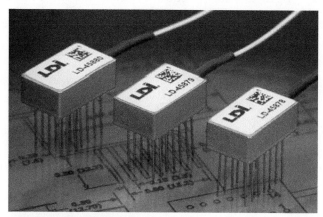

302F18.EPS

Figure 18 Edge-emitting LEDs.

- LEDs have a wide spectral output, typically 25 to 50nm or greater.
- The emission pattern is broad, and coupling losses into fibers may be in the range of 10dB and higher.
- LEDs typically have a longer operating life than lasers.
- LEDs are generally inexpensive.
- LEDs have a rather large active area where light is generated and have characteristically high capacitance. This makes them difficult to digitally modulate at higher transmission rates.
- Since LED light power output is relatively linear over a wide range of input drive current, LEDs are particularly well-suited to analog modulation systems.
- The light output of LEDs is typically less powerful than that of lasers.

For all of the preceding reasons, LED sources are best-suited for use in lower bit-rate applications or shorter wavelength systems where many of the deficiencies are not significant.

4.3.2 Laser Diodes

The term *laser* is an acronym derived from the descriptive phrase Light Amplification by Stimulated Emission of Radiation. A laser (*Figure 19*) has three key qualities. First, its light is monochromatic (of one color). Second, laser light is coherent, meaning all of its radiation is of the same wavelength and the waveforms are all in phase with each other. Third, the light generated by a laser is highly focused and does not spread.

There are two mirrored surfaces, one at each end of the lasing tube. One is highly reflective, the other half-reflective, which allows some of the light to exit the tube. Because the light is of the same wavelength and phase, it stays tightly grouped, with minimal dispersion, creating an intense beam. The light that does not escape keeps

bouncing end to end, stimulating the release of additional light.

As the device temperature increases, the threshold current increases and the optical output power changes. The wavelength of the optical output is temperature dependent. Thus, lasers are far more complex than LEDs in that both temperature sensing and compensation mechanisms are required.

Lasers have short rise times, making them an ideal light source for high speed data transmission systems.

4.3.3 VCSEL

The VCSEL (*Figure 20*) is a type of laser emitter, but it is not the same as the laser diode. One major difference is that the VCSEL is a surface-emitting device, unlike the laser diode, which is edge-emitting. VCSELs are somewhat less powerful than edge-emitting laser diodes, but can be manufactured efficiently at lower costs. For fiber optic systems needing a narrow, focused light source with high transmission capabilities, VCSELs are often a good compromise between more complex lasers and LEDs. VCSELs have largely replaced LEDs on new installations.

A VCSEL has a vertical laser cavity. The active region of the VCSEL emits light. Mirrors created by layers of semiconductor material reflect the light back into the cavity at different wavelengths, but the resulting output is a narrow, single-wavelength beam. In contrast to an edge-emitting device, the light emitted by the VCSEL is a narrow, circular beam. This beam configuration offers highly efficient transfer to an optical fiber. VCSELs are highly suited to applications in which

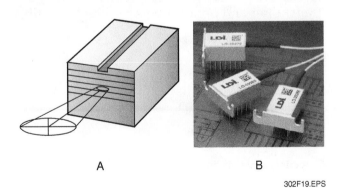

A   B

302F19.EPS

Figure 19 Laser diode.

302F20.EPS

Figure 20 VCSEL.

Laser Hazards

The Occupational Safety and Health Administration (OSHA) classifies lasers according to their hazard potential. The various classifications are summarized as follows:

- *Class 1* – Cannot emit laser radiation at known hazard levels. The least hazardous class, typically continuous wave: cw 0.4mW at visible wavelengths.
- *Class 1A* – A special designation that is based upon a 1,000-second exposure and applies only to lasers that are not intended for viewing, such as a supermarket laser scanner. The upper power limit of Class 1A is 4.0mW. The emission from a Class 1A laser is defined such that the emission does not exceed the Class 1 limit for an emission duration of 1,000 seconds.
- *Class 2* – Applies only to visible laser emissions and may be viewed directly for time periods of less than or equal to 0.25 seconds, which is the aversion response time. 0.25 seconds is the amount of time it takes to blink or turn your head away.
- *Class 3A* – Dangerous under direct or reflected vision. These lasers are restricted to the visible electromagnetic spectrum. Intermediate power lasers (cw: 1 to 5mW). Laser pointers generally fall into this class.
- *Class 3B* – May extend across the whole electromagnetic spectrum and are hazardous when viewed intrabeam. Moderate power lasers (cw: 5 to 500mW, pulsed: 10 J/cm^2 or the diffuse reflection limit, whichever is lower).
- *Class 4* – The highest energy class of lasers, also extending across the electromagnetic spectrum. This class of laser presents significant fire, skin, and eye hazards. High-power lasers (cw: 500mW, pulsed: 10 J/cm^2 or the diffuse reflection limit).

ribbon fiber is used to obtain serial data transmission to support high data rates.

To summarize, fiber optic light sources have the following characteristics:

- An LED is not as bright as a VCSEL.
- A VCSEL is not as bright as an edge-emitting laser diode.
- An VCSEL is only slightly higher in price, and almost as easy to produce as an LED.
- An edge emitting laser diode is much more expense than a VCSEL.
- An LED or VCSEL does not require as much temperature compensation as an edge-emitting laser diode.
- The light spectrum, or frequency bandwidth, of a VCSEL is almost identical to that of an edge-emitting laser diode.
- The light spectrum of an LED is much too broad for longer distance transmissions of data.
- For analog transmissions, the LED varies in brightness as the voltage is varied, where the VCSEL and edge-emitting laser diode vary in wavelength rather than luminance, or radiance.
- The rays of light emitted from an edge-emitting laser diode are almost perfectly parallel to each other.
- The rays of light emitted from a VCSEL are not quite parallel, but are very tightly bundled.
- The rays of light emitted from an LED are splayed in a wide direction from the exit window of the LED, and overfill the core of a fiber optic strand.
- Lasers and VCSELs typically have a more powerful output than LEDs.

Lasers and VCSELs normally use alternate-phase return-to-zero (APRZ) coding where the laser and VCSEL are kept active as long as power is applied, which aids in maintaining nominal operating temperature for these devices. Plus, timing signals are always present to help maintain synchrony between transmitter and receiver.

5.0.0 RECEIVERS

The function of a receiver is opposite that of the transmitter; it converts optical signals back into electrical signals. The receiver combines with the transmitter and optical fibers to complete a fiber optic system.

5.1.0 Basic Receiver Elements

Receivers vary greatly in cost and capabilities. Some receivers are simple light/dark photodetectors, while others are intricate, sensitive systems that accurately receive weak, high-speed signals. The basic functional components of a receiver are as follows:

- *The housing* – Encloses and protects receiver circuitry

Fiber Optics

- *The detector* – Converts the received optical signal into an electrical form
- *Amplification stages* – Amplifies the signal and converts it into a form ready for processing
- *Demodulation or decision circuits* – Replicates the original electronic signal

Photodiodes and photodetectors are used in an optic system as a receiver. Detectors are much faster and more sensitive if electrically reverse-biased. This is the opposite of LEDs, which are forward-biased.

Photodetectors can be made from a variety of materials. The composition determines the wavelengths to which the detectors are sensitive.

5.2.0 Speed

The receive section must operate at least at the same speed as the transmitter. There is an inherent time delay in a detector, dependent on the material and the design of the device. The delay is called response time, and includes the rise time. You will recall that rise time is the time required for an output signal to rise from 10 to 90 percent of the final level after the input is turned on abruptly. Fall time is the opposite—the amount of time required for an output signal to fall from 90 to 10 percent after the input is turned off.

The rise and fall times of a receiver have a significant impact on bandwidth, frequency response, and maximum bit rate. As rise and fall times increase, the response time also increases. Because the detector can only detect one signal at a time, as the response time increases, the limiting bandwidth, frequency response, and maximum bit rate all decrease.

6.0.0 CONNECTORS, SPLICES, AND SPLITTERS

Connectors, splices, and splitters are fiber optic components designed to link components such as transmitters, fibers, and receivers, or to direct or route signals along various paths between components.

A connector is a device attached to the end of a fiber optic cable that mates to another cable or device. Generally, connectors can be unplugged and reconnected repeatedly. When connected, the connector couples light optically into and out of its optical fiber.

A splice is a specialized connector that joins two fiber ends permanently. Splices are not designed to be unplugged and reconnected like regular connectors.

A splitter is a routing device that connects three or more fiber ends, dividing one input between two or more outputs, or combining two or more inputs into one output.

6.1.0 Connectors and Splices

In this discussion, connectors and splices are treated as similar devices because their requirements and losses are similar. However, recall that a splice is permanent, while a connector allows for disconnection.

In a long link, fibers have to be connected or spliced end to end because cables are available in limited lengths (typically 1 to 6 kilometers). For example, if a 30-kilometer span of cable is required and the longest available length is 6 kilometers, five sections will be needed. This will require four connections or splices, plus a connection at each end for the transmitter and receiver.

Connectors may also be required at building entrances, wiring closets, splitters, and other intermediate points between the transmitter and receiver. These allow for transitions between outdoor and indoor cables, rearrangement of circuits, and the division of optical power from one fiber into several fibers.

> **NOTE**
>
> The terms *splitter* and *coupler* are often used interchangeably in fiber optic work to describe devices that divide a signal into two or more paths. In this module, the term *splitter* is used for this purpose.

Dividing a fiber optic system into several subsystems connected together by splices and connectors also simplifies component selection, installation, and maintenance. Components may be selected from different suppliers, and can be installed at different times by different vendors or contractors. Maintenance of the system is simplified when a faulty or outdated part can be disconnected and a new part installed. Transmitters and receivers, for instance, can be upgraded without disturbing the rest of the system.

6.2.0 Connector Requirements

The following is a list of desirable features for a fiber optic connector or splice:

- *Low loss* – The connector or splice should minimize loss of optical power across the junction.
- *Easy installation* – The connector or splice should be easily and rapidly installed without the need for extensive training or special tools.

- *Repeatability* – A connector should be able to be connected and disconnected many times without changes in loss.
- *Consistency* – There should be no variation in loss; loss should be consistent whenever a new connector is applied to a fiber.
- *Economical* – The connector or splice and its required tooling should be inexpensive.

It can be difficult to find a connector or splice that meets all of these requirements. A low-loss connector may be more expensive than a high-loss connector, or it may require expensive installation tooling. The lowest losses are desirable, but the other factors clearly influence the selection of a connector or splice.

In general, loss requirements for splices and connectors range as follows:

- Less than 0.5dB for telecommunication and other splices in long-distance systems
- 0.5 to 1dB for connectors used in intra-building systems, such as local area networks or automated factories
- 1 to 3dB for connectors and splices used with plastic fiber or other applications where higher losses are acceptable and low cost is more important than low loss

6.3.0 Causes of Connection Losses

Three types of factors cause signal loss in fiber optic connections:

- *Intrinsic factors* – Caused by variations in the fiber itself
- *Extrinsic factors* – Contributed by the connector itself
- *System factors* – Contributed by the system

6.3.1 Intrinsic Factors

When joining two fibers to one another, you cannot assume that the two fibers are identical. Optical fibers are manufactured within certain tolerances, and fibers of the same specification vary within those stated limits. *Figure 21* shows the most common intrinsic connection factors.

A numerical aperture (NA) mismatch occurs when the NA of the transmitting fiber is larger than that of the receiving fiber. Core diameter mismatch occurs when the core or diameter of the transmitting fiber is larger than that of the receiving fiber. Cladding diameter mismatch occurs when the claddings of the two fiber ends differ and the cores do not align.

Concentricity loss occurs when fiber cores are not perfectly centered in their cladding. Ide-

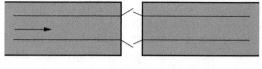

NA MISMATCH

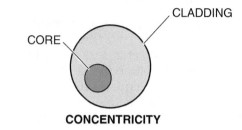

CONCENTRICITY

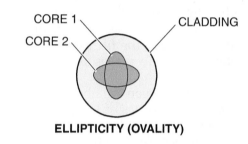

CORE DIAMETER MISMATCH

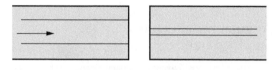

ELLIPTICITY (OVALITY)

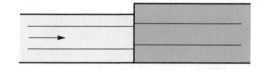

CLADDING DIAMETER MISMATCH

302F21.EPS

Figure 21 Intrinsic connection losses.

ally, the geometric axes of the core and cladding should coincide. The concentricity tolerance is the distance between the core center and the cladding center.

If the core or cladding is elliptical rather than circular, ellipticity losses may occur. Notice that the alignment of two elliptical cores varies depending on the angular dispositions. *Figure 21* shows two elliptical fibers joined with a maximum mismatch at 90 degrees. The ellipticity tolerance of the core and cladding equals the minimum diameter divided by the maximum diameter.

These fiber-related misalignment scenarios represent worst-case connections, and losses in general are not so severe. The probability of joining two fibers at the point of greatest diameter mismatch is small.

6.3.2 Extrinsic Factors

Connectors and splices also contribute losses to a system. When two fibers are not perfectly aligned by a connector, loss occurs even if there is no intrinsic variation in the fibers. Many different alignment mechanisms have been devised for joining two fibers.

A connector or splice must control the following four main causes of loss:

- Lateral displacement
- End separation
- Angular misalignment
- Surface roughness

To minimize lateral displacement, a connector should align the fibers on their center axes. When one fiber's axis does not meet with that of the other, loss occurs. With lateral displacement, the amount of signal loss depends on the ratio of the lateral offset to the fiber diameter. Thus, the acceptable offset decreases as the fiber diameter decreases.

An air gap between the fiber ends (end separation) causes two types of loss. Fresnel reflection loss is caused by the difference between the refractive indices of the two fibers and the intervening air gap. Fresnel reflection occurs at both the exit from the first fiber and at the entrance to the second fiber. Fresnel losses can be greatly reduced by using an index-matching fluid, such as an optically transparent gel with a refractive index the same as or near that of the fiber.

The second type of loss from end separation in multi-mode fibers results from the failure of high-order modes to cross the gap and enter the core of the second fiber. High-order rays exiting the first fiber miss the acceptance cone of the second fiber. Light exiting a fiber spreads out conically. The amount of separation loss, then, depends on the NA of the receiving fiber. A fiber with a high NA does not tolerate separation loss as well as a fiber with a lower NA.

Ideally, fibers should butt together with physical contact to minimize losses from end separation. However, a very small gap in connectors is sometimes desirable to prevent ends from rubbing and causing abrasion during reconnection. Fibers brought together with too much force may fracture. Therefore, some connectors are designed to maintain a very small gap between fibers. Others use spring pressure to bring fiber ends together gently without damage.

To prevent angular misalignment, the ends of mated fibers should be perpendicular to the fiber axes and parallel to each other during engagement. Angular misalignment losses result when one fiber is cocked in relation to the other. Again, the degree of loss depends on the fiber NA. The fiber face must be smooth and free of defects, such as hackles, burrs, and fractures. Surface irregularities can disrupt light rays and prevent them from entering the second fiber.

6.3.3 System-Related Factors

The loss at a fiber-to-fiber joint depends not only on the losses contributed by the fibers and connector, but on system-related factors as well. The performance of a connector in a system depends on modal conditions and the position of the connector in the system, because modal conditions vary along the length of the fiber.

In general, a connector inserted closer to the light source experiences higher signal losses than a connector farther down the link. This is due to the greater amount of light in high-order modes near to the light source. As light moves farther from the light source, mode-related losses tend to decrease.

6.4.0 Splices

Splices are used to make a permanent mechanical connection of optical fibers. A common and reliable technique for making these connections is called fusion splicing. A fusion splicer uses an electric arc to weld two glass fibers together. The fiber ends are prepared and placed on an alignment block in the splice machine. Micromanipulators bring the fibers together both axially and laterally. Alignment is observed through a built-in microscope or by monitoring the strength of the light signal across the junction during manipulation. Once aligned, the fiber ends are fused with an electric arc. Surface tension tends to align the fiber axes and minimize lateral displacement. The finished splice is then encased in epoxy or heat-shrink tubing to protect it.

Splicing techniques are examined in more detail later in this module.

6.5.0 Splitters

Thus far, the fiber optic link has been viewed as a point-to-point system (one transmitter linked to one receiver by an optical fiber). In many applications, however, it is desirable or necessary to divide light from one fiber into several fibers, or to couple light from several fibers into one fiber. This is accomplished with splitters, also known as passive optical nodes (PONs). A PON, or splitter, is a device that distributes light from one fiber to several other fibers. *Figure 22* shows the basic operation of a splitter.

6.5.1 Basic Splitter Theory

A splitter has several ports. Each port is an input or output point for light. *Figure 23* shows a schematic representation of a four-port directional splitter that will be used to discuss several ideas important to understanding splitters. Arrows indicate the possible direction of flow of optical power through the splitter. Light injected into port 1 will exit through ports 2 and 3. Ideally, no light will appear at port 4. Similarly, light injected into port 4 will also appear at ports 2 and 3 but not at port 1.

A splitter is passive and bidirectional. Ports 1 and 4 can serve as input ports, and ports 2 and 3 as output ports. Reversing the direction of the signal flow allows ports 2 and 3 to serve as input ports and ports 1 and 4 to serve as output ports.

Port 2 is called the throughput port and port 3 is known as the tap port. These terms are used to suggest that a path containing the greater part of the power is directed to the throughput port, whereas a path containing the lesser part is directed to the tapped port. Directionality is the ra-

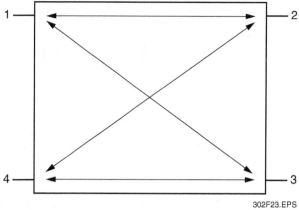

302F23.EPS

Figure 23 Four-port directional splitter.

tio between the unwanted power at port 4 and the input power at port 1. Directionality is sometimes called isolation. Ideally, no power appears at port 4, so the loss would equal zero. However, some power does appear at port 4, as a result of leakage or reflection.

Excess losses are losses that occur because the splitter is not a perfect device. Excess loss is the ratio between the output power at ports 2 and 3 to the input power at port 1. Losses can occur within the splitter from scattering, absorption, reflections, misalignment, and poor isolation. In a perfect splitter, the sum of the output power equals the input power (P2 + P3 = P1). In reality, the sum of the output power is always somewhat less than the input power because of the excess loss (P2 + P3 < P1).

The input power must obviously be divided between the two output ports. The splitting ratio is simply the ratio of signal division between the throughput port and the tap port: P2/P3. Typical ratios are 1:1, 2:1, 3:1, 6:1, and 10:1.

6.5.2 Splitter Configurations

This section discusses the physical and operational characteristics of splitters used to connect devices.

A tee splitter is a three-port device. *Figure 24* shows its application in a typical bus network. A splitter at each node splits off part of the power from the bus and carries it to a transceiver in the attached equipment. The throughput power at each splitter is much greater than the tap power.

As the number of terminals using tee splitters increases, losses mount quickly. As a result, tee splitters are only useful when a small number of terminals are involved.

Some networks are single-directional. A transmitter at one end of the bus communicates with a receiver at the other end. Each terminal also contains a receiver. A duplex network can be created either by adding a second fiber bus or by using

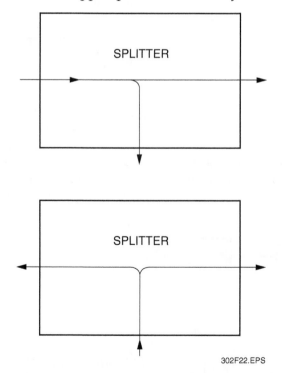

302F22.EPS

Figure 22 Operation of a basic fiber optic splitter.

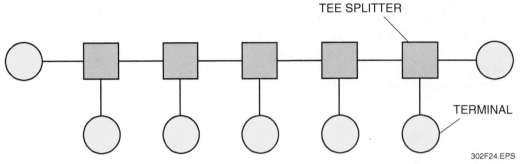

Figure 24 Tee splitter network.

an additional directional splitter at each end of a terminal. Such additions allow signals to flow in both directions. *Figure 25* shows an example. The failure of a single splitter does not shut down the entire network; it simply divides the network into smaller networks, one on each side of the failed splitter.

A star splitter is an alternative to a tee splitter and has additional benefits. *Figure 26* shows a transmissive star splitter that has an equal number of input and output ports. Light into any input port is equally divided among all the output ports.

The insertion loss of a star splitter is the ratio of the power appearing at a given output port to the power appearing at an input port. Thus, the insertion loss varies inversely with the number of terminals and does not increase linearly with the number of terminals. Star splitters are therefore more useful for connecting a large number of terminals to a network.

The actual amount of power into each output port varies somewhat from the ideal as determined from the insertion loss. The term *uniformity* is used to specify the variation in output power at an output port. Uniformity is expressed either as a percentage or in decibels.

A reflective star splitter contains a certain number of ports. Each port can serve as an input or an output. Light injected into any one port also appears at all other ports.

6.5.3 *Splitter Construction*

The input/output ports of most splitters are either fiber pigtails or connector bushings. The advantage of a pigtail is that it permits greater flexibility of application, because connections to pigtails can be made with any compatible splice or connector.

A fused star splitter is made by wrapping fibers together at a central point and heating the point. The glass will melt into a unified mass so that light from any single fiber passing through the fused point will enter all the fibers on the other side.

A transmissive star splitter results when an end of each fiber is on each side of the fused section. A reflective star has fibers that loop back so that each fiber is fused twice in the common section. Both types of fused star splitters are shown in *Figure 27*.

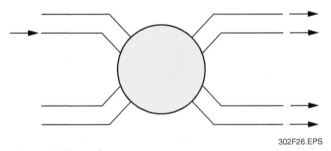

Figure 26 Star splitter.

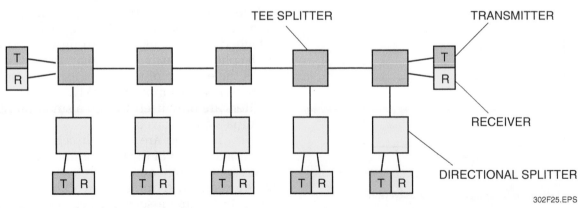

Figure 25 Tee network with directional splitter.

NCCER – Contren® Learning Series 33302-11

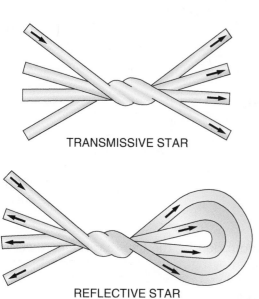

TRANSMISSIVE STAR

REFLECTIVE STAR

302F27.EPS

Figure 27 Fused star splitters.

A mixing rod splitter employs a relatively large rectangular or circular waveguide. The rectangular mixer resembles an optical fiber in that its top and bottom typically have a layer of material with a lower refractive index. A short length of large-core fiber can also serve as a mixer. Light from the core of an attached fiber mixes throughout the mixer.

In a transmissive mixing rod splitter, input fibers butt against one end of the rod and output fibers butt against the other end. Light injected into the mixing rod is mixed throughout the rod so that it enters all the fibers at the output end. In a reflective mixing rod splitter, all fibers butt against one end of the mixing rod. The other end is coated with a highly reflective material. Light from one fiber into the rod reflects back into all of the fibers. Both types of mixing rod splitters are shown in *Figure 28.*

6.5.4 Wavelength-Division Multiplexing Splitters

Multiplexing splitters are specialized components that send several different signals over a line simultaneously. Multiplexing allows telephone companies to send hundreds of telephone calls over a single fiber optic line.

Wavelength-division multiplexing (WDM) uses different wavelengths to combine two or more signals. Transmitters operating at different wavelengths can each inject their light signals into an optical fiber. At the other end of the link, the signals can be distinguished and separated by wavelength. A WDM splitter is used to combine separate wavelengths onto a single fiber or split combined wavelengths back into their component signals. *Figure 29* shows an example of a WDM system.

Crosstalk is a term for how well the de-multiplexed channels are separated. Each channel should appear only at its intended port and not at any other output port. The crosstalk specification expresses how well a splitter maintains this port-to-port separation.

Channel separation refers to the gap between the different wavelengths used in a fiber. In most splitters, the wavelengths must be widely separated, such as 850nm (nanometers) and 1,300nm. WDM devices do not distinguish between wavelengths as close as 1,140nm and 1,160nm.

6.5.5 Active Splitters

The splitters discussed so far have all been passive. A passive device is not powered, usually causes loss, and normally works bidirectionally. An active device, on the other hand, usually depends on some power or actuator to cause a shift in the device's state.

An active splitter consists of a light source, a photodetector, a beam divider, a focusing mirror, and a pigtailed fiber. Light from the source is coupled into the fiber. Light from the fiber passes through the beam divider and is directed by the focusing mirror onto the photodetector.

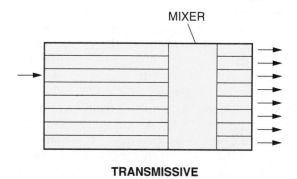

MIXER

TRANSMISSIVE

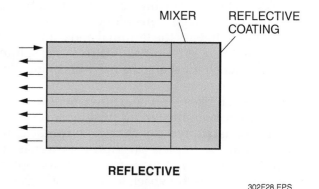

MIXER

REFLECTIVE COATING

REFLECTIVE

302F28.EPS

Figure 28 Mixing rod splitters.

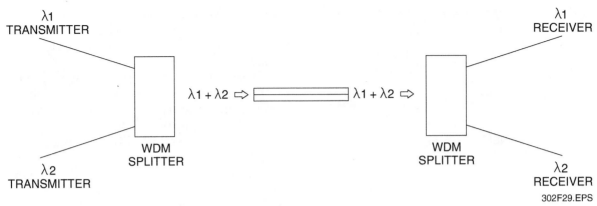

Figure 29 Wavelength-division multiplexing.

302F29.EPS

Incorporating both a source and a detector along with coupling mechanisms in a single package allows transmission in both directions along a single fiber. Operation is fully duplex, because both the source and detector can operate simultaneously. Electrical leads for the optoelectronic device permit mounting on a printed circuit board.

6.5.6 Optical Switches

It is sometimes necessary to couple light from one fiber to either of two end fibers, but not to both. This is not possible with a passive splitter because the division of light is always the same. An optical switch can be used in these applications. Like an electric switch, it connects one of two circuit paths, depending on the switch setting.

Some computer networks are configured as a ring in which the terminals are daisy-chained. In other words, the transmitter of the first terminal connects to the receiver of the second terminal. The second terminal's receiver feeds the transmitter, which connects to the receiver of the third terminal. This arrangement continues from terminal to terminal around the ring. A ring network has the advantage of a simple arrangement. The main disadvantage of this ring arrangement is that failure of a single terminal shuts down the entire network.

A fiber optic bypass switch overcomes this problem. The two settings of the switch permit the light signal to be transmitted to the terminal receiver or to bypass the terminal and continue on the ring to the next terminal. A directional splitter after the switch must also be used in conjunction with the switch.

The switch uses a relay arrangement to move the fiber between the circuit positions. A switch can be constructed so that it automatically switches to the bypass position if power is removed. The result is a measure of fail-safe operation. The general drawback to optical switches is the difficulty in minimizing signal losses. Normal problems with maintaining alignment are compounded by the fact that the connection involves moving parts.

7.0.0 INSTALLATION

> **WARNING**
>
> Working with fiber optic systems can be hazardous. The glass of optical fibers can damage eyes and skin, and can cause internal injury. Always wear appropriate protective equipment when handling and installing fiber optic systems. Avoid touching optical fiber with your fingers. Never eat or drink in an area where optical fiber is being handled.

Due to their light weight and extreme flexibility, fiber optic cables are often easier to install than copper cables. They are relatively easy to handle and can be pulled over greater distances.

The minimum bend radius and maximum tensile rating are the critical specifications for any fiber optic cable. The minimum bend radius is specified by the manufacturer, and depends on whether the cable is in tension. A typical minimum bend radius is 10 times the diameter for cables not in tension and 20 times the diameter for cables in tension. Careful planning of the layout and installation will ensure that these specifications are not exceeded. This section discusses the factors involved in planning an installation.

> **NOTE**
>
> The maximum pulling tension allowed during installation is higher than the allowable static load after installation. Likewise, the minimum bend radius allowed during installation is larger than the static allowance. One reason is that the minimum bend radius increases with pulling tension. Because the fiber is under load during installation, the minimum bend radius must be larger. The allowable bend radius after installation depends on the static tensile load.

Figure 30 shows cross sections of both simplex and duplex cables, both of which are commonly used in indoor applications. As discussed earlier, outdoor cables are typically multi-fiber cables with a much more rugged construction than indoor cables.

7.1.0 Direct and Indirect Burial Installation

Cables can be buried in the ground through plowing, trenching, or directional drilling. The plowing method uses a cable-laying plow, which opens the ground, lays the cable, and buries it in a single operation. In the trenching method, a trench is dug with a machine such as a backhoe. Then, the cable is laid, and the trench is filled. The trench method is better suited to short-distance installations.

Horizontal directional drilling is a technique of drilling a horizontal passageway for conduit with minimal disturbance of the surrounding environment. This method can be used for burials of a few feet to more than 1,500 meters, and is becoming increasingly popular for cabling residential and scenic areas. *Figure 31* shows a horizontal directional drilling system and conduits.

Buried cables must be protected against frost, water seepage, attack by burrowing and gnawing animals, and mechanical stresses that could result from earth movement. Cables should be buried at least 30" deep so they are below the frost line, and should be either specially designed for direct burial (without conduit), or enclosed in sturdy polyurethane or PVC conduit (indirect burial). The conduit should have an inside diameter that is several times the outside diameter of the cable to protect against earth movement. A small amount of excess cable length in the conduit helps reduce pulling tension on the cable.

7.2.0 Aerial Installation

Aerial installation includes stringing cables between utility poles. Unlike copper cables, fiber optic cables may be run along power lines with no danger of inductive interference.

Aerial fiber optic cables must be able to withstand the forces of high winds, extreme temperatures, and ice loading. Self-supporting aerial cables can be strung directly from pole to pole. Other cables must be lashed to a high-strength steel wire, which provides necessary support, or they must be used with a separate support structure.

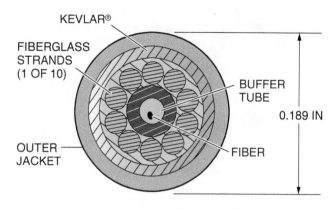

SIMPLEX CABLE CROSS-SECTION

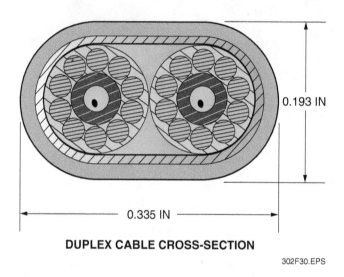

DUPLEX CABLE CROSS-SECTION

302F30.EPS

Figure 30 Simplex and duplex cable cross sections.

7.3.0 Indoor Installation

Most indoor cables are installed in conduit or trays. Because standard fiber optic cables are electrically nonconductive, they may be placed in the same ducts as high-voltage cables without the special insulation required by copper wire. Many cables cannot, however, be placed inside air conditioning or ventilation ducts for the same reason that PVC-insulated wire should not be placed in these areas. A fire inside these ducts could burn the cable covering, sending toxic gases throughout the structure.

Plenum cables, however, are specified to be run in any plenum area within a building without special restrictions. The material used in these cables produces less smoke, lower toxic fumes, and burns more slowly than non-plenum cable when it burns.

SMALL RIG

MEDIUM RIG

302F31.EPS

Figure 31 Horizontal directional drilling system and conduits.

7.4.0 Tray and Duct Installation

The first mechanical property of the cable that must be considered when planning an installation is the outside diameter of the cable and connectors. If the cable must be pulled through a conduit or duct, the minimum cross-sectional area required is $0.769" \times 0.43"$ ($20mm \times 11mm$) for duplex cable. For simplex cable, the minimum cross-sectional area is determined by the pulling grip used.

The primary consideration in selecting a route for fiber optic cable through trays and ducts is to avoid potential cutting edges and sharp bends. Areas where special caution must be observed are corners and exit slots in the sides of trays, as shown in *Figure 32*.

If a fiber optic cable is in the same tray or duct with heavy electrical cables, care must be taken to avoid crushing the fiber optic cable where the heavy cables cross over. In general, cables in trays and ducts are not subjected to pull tension; however, keep in mind that in long vertical runs the weight of the cable itself will create a tensile load of approximately 0.16 lb/ft (0.25N/m) for simplex cable and 0.27 lb/ft (0.44N/m) for duplex cable. This load must be considered when determining the minimum bend radius at the top of the vertical run.

Long vertical runs should be clamped at intermediate points, preferably every 3' to 6', to prevent excessive tension on the cable.

7.5.0 Conduit Installation

Fibers are pulled through conduit by a wire or synthetic rope attached to the cable. Any pulling forces must be applied to the cable members and not to the fiber or jacket. For cables without connectors, pull wire can be tied to Kevlar® strength members, or a pulling grip can be taped to the cable jacket or sheath if approved by the cable manufacturer. Whenever possible, use an automated puller with tension control and a swivel pulling eye. Use extra caution when pulling cables with connectors installed.

Enough clearance must be available in a conduit to allow the fiber optic cable to be pulled through without excess friction or binding. As a general rule, the pulling force should not exceed 90 pounds (400.5N). Note that this is a general rule. Manufacturers will specify the pulling tension that may be used on their cables. Because the minimum bend radius increases with increasing pulling force, bends in the conduit should not have a radius of less than 5.9" (150mm). Fittings in particular should be checked carefully to ensure that they will not cause the cable to make sharp bends or be pressed against corners. If the conduit must make a 90-degree turn, a fitting like the one shown in *Figure 33* must be used to allow the cable to be pulled in a straight line and to avoid sharp bends in the cable.

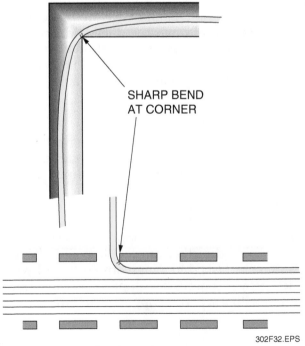

SHARP BEND AT CORNER

302F32.EPS

Figure 32 Sharp radius hazards.

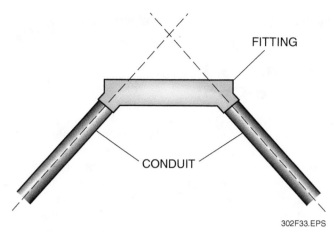

Figure 33 Turn fitting.

For outdoor underground installations, pull boxes should be used on straight runs at intervals of 250' to 300' for outside plant. Inside plant runs are generally limited to 100' to 150' between pull points for straight runs. This reduces the length of cable that must be pulled at any one time, thus reducing the pulling force. Also, pull boxes should be located in any area where the conduit makes several bends that total more than 180 degrees. To guarantee that the cable will not be bent too tightly while pulling the slack into the pull-box, use one with an opening that is equal to at least four times the minimum bend radius (4.75" or 120mm). *Figure 34* shows the correct form for pulling the last length of slack into the box.

The tensile loading effect of vertical runs discussed in the section on tray and duct installations is also applicable to conduit installations. Because it is more difficult to properly clamp fiber optic cables in a conduit than in a duct or tray, long vertical runs should be avoided, if possible.

7.6.0 Blown Fiber

Another method of installing fiber optic cables uses pressurized air, or another pressurized gas, to carry the fiber optic cable through small-

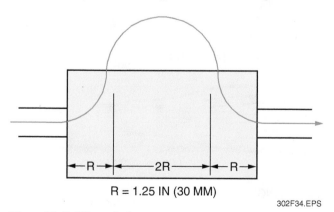

R = 1.25 IN (30 MM)

Figure 34 Pull box slack.

diameter tubes. Individual tubes may be spliced or joined together to route or re-route the fiber optic cables to particular points. The individual tubes may be bundled and then separated at distribution points along the route to install the fiber optic cable at the proper points. Only one cable may be installed in a single tube, but there can be 24 or more fiber optic strands in a single cable. Additionally, there may be as many as 19 tubes in a single bundle and as many as eight 19-tube bundles in a single 4" conduit or duct.

The major disadvantage with blown fiber is that special equipment and training are usually required for the installation of both the tubes and the fiber optic cables. The distance that these special fiber optic cables can be blown normally does not exceed about 2,000 feet, according to most manufacturers' specifications.

Unlike conventional fiber optic cable, fiber bundles can be blown out of the fiber pathway undamaged and may be immediately recycled and reused in the network, creating a continuously renewable and sustainable network infrastructure with no end to its life cycle. Network upgrades, expansions, and reconfigurations require no construction work, thereby eliminating waste and debris, as well as hazardous abandoned cable, unused dark fiber, and other environmentally compromising materials. Moreover, a blown fiber infrastructure takes up less building space and provides greater capacity, thereby allowing HVAC and other energy systems to operate with unobstructed air flow.

7.7.0 Pulling Fiber Optic Cables

Fiber optic cables are pulled using many of the same tools and techniques used in pulling copper cable. The departures from standard methods are due to three factors: the connectors are usually pre-installed on the cable; smaller pulling forces are allowed; and the minimum bend radius requirements differ.

The pulling tape or line must be attached to the strength members of the cable and not the connectors or fibers themselves. The recommended attachment method for simplex cable is the finger trap cable grip shown in *Figure 35*. The connector should be wrapped in a thin layer of foam rubber and inserted in a stiff polymer sleeve for abrasion and shock protection.

Duplex cable is typically supplied with Kevlar® strength members extending beyond the outer jacket to provide a means of attaching the pulling tape, as shown in *Figure 36*. The strength members are epoxied to the outer jacket and inner layers to prevent twisting. The free ends of

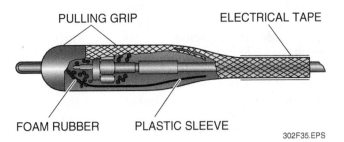

Figure 35 Simplex pulling grip.

the Kevlar® fibers are inserted into a loop at the end of the pulling tape and then epoxied back to themselves. The connectors are protected by foam rubber and a heat-shrink sleeve. The heat-shrink sleeve is clamped in front of the steel ring in the pulling tape to prevent pushing the connectors back into the rest of the cable.

> NOTE
>
> When pulling cables with connectors attached, be sure the link between the pull tape and the cable protects the connectors, and that pulling loads are applied to the cable's strength members and not the fibers or connectors themselves.

Constantly monitor the pulling force with a gauge while the cable is being pulled. If any increase in pulling force is noticed, immediately stop pulling and determine the cause of the increase. The pulling tension can be monitored by a running line tension meter or a dynamometer and pulley arrangement. If a power winch is used to assist the pulling, a power capstan with an adjustable slip clutch is recommended. The clutch, which is set for the maximum load, disengages if the set load is reached.

On Site

Coiling a Fiber Optic Cable

To find the natural coil of a cable that you are going to twist into a figure eight, grasp the cable with both hands, with your hands as far apart as you can reach. With the cable firmly held in each hand, bring your hands together. The cable will naturally coil. Now, start to pull your hands apart, making the coil tighter and tighter. At some point, this natural coil will start to open. This diameter is the natural coil of this cable and the cable will not kink or be deformed if the figure-eight diameters do not fall below this natural diameter.

If necessary, the cable should be continuously lubricated. At points such as pull boxes and man-holes, where the cable enters the conduit at an angle, a pulley, sheave, or bull wheel should be used. This will help ensure that the cable does not scrape against the end of the conduit or make sharp bends.

As the cable emerges from intermediate-point pull boxes, it should be coiled in a figure-eight pattern with loops at least 1' in diameter for simplex and duplex cable and proportionally larger diameters for larger cables. This prevents tangling or kinking of the cable. When all the cable is coiled and the next pull is to be started, the coil can be turned over and the cable laid out from the top. This will eliminate twisting of the cable. The amount of cable that has to be pulled at a pull box can be reduced by starting the pull at a pull box as close as possible to the center of the run. Cable can then be pulled from one spool at one end of the run; then the remainder of the cable can be unspooled, coiled in a figure-eight pattern, and pulled to the other end of the run.

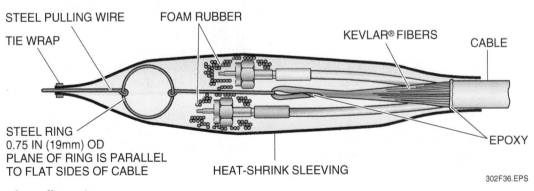

Figure 36 Duplex pulling grip.

7.8.0 Enclosures and Organizers

There are several types of enclosures, racks, and organizers used in fiber optic installations. Each is designed to protect system components and connections, and to make the job of installing a fiber optic system somewhat easier.

A splice closure is a standard piece of hardware used to protect cable splices. They can be seen on aerial telephone runs, although they are also used in underground applications. Splices are protected mechanically and environmentally within the sealed closure. The body of the closure serves to join the outer sheaths of the two or more cables being joined.

Organizers are designed to secure splices within a closure. In a typical application, the cable jacket is removed to expose the fibers at the point where they enter the closure. The exposed fiber should loop one or more times around the organizer, as shown in *Figure 37*. Such routing provides extra fiber for resplicing or rearrangement of splices. Closures typically hold one or more organizers to accommodate 12 to 144 splices.

Closures have two main applications. The first is to protect the splicing of two or more cables when an installation span is longer than a single cable length. The second is to protect a point where cable types have been switched for various reasons. For example, a 48-fiber cable can be brought into one opening of the closure. Four 12-fiber cables, all going to different locations, can be spliced to the first cable and brought out another opening of the closure.

7.9.0 Distribution Hardware

When installing a large fiber optic system, the individual fibers eventually must be routed to individual terminal locations. There are various types of hardware that facilitate transitions from one fiber type to another or the distribution of fibers to different points.

In indoor applications, a wiring center is used as a central distribution point. From this point, fibers can be routed to their destinations. A typical example is an outdoor multi-fiber cable brought into a building to the distribution point. At the wiring center, each fiber in the outdoor cable is spliced to a simplex, duplex, or multi-fiber cable that is routed to different locations.

A rack box containing an organizer serves such distribution needs. The boxes are mounted in 19" and 23" equipment racks, which are standard sizes in the telephone and electronics industries (*Figure 38*). Besides the organizer, rack boxes contain provisions for securing fibers and strain-relieving cables as well as storage space for service loops.

302F37.EPS

Figure 37 A splice organizer.

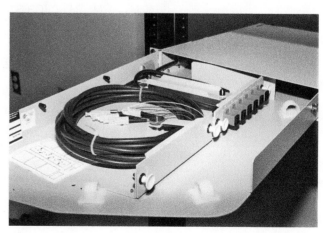

302F38.EPS

Figure 38 A rack-mounted fiber termination tray.

Patch panels provide a convenient way to rearrange fiber connections and circuits. A simple patch panel is a metal or plastic frame containing bushings on either side. The fiber optic connectors are plugged into the bushings. One side of the panel is usually fixed, meaning that the fibers are not intended to be disconnected. On the other side of the panel, fibers can be connected and disconnected to arrange the circuits as required (*Figure 39*). Like rack boxes, patch panels are compatible with 19" and 23" equipment racks. A variety of factory-assembled patch cords are available for use in fiber optic systems. For testing purposes, there are many types of loopback devices available that are fitted with all the common connector types (*Figure 40*). Additionally, there are adapter-type patch cords with different connectors on each end (*Figure 41*).

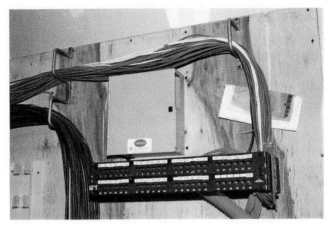

302F39.EPS

Figure 39 A rack-mounted fiber patch panel.

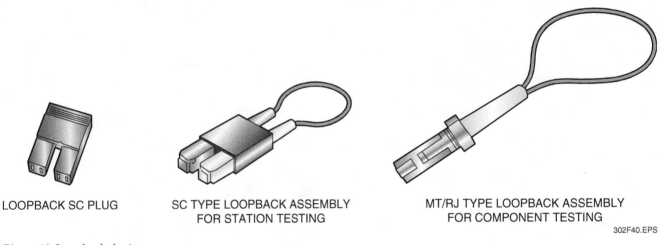

LOOPBACK SC PLUG

SC TYPE LOOPBACK ASSEMBLY
FOR STATION TESTING

MT/RJ TYPE LOOPBACK ASSEMBLY
FOR COMPONENT TESTING

302F40.EPS

Figure 40 Loopback devices.

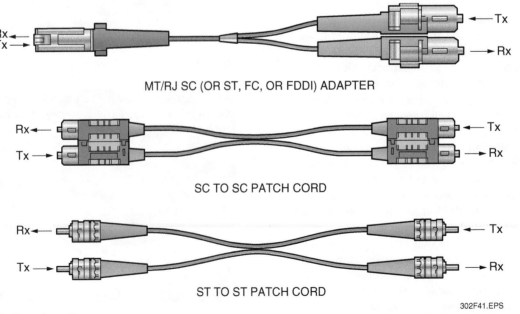

MT/RJ SC (OR ST, FC, OR FDDI) ADAPTER

SC TO SC PATCH CORD

ST TO ST PATCH CORD

302F41.EPS

Figure 41 Typical patch cords.

A junction box for fiber optic systems can be thought of as similar to an electrical outlet. A home or office is wired for electrical power with cable running between walls or under floors. Electrical equipment plugs into this power at a junction box containing an electrical outlet. *Figure 42* shows one type of fiber optic junction box. The box's cover plate contains two bushings for fiber optic connectors. One side accepts the building's fiber optic cable; the other side accepts the drop cable from the room. When installed, the room side of the bushing faces down to help prevent dust contamination when not in use.

8.0.0 TERMINATING OPTICAL FIBER CABLE

The quality of the fiber terminations in a system is critical. The job of a terminal connector is to provide a low-loss coupling of light energy with a terminal device or connection. To achieve low loss, the core of a fiber must be precisely aligned with another fiber or with the active area of a source or detector. An appreciable loss of power can occur at a cable connection if it is not properly assembled. For that reason, every fiber optic cable interconnection must be made using the correct tools and techniques.

8.1.0 Mechanical Considerations

Durability is a concern with any kind of connector. Repeated mating and unmating of fiber connectors can wear mechanical components, introduce dirt into the optics, strain the fiber and other cable components, and even damage exposed fiber ends. Typical connectors for indoor use are specified for 500 to 1,000 mating cycles, which should be adequate for most use.

302F42.EPS

Figure 42 Junction box.

Connectors are attached to cables by forming mechanical and/or epoxy bonds to the fiber, cable sheath, and strength members. The connection is adequate for normal wear and tear, but not for sudden, sharp loading, as when someone trips over a cable. A sharp tug can detach a cable from a mounted connector because the bond between connector and fiber is the weakest point. The same is true for electrical cords; the best way to address the problem is to be careful with the cables.

Care should be taken to avoid sharp kinks in cables at the connector. Fibers are particularly vulnerable if they have been nicked during connector installation. Care should also be taken to ensure that fiber ends do not protrude from the ends of connectors. If fiber ends hit other fibers or objects, they can easily be damaged, increasing attenuation.

Most fiber optic connectors are designed for use indoors, where they are protected from environmental extremes. Connectors and patch panels come with protective caps for use when they are not mated. Be sure to use the caps at all times to prevent contamination of the connectors.

Special hermetically sealed connectors are required for outdoor use. As you might expect, those designed for military field use are by far the most durable. Military field connectors are bulky and expensive, but when sealed they can be left on the ground, even exposed to mud and moisture. Generally it is best to avoid outdoor connectors wherever possible, or house them in enclosures that are sealed against dirt and moisture.

8.2.0 Basic Connector Structure

Most fiber optic connectors in use today have some common elements, as shown in *Figure 43*. The fiber is mounted in a long, thin cylinder called a ferrule, with a hole sized to match the fiber cladding diameter. The ferrule centers and aligns the fiber and protects it from mechanical damage. The end of the fiber is polished flush with the end of the ferrule. The ferrule extends from the connector body, which is attached to the cable structure. A strain-relief boot shields the junction of the connector body and the cable.

Standard fiber connectors lack the male-female polarity common in electronic connectors. Instead, fiber connectors mate in adapters, often called coupling receptacles or sleeves, that fit between and join the two fiber connectors.

Ferrules are typically made of metal, ceramic, or plastic. The hole through the ferrule must be large enough to accept the clad fiber but tight enough to hold it in a fixed position. Adhesive

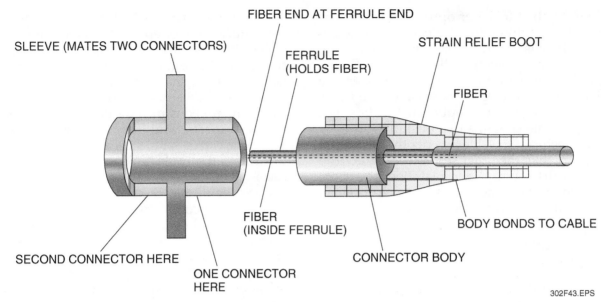

FIBER END AT FERRULE END

SLEEVE (MATES TWO CONNECTORS)

FERRULE (HOLDS FIBER)

STRAIN RELIEF BOOT

FIBER

SECOND CONNECTOR HERE

ONE CONNECTOR HERE

FIBER (INSIDE FERRULE)

CONNECTOR BODY

BODY BONDS TO CABLE

302F43.EPS

Figure 43 Parts of a fiber connector.

is typically put into the hole before the fiber is pushed in. The fiber tip may be pushed slightly past the end of the ferrule, then polished down to a smooth face.

The ferrule may be slipped inside another hollow cylinder, also called a sleeve, before it is mounted in the connector body. The body, typically made of metal or plastic, includes one or more pieces that are assembled to hold the cable and fiber in place. Details of assembly vary among connectors; cable bonding usually bonds the connector with strength members and the jacket. A strain-relief boot is slipped over the cable end of the connector to protect the cable connector junction.

8.3.0 Connector Installation: Field versus Factory

Fiber users face an important tradeoff in deciding where and how to install connectors. The tight dimensional tolerances needed for minimizing losses are easier to achieve in a factory environment. However, field installation gives installers much more flexibility, allowing on-the-spot repairs and modifications. Each approach has its advantages, and connector manufacturers have taken some steps to offer users the best of both worlds.

A big advantage of factory installation is that it is the cable supplier's responsibility to install the connector correctly. Factory technicians mount and test the connectors in a controlled environment using sophisticated test equipment. Generally, they can mass-produce standard lengths of connectorized cable economically. That is fine for short jumper cables used in patch panels, but

it is more difficult to supply the many different lengths needed for intra-building cable runs.

An intermediate step is to supply cable segments with factory-mounted connectors on one end and fiber pigtails on the other. These pigtails can be spliced to cables in the field using mechanical or fusion splices. This is a quick and easy approach using splicing equipment that many field technicians already have. Factory polishing makes connector losses low, and many types of connectors can be used. However, it requires additional splicing hardware and can add to cable costs.

Field installation of the complete connector enhances flexibility and has low consumable costs. Labor costs may also be low, depending on the location, but installation results depend on both the skill of the technician and the connector design. Some manufacturers offer field connector kits with some of the most sensitive alignments already completed.

8.4.0 Fiber Connectors

A variety of connectors emerged within a few years of the introduction of fiber optics. Many of the early designs have faded from the scene, but some have survived, and remain in production today. Standardization narrowed the field considerably. The International Standards Organization (ISO), the International Electrotechnical Commission (IEC), and the Telecommunications Industry Association (TIA) endorsed a design called the subscriber connector (SC). A second type, the straight terminus (ST) connector, has been widely used in data communication applications and the telephone industry. These connectors remained the industry standard until the introduction of small form factor dual-fiber connectors such as LC connectors. These connector ferrules are half the diameter of an SC or ST connector and allow doubling the density of fibers in a patch panel or junction box, which means reduced equipment size. A duplex connector developed for local area networks, called the fiber distributed data interface (FDDI), has also become an accepted standard. The small form factor connector is often used in designs where the ST or SC connector would otherwise have been the choice. *Figure 44* shows the connectors in common use.

The following is an overview of common connector types and their characteristics. The alpha designations are interpreted as follows:

- ST – Straight terminus
- SC – Subscriber connector
- LC – Lightwave connector
- MT-RJ – Mass termination RJ-45 style
- MTP – Multiple termination-preterminated (used with ribbon fiber to adapt to individual connectors)
- MPO – Multiple preterminated optics (used with ribbon fiber)

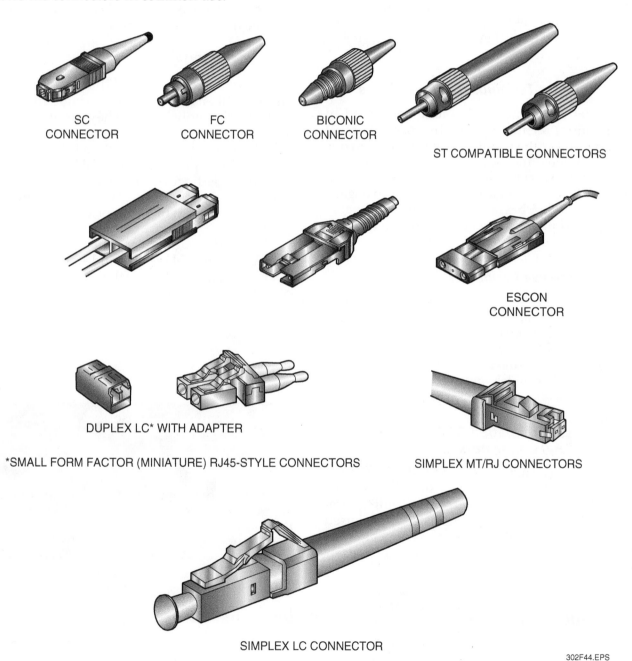

302F44.EPS

Figure 44 Standard connector types.

8.4.1 Subscriber Connectors (SCs)

Like most modern connectors, the SC is built around a 2.5mm diameter cylindrical ferrule that holds the fiber and mates with an interconnection adapter or coupling receptacle. It has a square cross section that allows high packing density on patch panels. The polarized duplex version has been adopted as a standard by ISO and TIA. Pushing the connector latches it in place, without any need to turn the connector in a tight space. It is better able to withstand pulling forces than ST connectors.

SCs have many variations. Most of the variation exists in the boot and ferrule to enable easy assembly in the field. When these connectors have an APC designation, they provide low return loss after polishing.

8.4.2 Straight Terminus (ST) Connectors

The older ST connector has long been used in data communications. Like the SC, it is built around a 2.5mm cylindrical ferrule and mates with an interconnection adapter or coupling receptacle. However, it has a round cross section and is latched into place by twisting it to engage a spring-loaded bayonet socket. Some variants on the ST connector can be latched into place by pushing them.

8.4.3 Lightwave Connectors (LC)

The lightwave connector (LC) is a push-and-latch-type connector, similar to an RJ-45 connector. It is like a miniature version of the SC. The small size is achieved by the use of a 1.25mm ferrule instead of the 2.50mm ferrule used on the SC, FC, and ST connectors. It is widely used in single-mode fiber applications.

8.4.4 Mass Termination RJ-45 (MT-RJ) Connectors

The mass termination RJ-45 (MT-RJ) style connector is a plug-and-jack-type connector like the RJ-45. It is used extensively in intra-building communication systems. It is a duplex connector used in multi-mode applications. The connector has two guide pins that help to precisely align the ferrules.

8.4.5 Fiber Distributed Data Interface (FDDI) Connectors

The fiber distributed data interface (FDDI) standard duplex connector has gained acceptance for duplex connections. Like the ST and SC, it is a dual 2.5mm ferrule-based connector that mates with a coupling receptacle or adapter. It is keyed so that it can be installed in only one polarity, which is critical for ensuring correct input/output connection.

8.4.6 Face Contact (FC) Connectors

The face contact (FC) connector is a screw-on type developed in the early 1980s that uses the same 2.5mm ferrule size as the ST and SC connectors. Its optical losses are similar to those of the other two types. Like the SC connector, it can resist pulling forces. However, the screw-on design cannot be mounted as easily and cannot be used as a module in duplex connectors.

8.4.7 MTP/MPO Connectors

These connectors (*Figure 45*) are made for use on multi-fiber ribbon cable in single-mode and multi-mode applications. They are used to terminate 6 to 48 strands in groups of six. Connectors used on e-mode connectors have an angled ferrule, which provides minimal back-reflection. Unlike single-fiber connectors, they are not glue and polish types. Rather, they are terminated using either a pigtail splice or cleave and crimp connection.

The multi-mode version usually has a flat ferrule. A ribbon cable is flat and has side-by-side fibers surrounded by a jacket (*Figure 46*). These connectors have an insertion loss for matched MTP/MPO connectors of about 0.25dB.

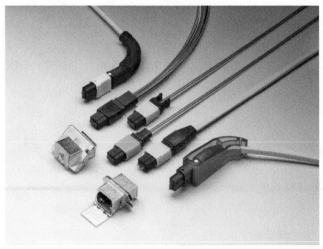

302F45.EPS

Figure 45 MTP- MPO connectors.

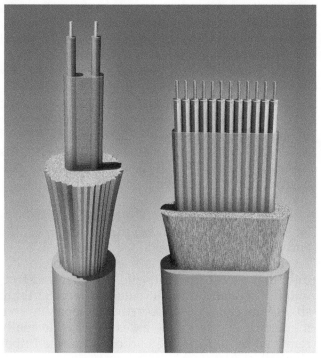

Figure 46 Ribbon cable.

302F46.EPS

8.4.8 Other Connector Formats

The catalogs of many fiber optic connector manufacturers list additional styles of connectors. Some, like the SMA, Mini-BNC, and D4, look like the SC, but are not necessarily compatible. Others types have their own distinct appearance. The use of coupling receptacles or adapters makes it possible to use these specialized connectors with standard types.

8.5.0 Connector Installation

The fiber termination process depends on whether the individual fibers or ribbon cables

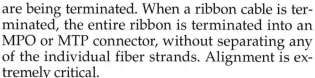

On Site

Hot-Melt Connector

One of the many innovative methods developed to reduce the time required to make a fiber connection in the field is the 3M™ hot melt connector. This connector arrives from the factory with adhesive pre-injected into it. The connector is placed into a hot oven, which melts the adhesive. The fiber is then inserted into the connector. Once it has cooled, the connection is ready to polish.

are being terminated. When a ribbon cable is terminated, the entire ribbon is terminated into an MPO or MTP connector, without separating any of the individual fiber strands. Alignment is extremely critical.

MPO/MTP connectors are impractical to cleave and polish in the field. It can be done, but it is very tedious and requires very exacting operations, especially for polishing. When and if MTP or MPO connectors must be terminated in the field, the crimp-on style is the most practical, because of the tricky and time-consuming polishing procedures. If any single fiber is out of alignment, the connector must be cut off and the process started all over again.

The primary advantage of the MPO/MTP connectors is density. These connectors have rows of 6 or 12 strands terminated without separating them. Some have as many as 24 strands in a single row, and may have as many as three rows of 24 strands. They are all connected at one time and disconnected at one time. Their primary interface is into a jack field that separates them into single strand connectors at a patch bay. The backbone cable can be moved from one patch bay to another in a matter of minutes, with a much lower chance of damage or broken fibers, compared to hours to disconnect and reconnect the same quantity of individual ST, SC, FC, or LC connectors.

There are a number of ways to terminate individual fibers in a connector. Each method has its own particular termination procedure:

- Hot-melt and polish connectors contain a heat-sensitive adhesive to attach the fiber optic strand into the connector.
- Anaerobic polish connectors use a two-part epoxy adhesive to attach the fiber optic strand to the connector.
- Epoxy polish connector adhesive uses ultraviolet light to cure the epoxy adhesive.

- There are various mechanical or crimp-on connectors that contain a very short piece of fiber already installed in the connector ferrule and a polished end-face. An index-matching gel allows the new fiber to be installed without polishing the end-face of the fiber optic strand. The individual fiber optic strands do require a length of the buffer removed, cleaned, and cleaved for proper mating in the connector.
- Pig-tail connectors are pre-polished with a short piece of a fiber optic strand (as short as a few inches or a few feet) which has been pre-installed and polished into a connector, with the new fiber optic strand spliced onto the fiber cable.

Physical contact (PC) connectors are the most commonly used connectors; however the angle physical contact (APC) connector has gained popularity, especially for long-distance and high-bandwidth applications. The APC is the primary connector used for broadband communications due to the two-way transmission of light on a single fiber and the fact that any reflected signal at a connector pair is sent off into the cladding and will not provide any appreciable interference in this type of signal. An ultra-polished connector (UPC) or super polish connector (SPC) is primarily manufacturer advertising to enhance their product line. UPC and SPC may provide a very slight lower loss at the mated pair, but their installation requires more polishing steps and is harder to achieve.

The termination of individual fibers begins with removal of the cable sheath using the proper stripping tool (*Figure 47*). The next step is to locate the proper buffer tube (if applicable), then locate the proper fiber optic strand or ribbon.

For all connectors that require insertion of a fiber optic strand into an open-ended connector ferrule, that fiber optic strand must be cleaved and polished as part of the termination technique. Before a fiber is cleaved, the strain relief is installed, and the outer coatings, strength members, and strand buffer are removed to a specific length. This length varies from one manufacturer to another. Once the bare clad fiber optic strand has been exposed and cleaned, the bare fiber is inserted into the fiber ferrule with whatever adhesive is required by the particular manufacturer and the connector type.

There should be a certain length of fiber optic strand (an inch or more) protruding from the end face of the ferrule. Once the adhesive has set, a scribe (*Figure 48*) is used to scribe, or cut into, that excess fiber close to the fiber end-face, without scratching the ferrule end-face. Scribing should not cause the excess fiber to separate from the ferrule end-face. The excess fiber is normally removed by pulling the end of the scribed fiber

302SA01.EPS

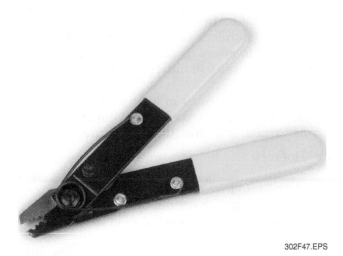

302F47.EPS

Figure 47 Fiber optic stripping tool.

Figure 48 Fiber optic scribe tool.

straight up from the ferrule end-face and discarding it into a proper waste receptacle.

After the scribing process, a very short piece of fiber, less than ¹⁄₁₆", should protrude from the fiber end-face. It is then polished so that the fiber end-face and ferrule end-face are smooth, flush, and free of scratches and dirt.

Once a fiber has been cleaved, it must be polished. Polishing grinds the fiber ends to create an acceptable finish on the fiber ends before they are mated. The hand polishing technique is the most popular for field polishing. The start of any polishing process is called air polishing. Air polishing is done in the open air with only a small amount of pressure applied onto the polishing paper by the fiber optic strand. Air polishing is used only to take the fiber optic strand-end down to very near the ferrule end-face. A few microns of excess fiber should still be visible at the ferrule end-face.

From that point, polishing of the end-face varies by manufacturer. The purpose of all polishing techniques is to remove any surface irregularities from the end-face of the fiber optic strand, and to make the end-face of the fiber optic strand and the ferrule end-face perfectly smooth and even. The ferrule is much harder than the fiber optic strand. Therefore, if the end-face is over-polished, the fiber optic strand may be polished below the ferrule end-face causing additional reflections at the mated pair. Under-polishing leaves the fiber optic strand protruding from the ferrule end-face, which may result in the fiber optic strand being crushed by the mating fiber during entry into a connector.

> **NOTE**
>
> APC connectors are seldom polished in the field by hand. Keeping the APC connector at the proper 7- to 9-degree angle, on the correct axis requires a special polishing puck and above average polishing techniques.

Upon completion of the polishing operation, the end-face of the ferrule should be examined using a 100- to 200-power microscope that uses both coaxial and then oblique lighting. Coaxial lighting shines around the ferrule and will show irregularities and dirt on the end-face. Oblique lighting shines at an angle across the end-face and will show scratches and polish of the end-face of the cladding and core of the fiber optic strand. The oblique lighting will also show dirt particles on the fiber optic end-face. If any dirt is present, nearly-dry, lint-free cleaning cloths should be used to clean the end-face to eliminate cleaner residue from the fiber, which can cause light distortion.

Another method of inspecting the fiber and ferrule end-face is the use of an interferometer (*Figure 49*). An interferometer is an electronic device that uses reflected light to examine the end-faces. It can show if the fiber is in fact at the center of the ferrule; that the ferrule end-face has the proper curvature; is flat at the mating of the fiber optic strand end-face; and the fiber does not protrude from, or is not recessed into the ferrule.

Once the end-face is polished and clean, the link should be examined with a light source/ light meter pair, and an OTDR. The OTDR is used first to display and record the light reflections along the entire path of the individual fiber optic strand. Although an OTDR does not give an actual reading of amount of light loss along the individual fiber optic strand, it estimates the amount of light loss, and it shows where excess losses are present. A trained eye can interpret mated connector pairs, splice points, and even points where there is a micro or macro bend in the fiber.

A light source/light meter pair shows the amount of light loss from end-to-end of a fiber optic strand. It does not show where a splice point is, or where there are mated pairs, but it measures and records the losses imposed (overall) by these points within the fiber optic strand length.

302F49.EPS

Figure 49 Interferometer.

9.0.0 SPLICING

Splices permanently weld, glue, or otherwise bond together the ends of two fibers. Like fiber optic connectors, splices are functionally similar to their wire counterparts. However, as with connectors, there are important differences between splicing wires and splicing optical fibers.

It is important for the electronic system technicians to understand when and why optical fibers are spliced, the types of splices, and the special equipment that is required.

9.1.0 Applications of Fiber Splices

Typically, splices are used to permanently join lengths of cable outside buildings, while connectors are used more commonly to terminate cables inside buildings. Splices may be incorporated in lengths of cable or housed in indoor or outdoor splice boxes. The functional characteristics of splices versus connectors are listed in *Table 3*.

The lower attenuation of splices is an advantage for long-distance fiber optic cable runs. Bare fiber normally comes on reels in standard lengths from 1 to 25km (from a little over half a mile to

Table 3 Fiber Splice Applications

Connectors	Splices
• Non-permanent	• Permanent
• Easy to use once mounted	• Lower attenuation
• Factory-installable on cables	• Lower back reflection
• Allow easy reconfiguring	• Easier to seal hermetically
• Provide standard interfaces	• Usually less expensive per splice
	• More compact

302T03.EPS

almost 15 miles). Cables are bulkier than bare fibers, particularly the heavy-duty types intended for outdoor use, and normally come in lengths of less than 12km (about 7 miles).

Longer cable runs are made by splicing cable segments together. If the cables are installed in underground ducts, the splices are made and installed in manholes, with the cable segment length dependent on the manhole spacing. Overhead cables are spliced in the field from segments that are typically well over a kilometer long.

To withstand hostile outdoor environments, splices are housed in protective enclosures. Although many splice enclosures are designed to be reopened if repairs or changes are needed, they can be hermetically sealed to protect against moisture and debris infiltration. This combines with the low signal loss of splices to make them the preferred method of joining lengths of fiber in long-distance telecommunication systems.

9.2.0 Types of Splicing

There are two basic approaches to fiber splicing: fusion and mechanical (*Figure 50*). Fusion splicing melts the ends of two fibers together so that they fuse, like welding metal. Mechanical splicing holds the two fiber ends together without welding them, using a mechanical clamp and/or glue.

9.2.1 Fusion Splicing

Fusion splicing is performed by butting the tips of two fibers and melting them together. Typical signal losses from fusion splices are 0.05 to 0.2dB, and most are below 0.1dB. The maximum loss listed as acceptable by *ANSI/TIA/EIA-568* is 0.3dB.

Fusion splicing machines are designed differently, but they all have the common goal of reliably producing good splices. Many are automated and feature expensive instruments, with prices in the thousands to tens of thousands of dollars. Major differences between available models include the degree of automation and the amount of instrumentation included.

A fusion splicer typically uses an electric arc. The electrode spacing and timing of the arc is user-adjustable. The discharge heats the fiber junction, fusing the ends. Portable versions are operated by batteries that carry enough charge for a few hundred splices before requiring recharging. Factory versions work from power outlets or batteries.

Fusion splicers feature a microscope and video camera for magnified viewing of fibers during the alignment process. Also, instruments are often

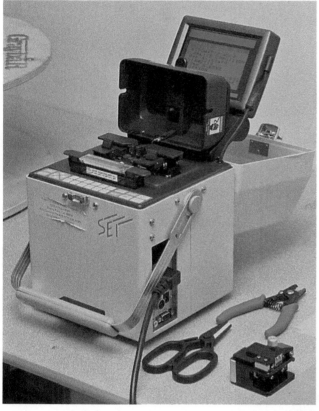

Figure 50 Splicing tools.

302F50.EPS

used to estimate the optical power transmitted through the fibers both before and after splicing.

Fusion splicing involves a series of steps. First, the fiber must be exposed by cutting open the cable. Then the protective plastic jacket must be stripped from the fiber at the ends to be spliced. The fiber ends must be cleaved to produce faces that are within 1 to 3 degrees of being perpendicular to the fiber axis. The ends must be kept clean until they are fused.

The next step is alignment of the fibers, which may be done manually or automatically, depending on the splicer model. After preliminary alignment, the ends may be pre-fused with a moderate arc that cleans their ends and rounds their edges.

These ends are then pushed together, allowing a test signal transmission to see how accurately they are aligned. The fiber ends are then aligned

both horizontally and vertically, with the ends brought together so they are just barely touching. After satisfactory alignment is achieved, the arc is fired to weld the two fiber ends together. Care must be taken to ensure proper time duration and temperature of the arc so the fiber ends are heated correctly and not overheated. After the joint cools, it can be coated with a polymer material to protect against environmental degradation. The spliced area can also be enclosed in a plastic or heat shrink-jacket. The entire splice assembly is then enclosed mechanically for protection. *Figure 51* shows a fusion splice in process.

9.2.2 Mechanical Splicing

Mechanical splicing is a simpler process whereby fibers are permanently joined and held within a mechanical device. There are many different types of mechanical splices, but most incorporate a sleeve or collar that aligns the fiber ends and secures them by means of adhesive or pressure. In creating a typical mechanical splice, the fiber ends are stripped, cleaned, and polished. Then the ends are inserted into opposite ends of the mechanical collar or sleeve. A tool is used to help maintain alignment and to apply pressure until the joint is fastened and any adhesive, if applicable, has cured.

In general, mechanical splicing requires less costly capital equipment, but has higher consumable costs than fusion splicing. This can tilt the balance in favor of mechanical splicing for systems requiring only occasional splicing, or for use in emergency on-site repairs. Mechanical splices tend to have slightly higher losses than fusion splices, but the difference is not dramatic. Back-reflections can occur in mechanical splices, but they can be reduced by using epoxy to connect the fibers, or by using a fluid or gel with a

302F51.EPS

Figure 51 Fusion splice.

matching refractive index. This index-matching gel suppresses the reflections that can occur at a glass-air interface.

An older kind of mechanical splice is the polished-ferrule splice, often called a rotary splice (*Figure 52*). As with other splices, the plastic coating is first removed from the fiber. Then, each fiber end is inserted into a separate ferrule, and its end is cleaved. The two ferrules are then mated within a jacket or tube. After the ferrules are inserted in the tube, they are rotated while the splice signal loss is monitored. The ferrules are fixed in place at the rotational position where the splice loss is at a minimum. Although this technique is more complex and time-consuming than conventional mechanical splicing, it offers a more precise way of mating fibers. Its sensitivity to rotation of the fiber around its axis makes it suitable for splicing polarization-sensitive fibers.

Fibers can also be mechanically spliced in V-grooved plates. V-groove splicing devices can take various forms. The two fiber ends can be slipped into the same groove on a single plate, and a matching plate applied on top. Or, the fiber ends can be put into separate grooved plates and have their tips polished before the plates are mechanically aligned and joined with each other. The V-groove mechanical splice is particularly suited to multi-fiber and ribbon cable applications. Kits such as the one shown in *Figure 53* contain the tools and materials needed to perform mechanical splices.

9.3.0 Splicing Issues

The three main concerns in splicing are the optical characteristics of the finished splice, its physical durability, and the ease of splicing. Choosing the appropriate splicing components and methods is a matter of prioritizing and reconciling these considerations.

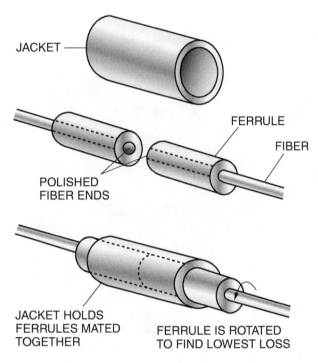

302F52.EPS

Figure 52 Rotary splice.

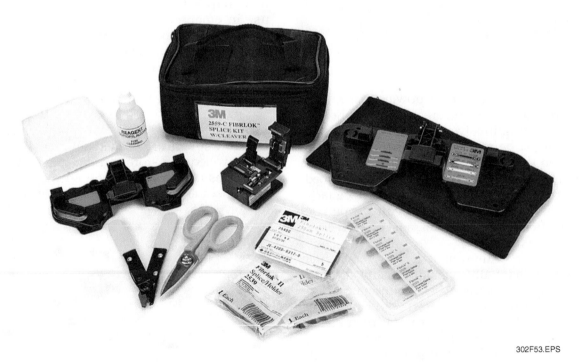

302F53.EPS

Figure 53 Mechanical splicing kit.

9.3.1 Attenuation

The sources of attenuation described for connectors also affect fiber optic splices. However, splicing tends to align fibers more accurately, resulting in lower attenuation, usually less than 0.075dB per splice. Some sources of signal loss are virtually eliminated in splices; others are greatly reduced.

Splice losses fall into two categories: intrinsic and extrinsic. Intrinsic losses arise from differences between the fibers in a splice connection. Splice-loss causes include variations in fiber core and outer diameter, differences in refractive index profile, and inconsistencies in the ellipticity and eccentricity of the fiber core. These losses can occur even in fibers with nominally identical specifications, because of inevitable imperfections in the manufacturing process.

Extrinsic losses arise from the nature of the splice itself. They depend on fiber end alignment, end quality, contamination, refractive-index matching between ends, spacing between ends, waveguide imperfections at the junction, and angular misalignment of bonded fibers.

Attenuation can be very low in properly made splices, but imperfect ones can suffer from high losses. A single dust particle between fiber ends can block the entire core of a single-mode fiber. However, with proper tools and techniques, splice attenuation is comparably low for single-mode and multi-mode fibers.

> **NOTE**
> The simple act of smoking while splicing can add smoke particles to a fusion or mechanical splice. These particles can cause unacceptable loss at the splice point.

9.3.2 Strength

If you pull hard on a spliced metal wire, it will normally break at the splice. Optical fibers are also more vulnerable at their splices.

Stripping the coating from fibers can damage them before splicing, causing microcracks that can result in a failure later on. In fusion splicing, contaminants can compromise the fusion zone itself, and thermal cycling from the welding process can weaken the surrounding area. When fusion splices fail, they typically fail near to, but not at, the splice boundary.

Mechanical splices also fail because the fibers have been damaged during preparation. The mechanical pressure and additional handling involved when making a mechanical splice can jeopardize the strength of the joint.

The lifetime of a splice can be enhanced by claddings or jackets that reinforce the splice and protect it from mechanical and environmental stresses.

9.3.3 Ease of Splicing

Because splices are often installed in the field, the convenience and portability of a splicing device are important considerations. This has led to the development of specialized equipment for field as well as factory use.

Fusion splicing produces good connections but is generally less practical in the field. Standard factory splicers are expensive units that measure up to 12" (30 centimeters) square, weighing in at 20 pounds (10 kilograms) and up. Special portable models can cost several thousand dollars, but weigh only a few pounds (a couple of kilograms), and measure 4" to 6" (10 to 15 centimeters) square. Field units have fewer features and less battery power, requiring more frequent charging.

The installation of mechanical splices requires special splice sleeves and tools. However, the sleeves are small and the tools are much less bulky and expensive than those used in fusion splicing. This makes mechanical splicing a practical option for splicing on a smaller scale. Emergency cable repair kits come with mechanical splicing equipment.

9.3.4 Splice Protection

Fiber optic splices require protection from hazards, whether they are indoors or outdoors. Splice closures help to organize spliced fibers from multi-fiber cables. They also protect the splices from strain, moisture, and debris.

Splice closures usually feature a rack or tray, as shown in *Figure 54*, which secures an array of individual splices. This rack is mounted in a case that provides protection from the elements. In case more splices are needed, an excess length of fiber is rolled and stored in the splice case. Like splice closures for telephone wires, fiber optic splice cases are mounted in strategic, accessible locations where splices are necessary, such as in manholes, on utility poles, or at points where fiber cables enter buildings.

Fiber splice closures should have the following characteristics:

- Hold a cable's strength members securely
- Be watertight
- Have a redundant level of seals in case one level fails

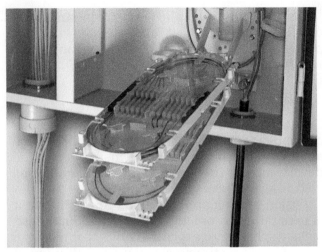

Figure 54 A splice organizer.

- Electrically bond and ground metallic components in the cable, such as strength members or wire sheaths
- Be re-enterable should the splice require maintenance
- Organize splices and fibers so they are quickly and clearly identified
- Reserve some space for more splicing and future service additions
- Accommodate minimum bend specifications for the fibers and cables to avoid signal losses and physical damage

10.0.0 FIBER OPTIC TESTING

Optical fibers are tested both at the factory and in the field for performance and dimensional compliance to specifications. As a quality control measure during the manufacture of fibers, the manufacturer must constantly test the fibers to ensure that they are supplying a quality product. These tests include the following:

- Core diameter tests
- Cladding diameter tests
- Numerical aperture (NA) tests
- Attenuation tests
- Refractive index profile examinations
- Tensile strength tests

Other tests performed on fibers or on fiber optic cables concern their mechanical and environmental characteristics. Mechanical tests, such as impact and crush-resistance tests, measure the cable's ability to withstand physical stresses. Environmental tests gauge the change in attenuation in cables under conditions of extreme temperature, repeated cycling of temperature, and humidity.

Fiber and cables are specified as the result of these tests. Using specifications, engineers choose cables with performance properties that are suited to a particular application. Manufacturers of other fiber optic components, such as sources, detectors, connectors, and splitters, also use testing to determine the performance specifications of their products.

Additional tests are performed during and after the installation of a fiber optic link. These tests ensure that the installed system will meet the performance requirements. Splices, for example, must not exceed certain loss values. During installation, each splice must be tested. If the loss is unacceptably high, the fibers must be respliced.

> **NOTE**
>
> Upon acceptance of a length of fiber optic cable, and before installing a length of fiber optic cable, it should be tested on-the-reel with an OTDR to ensure that none of the individual fiber optic strands were damaged in shipping and moving. It is extremely embarrassing and expensive to install a defective fiber optical cable that has to be replaced before it is ever put into service.

10.1.0 Optical Power Meter

An optical power meter is used to measure the strength of a fiber optic light signal. Sensors that plug into the unit contain the detector and perform the optical-to-electrical conversion. Different sensors are available for use at different power levels and operating wavelengths from 400 to 1,800nm. Adapters permit bare fibers or a variety of popular connectors to be used for input. *Figure 55* shows an example of a fiber optic test set.

The range of a typical meter, using different sensor heads, is –80dBm (10μW) to +33dBm (2W). The resolution is variable at either 0.1 or 0.01dB. An output jack, which provides an adjustable regulated current, allows plug-in LEDs, VCSELs, or lasers to provide the light source for measurements.

Figure 55 Fiber optic test set.

The meter can be used for a number of measurements and tests.

The power transmitted through a fiber optic link, as well as the loss through the link, can be tested with a light source and power meter (*Figure 56*). The light source is connected to one end of the link using a jumper with connectors that match the connectors on the link to be tested. The power meter is connected to the other end of the link. Single-mode light sources use a laser as the light source, while multi-mode sources use an LED or VCSEL.

10.2.0 Insertion Loss Testing and Mode Control

Insertion loss testing measures the power through a length of spliced or connected fiber. As discussed earlier, a certain amount of light can be lost through imperfections in splices and connectors. It can be difficult to obtain accurate and repeatable results when testing for insertion loss. Reliable testing is crucial in making high quality fiber connections and splices.

Signal launch conditions have an effect on the distribution of high- and low-order modes (the longer and shorter paths taken by light) within a multi-mode fiber. Since light in high-order modes is more likely to be lost through fiber or connector imperfections, the mode distribution has an effect on loss measurements. For example, if a test light signal populates most of the vulnerable,

high-order modes, an artificially high percentage of signal loss may be indicated in the test results. Conversely, if the test signal fills mostly the more stable, low-order modes, the percentage of signal loss may seem artificially low.

For repeatable, meaningful measurements in loss tests of multi-mode fibers, launch conditions must be controlled. In the past, the Electronics Industry Association (EIA) recommended a 70/70 launch. This condition was created by a highly tuned light source filling 70 percent of the fiber core diameter and 70 percent of the fiber NA. This precise launch condition was thought to result in equilibrium of mode distribution (EMD) in a graded-index fiber, meaning that the modes (paths) have blended within the fiber to the point where their high/low distribution no longer changes. However, the difficulty of creating such precise signal launch conditions, along with advances in transmitter and fiber technology, prompted new recommendations for loss testing and measurement.

The newer recommendation for testing and measuring signal insertion loss of multi-mode fibers is to overfill a fiber with light (the emitter coverage more than 70 percent of the fiber core diameter and numerical aperture), and to simply filter out the excessive higher order modes that result in artificially high loss readings. A fiber link's

302F56.EPS

Figure 56 Light source and power meter for testing fiber optic links.

On Site

Laser Light Source

The laser light source shown here is used as a visual fault locator (VFL) to find breaks or damage in a fiber, a patch cord, or a splice. The light source is intense or powerful enough that it will be visible through the buffer at the point where the fiber is broken. The VFL is also used to identify fibers in a cable. The VFL can insert a light in a fiber at one end of a cable and see it at the other end to identify the fiber. The range of the VFL is about 3 miles.

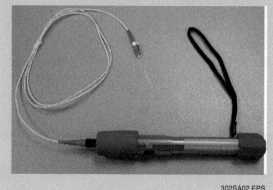

302SA02.EPS

high order mode power (HOMP) indicates how much of a light signal exists in those problematic modes. The HOMP value can be found by first measuring the power of a light signal through the fiber, and then measuring it with a mandrel-wrap filter in place. The filter is typically five or six wraps of bare fiber around a mandrel of a certain diameter (approximately 1"). This coiling of optical fiber effectively filters out the higher order modes. The unfiltered power reading minus the filtered reading gives the HOMP value for that link.

If the HOMP value is high, then a mode filter should be used when testing loss through connectors or splices in the link.

10.3.0 Fiber Loss Measurements

Losses in a fiber optic cable can be measured in two ways: the cutback method and the known reference method. The cutback method uses a single fiber. First, power through the fiber is measured. A section of fiber is then cut off and the power through the remaining length of fiber is measured. The fiber loss in decibels per kilometer can then be determined using the following formula:

$$Loss = P2 - P1/L$$

Where:

P1 = dBm reading from
   the first measurement
P2 = dBm reading from the
   second measurement
L = difference (in kilometers) of
   the two cable lengths

A second method of measuring attenuation in an optical fiber is to compare the power through the cable under test to the power through a known reference cable. The power through the cable under test is measured in absolute units on the meter. Next, the power through the reference cable is measured. Loss is quantified by comparing the two results.

The light loss measurements accurately determine if a particular fiber optic strand will pass enough light to operate within the defined application. It does not show where a problem is within a fiber, but shows that there is a problem in that the loss of light is too great for proper operation of the connected equipment.

10.4.0 Time and Frequency Domains

A signal can be described in terms of either the time domain or the frequency domain. This module has used both domains without bothering to distinguish between them. We described the limitations on a fiber's information-carrying capacity in terms of both bandwidth and dispersion. Bandwidth is in the frequency domain; rise-time dispersion is in the time domain. Analog engineers, dealing with analog signals in the frequency domain, talk in terms of frequencies. Digital engineers, dealing with pulses, use the time domain and talk in terms of rise times and pulse widths.

10.5.0 Optical Time-Domain Reflectometry

As its name implies, an optical time-domain reflectometer (OTDR) allows evaluation of an optical fiber in the time domain (*Figure 57*). The OTDR is a useful tool in analyzing the performance of a fiber optic system. This section describes the basic principles of optical time-domain reflectometry.

The OTDR relies on the backscattering of light that occurs in an optical fiber. Backscattering results from Rayleigh scattering and Fresnel reflections. Rayleigh scattering is caused by the refractive displacement due to density and compositional variations in the fiber. In a quality fiber, this scattered light can be assumed to be evenly distributed with length. Fresnel reflections occur because of abrupt changes in the refractive index at connections, splices, and fiber ends. A portion of the Rayleigh scattered light and Fresnel reflected light returns to the input end as backscattered light.

An OTDR requires access to only one end of the fiber in cable segments that may be tens of miles long. However, light that may be injected at the far end should be eliminated by covering the far

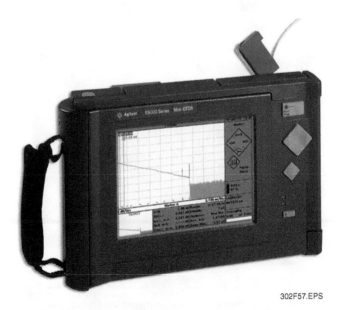

302F57.EPS

Figure 57 Optical time-domain reflectometer.

end with a cap or other opaque device to prevent stray light from appearing as a reflection and providing a false reading. Timing how long it takes light to travel from the instrument to a point in the fiber and back can locate flaws and junctions in the fiber. In this way, an OTDR is much like an optical version of radar.

The OTDR screen displays time horizontally and power vertically. Fiber attenuation appears as a line decreasing from the left (the input end of the fiber) to the right (the output end). Both the input and the backscattered light attenuate over distance, so the detected signal becomes smaller over time. A connector, splice, fiber end, or abnormality in the fiber appears as an increase in power on the screen, because backscattering from Fresnel reflections will be greater than backscattering from Rayleigh scattering. The quality of a splice can be evaluated by the amount of backscattering; greater backscattering means a higher-loss splice. A connector shows both a power increase from reflection and a power drop from loss. The degree of loss indicates the quality of the connection. *Figure 58* shows a representation of a typical OTDR display.

Light travels through a fiber at a speed of about 5ns/m, depending on the refractive index of the core. If the launch cable used is too short, the OTDR may not actually see the first connec-

tor, so a launch cable of between 50m and 150m should be used. To be able to see the last connector on the fiber optic span, a launch cable should also be used at the far end. The far-end launch doesn't need to be more than about 10m to 20m, but it should have an opaque cap on the very end to prevent ambient light from giving false readings on the OTDR. False readings show up as a reflection near the middle of the cable. Most OTDRs use a cursor to mark a horizontal point on the trace and display the distance in terms of both time and physical distance. One can, for example, determine the distance to a splice with a great degree of accuracy, typically to within a foot. However, an OTDR measures the distance along the fiber, not necessarily the length of cable. If the fiber is wound around a central core within the cable (most are), the actual fiber length will be somewhat longer than the cable length.

10.5.1 *Uses of an OTDR*

OTDR technology has many uses in fiber optic science. Three of the major field uses are the following:

- System loss evaluation
- Splice and connector evaluation
- Fault location

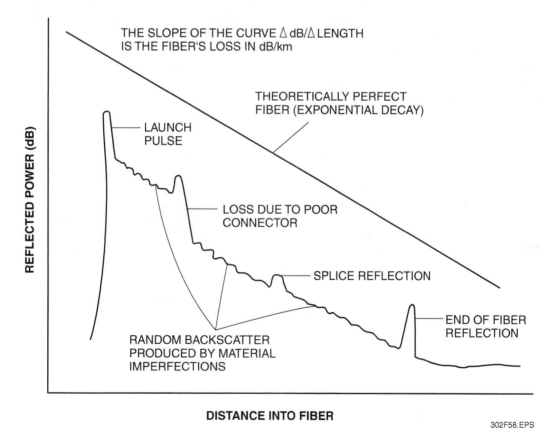

Figure 58 Typical OTDR display.

302F58.EPS

The system loss budget of a fiber optic installation assumes a certain amount of signal loss per unit length. OTDR can measure attenuation in a fiber system before and after installation. Measurements before installation ensure that all fibers meet specified limits. Measurements after installation check for an increase in attenuation resulting from bends, installation stresses, or unexpected static loads.

An OTDR can be used during system installation to verify that all splices and connector losses are within acceptable limits. After a splice or connector is inserted into the line, an OTDR checks the amount of signal loss. If the amount of loss is unacceptable, the connection is remade. Some connections, like rotary splices, are tunable; that is, by rotating one half of the splice in relation to the other, losses from lack of symmetry in the fiber or splice can be minimized. As the splice is rotated, an OTDR can measure the signal to determine the position of lowest loss, thereby maximizing transmission through the splice.

Faults, such as broken fibers or splices, may weaken a system during or after installation. The location of a fault may not be obvious. It may be impossible to locate visually if the cable is buried or located in a conduit, or if the break is only in the fiber and not the outer cable. An OTDR provides an invaluable method of locating hidden faults accurately, saving time and expense.

SUMMARY

At the center of a fiber optic cable is the optical fiber. The core of this fiber is surrounded by a cladding layer that prevents light from escaping. Additional strength members and a buffer may be included, as well as an outer jacket or sheath. Among the advantages of fiber optic cable systems are high transmission speed and low signal loss. Fiber optic cables are available for indoor or outdoor use.

A fiber optic system requires a transmitter, a conductor (the fiber), a receiver, and a means of connection, such as connectors or splices. Various imperfections in connectors, splices, or in the fibers themselves can cause a loss of light called attenuation. Such losses are minimized in a quality fiber optic system.

There are three main types of light sources: laser diode, VCSEL, and LED. LEDs are used in low bit-rate and analog applications of shorter wavelength systems. VCSELs are lower-cost laser diodes, and have replaced LEDs in many applications. Lasers have shorter rise times and produce higher output power than LEDs, and are therefore useful for long-distance and higher-speed transmission.

Bend radius and tensile loading are critical factors in fiber optic installation because the fiber can easily be damaged if overstressed. Pulling forces must be carefully monitored when the cable is being pulled.

1. The device used to convert a light signal into an electrical signal at the receiver is known as a(n) _____.
   a. refractive index reflectometer
   b. numerical aperture detector
   c. multiplexer
   d. photosensitive detector

2. Signal propagation is defined as _____.
   a. the scattering of light due to imperfections in the fiber
   b. the characteristic movement of a light signal through a medium
   c. a measure of a detector's output current or output voltage in relation to its optical input power
   d. a diode's sensitivity expressed as a ratio of the number of photons impinging on an external circuit to the number of electrons flowing in an external circuit

3. The value that represents the amount of light lost within a fiber optic system is _____.
   a. attenuation
   b. refractive index
   c. propagation
   d. responsivity

4. The two basic components of an optical fiber are the _____.
   a. insulator and splitter
   b. core and cladding
   c. jacket and connector
   d. connector and splitter

5. The term used when there are several paths through which light can travel in an optical fiber is _____.
   a. simplex
   b. modal dispersion
   c. attenuation
   d. multi-mode

6. A mode is defined as _____.
   a. one path a light ray can travel through a material
   b. the frequency at which light travels through a medium
   c. a measure of a detector's output current or output voltage in relation to its optical input power
   d. a diode's sensitivity expressed as a ratio of the number of photons impinging on an external circuit to the number of electrons flowing in an external circuit

7. Optical fibers are classified by _____.
   a. the thickness of the fiber itself and the refractive index of the core
   b. their material makeup, the refractive index of the core, and the attenuation factor of the cladding
   c. their material makeup, the refractive index of the core, and the modes that the fiber propagates
   d. their material thickness, the refractive index of the core, and the bandwidth factor of the cladding

8. The four common components in a fiber optic cable are _____.
   a. optical fiber, jacket, transmitter, and receiver
   b. optical fiber, jacket, connector, and splitter
   c. optical fiber, buffer, strength members, and jacket
   d. optical fiber, jacket, multiplexer, and cladding

9. A VCSEL is a type of _____.
   a. receiver
   b. emitter
   c. splitter
   d. connector

10. In an analog system, transmission speed is measured as _____.

   a. bandwidth
   b. data rate
   c. frequency
   d. reflectivity

11. The term *laser* was derived from light _____.

   a. attenuation by standard emission and radiation
   b. amplification by stimulated emission of radiation
   c. attenuation by systematic emission or reduction
   d. amplification by systematic emission or reproduction

12. In a fiber optic system, the losses caused by variations in the fiber itself are called _____.

   a. responsivity losses
   b. extrinsic influences
   c. modal dispersion
   d. intrinsic factors

13. If a connection leaves a gap between the fiber ends, the result can be _____.

   a. misalignment losses
   b. end-separation losses
   c. ellipticity losses
   d. modal dispersion

14. A device used to couple light from one fiber to one of two end fibers but not to both is a(n) _____.

   a. splicer
   b. optical switch
   c. connector
   d. tee splitter

15. Which of the following is a push-and-latch connector?

   a. FC
   b. ST
   c. LC
   d. SMA

16. The type of fiber connection that requires injection of a two-part epoxy into the connector is _____.

   a. anaerobic
   b. hot melt
   c. epoxy polish
   d. pigtail

17. Splices generally result in less attenuation than connectors.

   a. True
   b. False

18. The two basic types of fiber splices are _____.

   a. connector and splitter
   b. fusion and mechanical
   c. double lap and V-groove
   d. simplex and duplex

19. Points within a fiber optic strand that induce losses can be identified using a(n) _____.

   a. optical time-domain reflectometer
   b. optical liquid-chromatology analyzer
   c. VTVM
   d. optical spectrum analyzer

20. A line on an OTDR display that descends from left to right represents _____.

   a. frequency
   b. reflectance
   c. Fresnel effect
   d. attenuation

Trade Terms Introduced in This Module

Attenuation: The degradation of a signal as it moves through a medium.

Cladding: A material that encloses the fiber in fiber optic cable, allowing the signal to be transmitted without escaping the cable.

Duplex: A term used to indicate two-way transmission.

Insertion loss: The amount of optical power lost through a connection or splice.

Laser: A source of high-power coherent light (derived from light amplification by stimulated emission of radiation).

Mode: A path that can be taken by a ray of light through a material.

Multi-mode: The term used when there is more than one way that light can travel through a material.

Multiplexing: The broadcast of several signals over a line simultaneously.

Numerical aperture: The range of angles over which a system can accept or emit light.

Propagation: The method of traveling through space or a specific material.

Receiver sensitivity: The term used to describe the minimum signal that a receiver can detect.

Return loss: Light reflected back through a fiber toward the source. Return losses are kept to less than –55dB to ensure speed and clarity of transmission.

Simplex: A term used to indicate one-way transmission.

Splitter: A device that connects three or more fiber ends to create one or more outputs.

Additional Resources

This module presents thorough resources for task training. The following resource material is suggested for further study.

Fiber Optic Reference Guide. David R Goff. Woburn, MA: Focal Press.

The Cabling Handbook. 2nd Edition, 2001. John R. Vacca. Upper Saddle River, NJ: Prentice Hall PTR.

The Fiber Optic Association website (www.thefoa.org) contains a variety of information and links for fiber optic equipment, cables, and termination procedures.

OSHA guidelines on laser hazards can be found at: www.osha.gov.

Figure Credits

The Fiber Optic Association, Inc., www.thefoa.org, Module opener, Figures 50 (bottom photo) and 51

Don Owens, Figure 1 and SA02

Wayne Adair, Table 2

Laser Diode Incorporated/Tyco Electronics, Figures 18 and 19 (photo)

BeamExpress SA, Figure 20

Vermeer Manufacturing Company, Figure 31

Topaz Publications, Inc., Figures 37 and 39

Mike Powers, Figure 38

Topstone Communication, Inc., Figure 42

Photo courtesy of Molex Incorporated, Figure 45

AFL, Figure 46

Greenlee Textron, Inc., a subsidiary of Textron Inc., SA01 and Figure 48

3M Company, Figures 47, 50 (top photo), and 53

Sumix Corporation, Figure 49

Tyco Electronics, Figure 54

Agilent Technologies, Inc., Figures 55 and 57

Advanced Fiber Solutions, Inc., Figure 56

CONTREN® LEARNING SERIES — USER UPDATE

NCCER makes every effort to keep its textbooks up-to-date and free of technical errors. We appreciate your help in this process. If you find an error, a typographical mistake, or an inaccuracy in NCCER's Contren® materials, please fill out this form (or a photocopy), or complete the online form at www.nccer.org/olf. Be sure to include the exact module number, page number, a detailed description, and your recommended correction. Your input will be brought to the attention of the Authoring Team. Thank you for your assistance.

Instructors – If you have an idea for improving this textbook, or have found that additional materials were necessary to teach this module effectively, please let us know so that we may present your suggestions to the Authoring Team.

NCCER Product Development and Revision
3600 NW 43rd Street, Building G, Gainesville, FL 32606

Fax: 352-334-0932
Email: curriculum@nccer.org
Online: www.nccer.org/olf

❏ Trainee Guide   ❏ AIG   ❏ Exam   ❏ PowerPoints   Other _____

Craft / Level: _____   Copyright Date: _____

Module Number / Title: _____

Section Number(s): _____

Description: _____

_____

_____

_____

_____

Recommended Correction: _____

_____

_____

_____

_____

Your Name: _____

Address: _____

_____

Email: _____   Phone: _____

Wireless Communication

33303-11

Trainees with successful module completions may be eligible for credentialing through NCCER's National Registry. To learn more, go to **www.nccer.org** or contact us at **1.888.622.3720**. Our website has information on the latest product releases and training, as well as online versions of our *Cornerstone* newsletter and Pearson's Contren® product catalog.

Your feedback is welcome. You may email your comments to **curriculum@nccer.org**, send general comments and inquiries to **info@nccer.org**, or use the User Update form at the back of this module.

V.1 4/11

WIRELESS COMMUNICATION

Objectives

When you have completed this module, you will be able to do the following:

1. Describe the different types of wireless communication.
2. Describe the limitations of wireless communication.
3. Identify the basic components used in wireless systems and explain the function of each.
4. Identify the equipment used when testing and troubleshooting wireless communication systems.
5. Identify interfering factors in wireless communication systems.
6. Describe the placement and function of an antenna.

Performance Tasks

This is a knowledge-based module; there are no performance tasks.

Trade Terms

Ad hoc	Micron	Packet sniffer	Voltage standing wave
Companding	Microwave	RS-232	ratio (VSWR)
Emitter	Oscillator	Transducer	

Contents ———————————————————

Topics to be presented in this module include:

Figures and Tables

1.0.0 INTRODUCTION

The term *wireless communication* simply means that signals containing voice and data are carried through the air instead of through cables. Some wireless communication is one-way; that is, the receiving station is not able to send a signal. A radio is a prime example of one-way wireless communication. A TV set, although it may receive its programs through cable, is capable of wireless reception through an antenna. In fact, it wasn't too many years ago that wireless was the only method of receiving TV signals. The number of channels you had available was determined by the number of TV stations in your area. Cellular phones are a good example of two-way wireless communication.

Wireless communication was conceived in the late 1800s when radio waves were generated in a laboratory. Wireless voice transmission was accomplished in 1900, and by 1920 the radio industry was in full swing. Since its invention, wireless communication has been the only means of communication with ships at sea.

The advent of radar in the 1940s made possible short-wavelength (microwave) transmission, which enabled large volumes of information to be transmitted at high speeds. Later, satellites became the vehicle for global wireless communication of entertainment, voice, and computer data.

2.0.0 WIRELESS COMMUNICATION PRINCIPLES

A wireless communication system must contain a transmitter, a receiver, and two antennas. The transmitter sends out a signal containing voice or data information at a specified frequency. The receiver, which is tuned to the same frequency as the transmitter, picks up the signal and extracts the information from it. As you know from using your radio, it is capable of receiving a wide band of frequencies. However, it will only play the station to which it is tuned at any given time. The receiver's frequency is adjusted to match the frequency of a particular sending station's transmitter.

Wireless systems operate in a range of frequencies specified by the Federal Communications Commission (FCC). Within that range, certain fre-

quency bands are reserved for AM and FM radio, TV, microwave, and other forms of communication.

Frequencies used for communications are very high, as shown in *Figure 1*. They are typically stated in thousands, millions, and even billions of cycles per second.

Special terms have been developed to define frequency ranges:

- *Kilohertz (kHz)* – Thousands of cycles per second
- *Megahertz (MHz)* – Millions of cycles per second
- *Gigahertz (GHz)* – Billions of cycles per second

Examples of uses of these frequencies are as follows:

- *Audible sound* – 40Hz to 16kHz
- *Voice telephone* – 300Hz to 3,400Hz
- *AM radio* – 530kHz to 1,600kHz
- *FM radio* – 87.5MHz to 108MHz
- *VHF broadcast TV* – 30MHz to 300MHz
- *UHF broadcast TV* – 300MHz to 3,000MHz
- *Cellular phone* – 800MHz to 4.1GHz
- *Infrared (IR)* – 300GHz to 428,000GHz

Frequencies above 1GHz are considered microwave frequencies. You may also hear the term *frequency band used*. For convenience, RF engineers have assigned band designations to ranges of frequencies. The industry standard frequency ranges for some of these bands are listed in *Table 1*. These designations differ from the band designations used in military applications. This fact creates some confusion in the industry.

Table 1 Industry Standard Frequency Ranges

Band Range	Frequency Range
HF	3 – 30MHz
VHF	30 – 300MHz
UHF	300 – 3,000MHz
L band	1,000 – 2,000MHz
S band	2,000 – 4,000MHz
C band	4,000 – 8,000MHz
X band	8,000 – 12,000MHz
Ku band	12 – 18GHz
K band	18 – 27GHz
Ka band	27 – 40GHz
Millimeter band	40 – 300GHz
Submillimeter band	300GHz and above

303T01.EPS

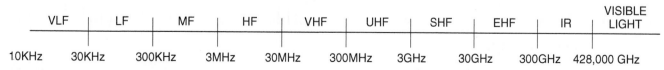

303F01.EPS

Figure 1 Radio frequency spectrum.

Wireless Communication

2.1.0 Modulation

The frequency at which a wireless communications system operates is known as the carrier frequency because it carries the information through the communications channel. The carrier frequency is necessary because the frequency of voice communication is too low for efficient transmission.

The carrier is a continuous, constant-amplitude sine wave that is varied (modulated) by superimposing the signal onto it. The general types of modulation in common use are as follows:

- *Amplitude modulation (AM)* – Amplitude modulation is accomplished by mixing the information signal with the carrier so that the amplitude of the carrier is modified to reflect the variations in the information signal. *Figure 2* shows an example.
- *Frequency modulation (FM)* – In frequency modulation, the time interval between sine waves is varied at a rate that reflects the variations in the information signal (see *Figure 3*). FM is less susceptible to noise than AM.
- *Phase modulation (PM)* – Phase modulation is similar to frequency modulation, except that in phase modulation the phase of the carrier is varied instead of the frequency. Phase modulation is used in digital communications.
- *Frequency shift keying (FSK)* – FSK is a form of frequency modulation in which the carrier is modulated with tones that represent binary codes. One tone represents a 0 and another represents a 1. The binary high and low are represented by different audio frequencies.

- *Pulse modulation* – In pulse modulation, the pulse amplitude, pulse duration, or the interval between pulses is varied to represent information. The modulated pulses are superimposed on the RF carrier.

2.2.0 Analog and Digital Signals

An analog signal is a continuously variable waveform in which each instant along the waveform represents a value such as the loudness and pitch of a human voice. When you observe analog frequencies with an oscilloscope, you will see a sine wave.

Analog telephone systems convert voice information into analog electrical signals. In this process, two conversions take place:

- A transducer is used to convert the sound waves from the speaker's voice into an electrical current with variations in frequency and amplitude that represent the tone and loudness of the speaker's voice. A transducer is a device that converts energy from one form into another. In this case, the transducer is a microphone containing a diaphragm that vibrates in response to the speaker's voice. The vibrations produce a corresponding electrical signal. On the receiving side, a device that is similar to a miniature loudspeaker converts the electrical signal into a reproduction of the speaker's voice.
- The electrical signal is transmitted over the airwaves to a distant receiver where it is converted back into sound waves.

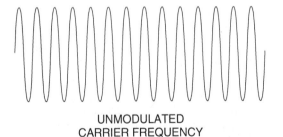

UNMODULATED
CARRIER FREQUENCY

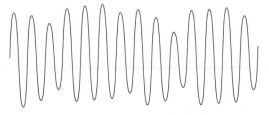

AMPLITUDE MODULATED
CARRIER FREQUENCY

303F02.EPS

Figure 2 Amplitude modulation.

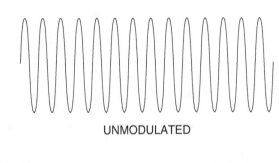

UNMODULATED

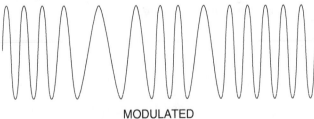

MODULATED

303F03.EPS

Figure 3 Frequency modulation.

A digital signal is a coded representation of the input signal. Unlike the analog signal, in which variations between the high and low points have value, digital signals have only two values: high and low.

Analog-to-digital (A/D) converters are used to convert analog voice signals into digital signals. The coded message is a series of pulses transmitted on an analog carrier at regular time intervals. The pulses are usually rectangular.

The analog signal is sampled at regular time intervals. In order to faithfully reproduce the signal content, the sampling rate must be at least twice the highest frequency component of the analog signal. For example, 8,000 samples per second are required for a voice signal of 4kHz. To obtain high-quality voice or music at 16kHz, the sampling rate would have to be 32,000 samples per second.

Each pulse within the sample is assigned a value, typically between 1 and 256, depending on its amplitude. The increments are not necessarily uniform. In the case of speech signals, for example, it is preferable to have more levels when the signal is weak (amplitude close to 0) than where the signal is strong. This process, called companding, helps improve the fidelity of the voice signal. The train of pulses that results from the companding process is known as a pulse amplitude modulated (PAM) digital signal.

The PAM signal is converted into a binary format (a series of ones and zeros). A pulse can be represented by an 8-bit binary number. For example, a pulse with an amplitude level of 137 (out of 256) can be represented by the binary number 10001001. The modulated PAM signal is called a pulse code modulated (PCM) signal.

Digital signals are less susceptible to interference than analog signals. The signal can be transmitted over a very long distance with much higher fidelity. One drawback of digital signals is that they require a much higher bandwidth (frequency range) than equivalent analog signals.

2.3.0 Multiplexing

The term *multiplexing* means to combine several low-capacity communications channels into one high-capacity channel that can carry signals from several sources. Multiplexing allows a single high-capacity channel to be used by multiple network devices.

A device known as a multiplexer (mux for short) is used for this process. A mux is typically designed so that it can be used at the transmitting end to do the multiplexing and at the receiving end to do the demultiplexing.

The two types of multiplexing commonly used in wireless communications are as follows:

- *Frequency division multiplexing (FDM)* – In FDM, each subscriber has a different frequency or channel on which to communicate. Because the channels are separate, a large number of signals can be sent at the same time without conflict. The channels can be used to send information in two directions.
- *Time division multiplexing (TDM)* – Instead of each subscriber having its own channel, each subscriber is allocated a time slot.

> **NOTE**
> Frequency division multiplexing and time division multiplexing are described in more detail later in this module.

3.0.0 RADIO FREQUENCY (RF) SYSTEMS

This section discusses radio frequency (RF) systems. The main components of an RF system are a transmitter, receiver, and antenna.

3.1.0 Transmitters

A transmitter is a device or subsystem that produces RF energy at a specific frequency. It modulates the signal if necessary, amplifies it to achieve the power level required for transmission, and then delivers it to an antenna or other load device.

There are many kinds of transmitters. The common denominator in all transmitters is an oscillator, which is an electronic circuit that produces a pure sine wave at a specified frequency called the carrier frequency. In the most basic transmitters, the oscillator output is sent directly to the antenna without modulation (*Figure 4*). The antenna transmits a continuous wave (CW) output. An example of such an application is the anti-shoplifting device that is attached to clothing. If the device is not removed or deactivated at checkout, the receiver at the door will detect the low-level (approximately 10 milliwatts) CW RF signal it emits and alert store personnel by means of an audible alarm or some other signal.

On Site

Other Multiplexing Methods

There are other multiplexing methods such as QAM or quadrature phased shift-key access multiplexing, which is used in satellite and CATV systems. CDMA, code division multiple access, is used in wireless telecommunications systems such as cellular PCS.

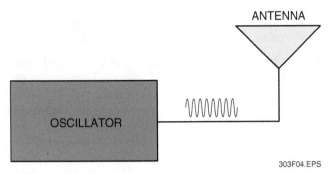

Figure 4 Basic transmitter.

At the next level, the oscillator RF output is modulated to provide information content (*Figure 5*). The receiver then decodes (demodulates) the signal to extract the information. Such a signal could simply be a series of pulses transmitted to a central security station to indicate the presence of an intruder.

As the level of communication increases, so does the level of sophistication of the transmitter. A voice communication device such as a two-way radio requires additional components, as shown in *Figure 6*. A typical fixed-station communications transmitter delivers RF output power between 25 and 200 watts. The signal that leaves the transmitter is a combination of two signals at different frequencies: one is the signal containing the modulated carrier frequency; the other is the output of the local oscillator (LO). The local oscillator is an electronic device that produces a pure sine wave at a fixed frequency ranging from 500 to 1,000MHz. In this case, assume that it is 500MHz.

Assume that the diagram in *Figure 6* represents a wireless phone. The low-frequency voice signal is amplified by an audio amplifier to increase the signal strength, then sent to the modulator where the audio is imposed on an intermediate carrier frequency (IF) generated by the local oscillator. In the mixer stage, the 400MHz modulated IF signal is mixed with the output of a high-frequency (HF) oscillator (500MHz) to produce the 900MHz signal that will be transmitted.

Actually, the mixer will produce two outputs: one at 900MHz and the other at 100MHz (500MHz – 400MHz). The unwanted frequency (sideband) is removed by a filter. A high-power RF amplifier is used to boost the RF output signal to the power level required for transmission.

3.2.0 Receivers

Receivers intercept electromagnetic waves, then extract and process the information carried by those waves. Like the transmitting system, the receiving system has an antenna selected for its ability to intercept the range of frequencies that the receiver is designed to process. Receivers range from simple to complex, depending on what kind of work they are doing. The most basic receiver consists of an antenna and detector that reacts to the presence or absence of a modulated signal.

A receiver that processes voice or data is somewhat more complex because it needs a means of demodulating (detecting) the information in the carrier (*Figure 7*). The earliest receiver consisted of a germanium crystal with a piece of fine wire attached. This type of detector is known as a cat's whisker and is basically a diode that converts the RF to DC voltage representing the voice waves. It can be used to detect AM radio signals.

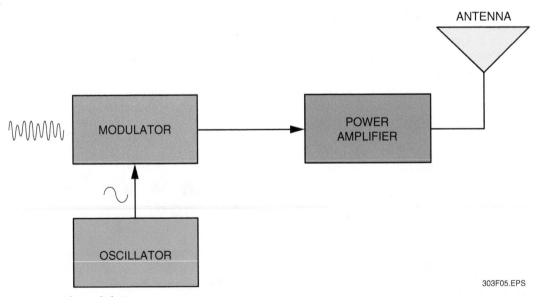

Figure 5 Transmitter with modulation.

The direct-conversion receiver uses a more sophisticated detection method (see *Figure 8*). The incoming RF signal is amplified to compensate for transmission losses. The amplified signal is then mixed with the output of a local oscillator, which is tuned to the carrier frequency. This process is known as beating, and the oscillator is sometimes called a beat frequency oscillator (BFO). In the mixing process, the carrier and LO frequency cancel each other out, leaving the audio.

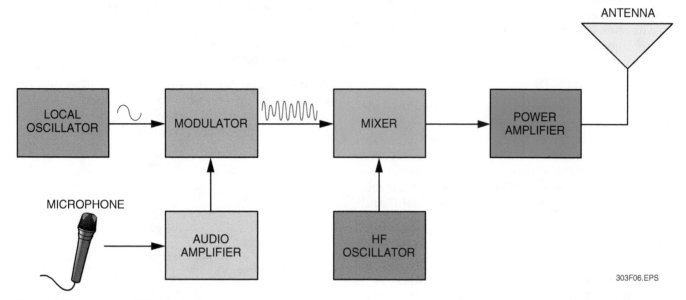

Figure 6 Voice communication transmitter.

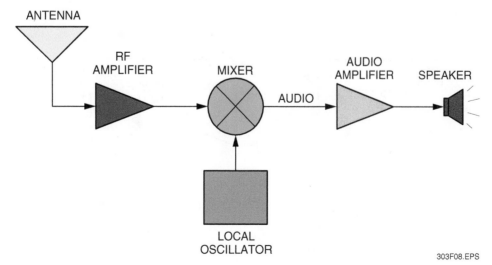

Figure 7 Basic receiver.

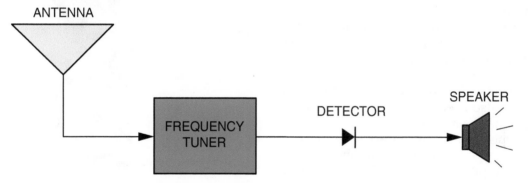

Figure 8 Direct-conversion receiver.

Most radio receivers currently in use are superheterodyne receivers (*Figure 9*). In a superheterodyne receiver, the LO operates at a different frequency than the carrier, typically 30MHz lower. When the two signals are mixed, the result is an IF which is amplified and filtered before the detection stage. Superheterodyne receivers have greater sensitivity to low-amplitude signals and better rejection of unwanted frequencies.

3.3.0 Transceivers

A transceiver is a single device that combines the functions of a transmitter and receiver (*Figure 10*). Familiar examples of transceivers are CB radios and cellular phones. A transceiver is likely to cost less than a separate transmitter and receiver because some of the components perform double duty. That is, they are used in both transmit and receive modes.

3.4.0 Repeaters

A repeater acts as a link between wireless systems. It links devices at the physical layer of the OSI model. Its primary function is to receive and regenerate the signal in the exact form in which it was received.

A repeater is sometimes used to extend the range of a cell in a cellular phone system or other wireless system. In such applications, it can be placed near the fringe of a cell to transfer data to a location beyond the effective range of the cell. Repeaters are also used to link dissimilar media. For example, a repeater can be used to connect a server equipped with a twisted-pair connection to a server equipped with an optical fiber connection. In these applications, the repeater is referred to as a media converter.

3.5.0 Waveguide

In very high power RF systems, coaxial cable cannot withstand the amount of energy flowing from the transmitter to the antenna. In those situations, special metal piping called waveguide is used to convey the RF. Waveguide is usually rectangular, but is sometimes round (*Figure 11*).

3.6.0 Antennas

As mentioned earlier, both the transmitter and receiver need an antenna. In some cases, the same antenna serves both purposes. One example is the short antenna on a cellular phone.

303F10.EPS

Figure 10 Portable transceiver.

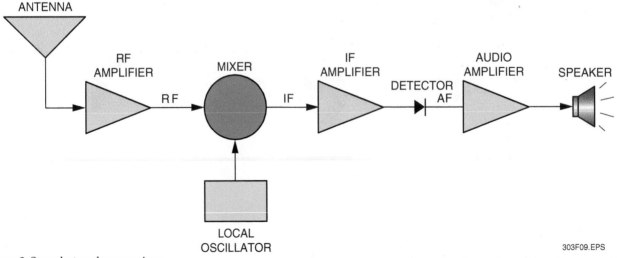

303F09.EPS

Figure 9 Superheterodyne receiver.

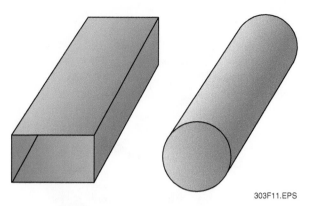

Figure 11 Waveguide.

The antenna can be anything from a loop of wire to a large dish, depending on the application. Antennas can be omni-directional (all directional) or directional. Some systems are line-of-sight, which means the transmitting antenna has to be able to see the receiving antenna. For that reason, transmitting antennas may be located on ground-level towers or on the tops of buildings. It is fairly common to locate transmitting antennas on top of mountains or hills, so that the transmitted signal is not obstructed.

The air is full of RF energy, so antennas are bombarded with signals of all types. The design of the antenna itself, along with the tuning/filtering system at the front end of the receiver, allows the receiving system to reject all but a limited band of frequencies. In order for an antenna to function in a given frequency band, its size must approximate the wavelength of the signal it is designed to process. The higher the frequency, the lower the wavelength will be. Therefore, the lower the frequency, the larger the antenna must be. This explains why antennas for AM radio stations (530 to 1,600kHz) can be more than a hundred feet high, while antennas for portable phones (900MHz range) may be only a few inches long.

The following provides a brief description of some of the different kinds of antennas.

- *Dipole antenna* – A dipole antenna is a half-wavelength antenna with two quarter-wavelength segments (*Figure 12*). Dipole antennas are common in short-wave and amateur radio. A folded dipole is a type of dipole made of two parallel wires with the ends connected. It is commonly used in FM receivers.
- *Parabolic antenna* – Parabolic antennas (*Figure 13A*), often called dishes, range in diameter from 18" to 15'. The size is a function of the frequency band(s) the antenna must handle. Dishes are used with UHF and microwave radio and TV receivers and transmitters, and for satellite TV.

Figure 12 Dipole antenna.

In transmitting mode, the transmitter RF is fed through the antenna's feed horn, then reflects off the surface of the dish and is radiated outward. In receive mode, the received RF signals are reflected off the dish into the feed horn. Some parabolic antennas use a different type of feed known as a Cassegrain feed. In this type of antenna, a small convex reflector is used to reflect the RF energy to or from the feedhorn at the center of the dish. The antenna in *Figure 13* has a Cassegrain feed.

It is interesting to note that satellite TV, because of its relatively low frequencies, used to require a very large antenna (several feet in diameter). As the technology evolved, the satellite frequencies got higher and the antennas got smaller. Now, you can get satellite TV with an 18" diameter antenna. The catch is that, unlike the original satellite TV, you have to subscribe to it in order to get the decoder that unscrambles the signal.

- *Helical antenna* – A helix is a coil shaped like a circular spring. In a helical antenna, the coil is attached lengthwise to the center of the antenna's reflecting surface, which can be circular, rectangular, cone-shaped, or horn-shaped (*Figure 13B*). Helical antennas are directional, and are used extensively in satellite communications.

- *Phased array antenna* – A phased array antenna (*Figure 14*) has two or more antenna elements connected to the feed line. Phased array antennas can be one- or two-dimensional. Most mobile communications phased array antennas are two-dimensional, which means the beam can be shaped both vertically and horizontally.

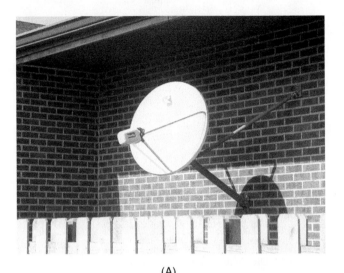

(A)

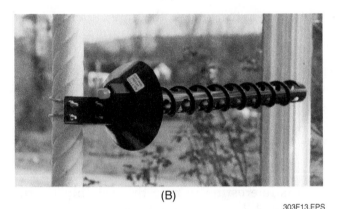

(B)

303F13.EPS

Figure 13 Parabolic and helical antennas.

303F14.EPS

Figure 14 Phased array antenna.

3.7.0 Voltage Standing Wave Ratio (VSWR)

Voltage standing wave ratio (VSWR), often pronounced vizwahr, is a measure of loss in a transmission line (such as coaxial cable or waveguide) caused by impedance mismatches between the antenna and the transmission line. VSWR is stated as a ratio. A perfect VSWR is 1 to 1, which is unattainable. A VSWR of 1.4 to 1 is considered pretty good. Later in the module, you will read about measuring VSWR.

4.0.0 INFRARED (IR) SYSTEMS

Infrared (IR) technology is widely used in residential and commercial security systems, remote lighting control, and remote control of audio and video systems. It is also used to implement wireless computer networks in an environment where RF or cabling would be ineffective. Wireless IR connections between computers and their peripherals, such as printers, have become fairly common.

Like RF systems, IR systems consist of a transmitter and receiver assembly used to operate a two-way electronic system in which IR is used to carry intelligence. The IR band is electromagnetic radiation at a frequency higher than RF and below that of visible light, having wavelengths ranging from 0.75 to 1,000 microns. The transmitter is required to generate and send the IR signal, and the receiver is required to receive the signal. Transmitter and receiver assemblies can be separate units, or they can be combined in a transceiver.

Two basic types of IR systems are direct and diffused. Direct (line-of-sight) systems require an obstruction-free transmission path for the IR signal between the transmitter and receiver. This requires that the transmitter and receiver units be accurately aligned. Direct systems are commonly used in applications such as security systems, home automation, and remote control of audio and video systems.

Unlike direct systems, diffused IR systems do not require a direct signal transmission path. They are capable of transmitting and receiving signals by reflections off walls, floors, and ceilings. This capability allows for more flexibility when placing equipment within an area. However, the diffused IR signal is more susceptible to multipath distortion, interference, and signal absorption caused by obstructions. *IEEE Standard 802.11* covers diffused IR transmission in the 300,000GHz to 428,000GHz band at data transmission rates of 1 and 2 megabytes per second (Mb/s). Such diffused IR systems are considered to be more secure from outside monitoring than when using RF wireless transmissions. Diffused systems are commonly used for data transmission in computer and data processing applications and wireless LANs.

Both the direct and diffused IR systems have a limited maximum range (typically about 200') which limits them to applications in an enclosed room or area. However, the direct IR system is sometimes used between buildings, providing there are no obstructions. The performance of an IR system is greatly affected by the location and environment in which it is installed.

Some points to consider concerning performance are as follows:

- Locations that have many obstructions reduce the effective range and create dead spots.

- IR signals are obscured by rain, fog, and air pollution (such as smog, soot, dust, and dirt).
- IR signals are relatively weak and cannot pass through solid or opaque objects.
- IR signals cannot bend around corners.

4.1.0 Basic IR Components

The basic components that make wireless transmission and reception of IR signals possible are the infrared light-emitting diode (IR LED) and the photosensitive detector, respectively. The IR signal source, which is contained in a transmitting assembly, is the IR LED (*Figure 15*), commonly called an IR emitter. It is a gallium arsenide

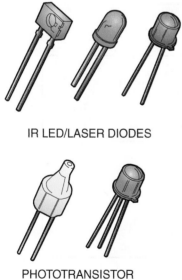

IR LED/LASER DIODES

PHOTOTRANSISTOR

IR LED/LASER DIODE
SCHEMATIC SYMBOL

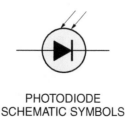

PHOTODIODE
SCHEMATIC SYMBOLS

PHOTOTRANSISTOR
SCHEMATIC SYMBOL

303F15.EPS

Figure 15 IR signal emission and detection.

(GaAs) semiconductor device encapsulated in a clear case and usually protected by a red plastic window. When forward-biased as a result of an applied voltage, it emits an IR signal just beyond the visible red wavelengths. This IR signal passes through the window in the transmitter assembly and is radiated into the surrounding space. Laser diodes are also used as IR emitters in some applications that require a higher output power level than that produced by an IR LED. Laser diodes and IR LEDs are similar in appearance. The schematic symbol is the same.

The photosensitive detector contained in the receiving assembly is either a phototransistor or photodiode, which are sensitive to IR radiation. The phototransistor is a semiconductor device that has a light-sensitive junction exposed to a received IR signal through a lens opening in the transistor package. When an IR signal strikes the phototransistor, it produces an output that is directly proportional to the intensity of the received IR signal. The photodiode is a semiconductor device similar in appearance to an IR LED. It has a small transparent window that allows light to strike its junction. It also produces an output that is directly proportional to the intensity of the received IR signal. The phototransistor or photodiode is usually mounted in a light shield called a collimator tube that is painted black to keep unwanted or stray light from reaching the detector. This tube also helps to guide the IR signals into the detector.

There are many kinds of IR circuits, each designed for a specific application. A few examples of common applications are described in the remainder of this section.

4.2.0 Remote Control Circuits

IR remote control circuits are commonly used to control audio and video equipment. The arrangement can be very simple, involving the control of a single device, or more complex, involving the control of several devices.

The use of a handheld remote control unit to control the operation of a TV set provides a good example of basic wireless IR transmitter and receiver operation (*Figure 16*).

4.3.0 Remote Control Distribution Systems

Remote control of several different audio/video components in an office or home is commonly done using an IR remote distribution system. There are many configurations of distribution

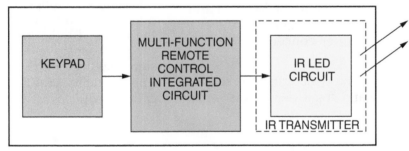

HANDHELD REMOTE

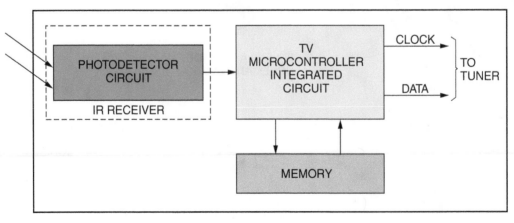

TV FRONT END

Figure 16 IR for audio and video systems.

303F16.EPS

systems. *Figure 17* shows a typical example of an IR distribution system used for controlling a satellite-fed TV/DVR system.

4.4.0 RS-232 Data Transmission Interface Systems

IR data transmission can be used in wireless LAN applications to interface computers and their peripherals. *Figure 18* shows a simplified block diagram of a basic high-speed RS-232 IR computer interface printed circuit board used for sending and receiving IR RS-232 data. Two boards are required per interface; one is located in each computer or other device being interfaced. The boards operate in an identical manner. As shown, a board consists of an IR LED, a photodiode, an IR transceiver integrated circuit chip, a serial interface chip, and an oscillator circuit.

4.5.0 IR Beam-Break Alarm Systems

Another application of IR is a wireless beam-break alarm system. There are many designs of beam-break circuits. A block diagram for one basic system is shown in *Figure 19*.

5.0.0 WIRELESS COMPUTER NETWORKS

The components of computer networks, such as LANs and WANs, are usually connected by cables. There are, however, situations in which it is not practical, or even possible, to use cables. That is where wireless networks (wireless LANs, or WLANs) come into play. The following are some applications in which a wireless LAN should be considered:

- A building in which it is not possible to run cabling above the ceiling or under the floor
- A remote area of the building where one person or a small group, such as the receptionist in the lobby of the building or a guard at a security checkpoint, needs to be connected to the network
- A remote building where it is not feasible to run cable
- A situation where buildings needing access to the network are separated by a highway, river, or other open area

A wireless network is created when two or more computers with wireless adapters operating on the same channel come close enough to communicate.

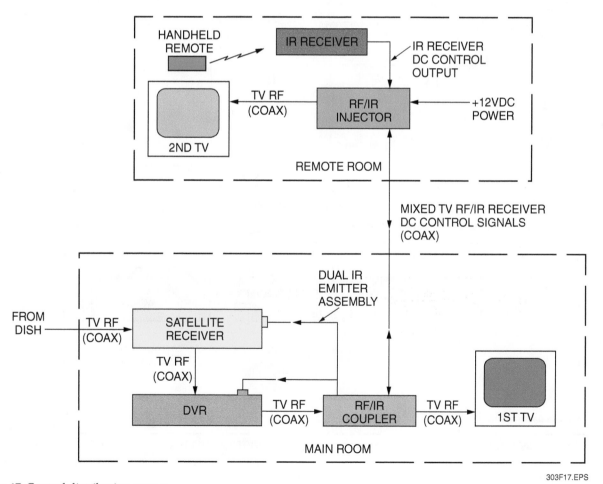

Figure 17 Control distribution system.

303F17.EPS

This is known as an **ad hoc** network. The more common application, usually found in businesses, uses a wireless access point (WAP) as the connection point for wireless devices. It also can be used to connect the wireless LAN to a hub of an Ethernet-cabled LAN, thereby giving authorized wireless users access to the company network.

An example of a wireless LAN application is a company that was expanding to another floor of an older office building and needed to do it quickly. When they learned that the building

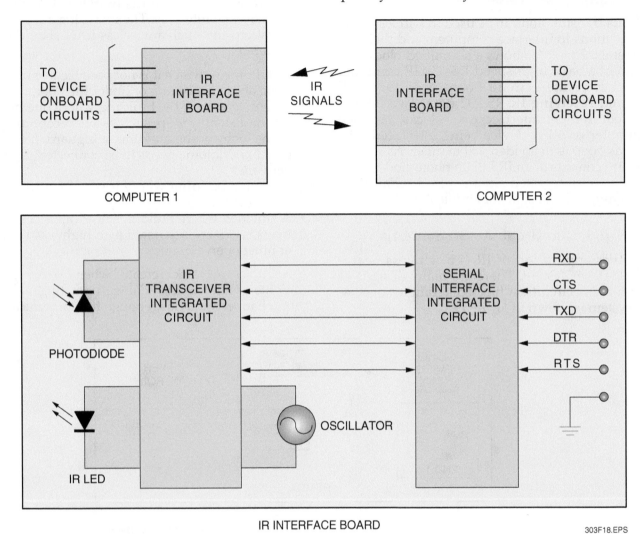

Figure 18 Overview of an RS-232 data transmission interface.

303F18.EPS

Figure 19 Overview of an IR beam-break alarm system.

303F19.EPS

had asbestos insulation between floors, they realized that they would have to go through an expensive and time-consuming asbestos removal process in order to run the cabling needed to expand their network. A wireless network extension was a perfect choice in this situation.

There are also situations in which a wireless LAN would not be a good choice such as the following:

- A factory or processing plant environment in which airborne particles could distort signals
- An environment in which signal security is a concern, unless the transmissions are encrypted
- A location in which other equipment, such as microwave ovens, could interfere with the RF signals

5.1.0 Background

The first wireless LANs operated in the 900MHz band and had data transfer speeds in the 1 to 2Mbps per second range. They were common in working environments such as warehouses and retail stores where employees used handheld devices such as scanners.

In the early 1990s, products in the 2.4GHz band began to hit the market. This band, which covers 83MHz of bandwidth from 2.4GHz to 2.483GHz, is sometimes known as the industrial, scientific,

and medical (ISM) band. With few exceptions, major manufacturers of wireless LAN equipment have adhered to this standard. The standard was formalized by *IEEE Standard 802.11*, which supports three physical transmission layers, two for RF and one for diffuse infrared. *IEEE Standard 802.11b* supports 5.5 and 11Mbps data transfer rates. Note that these transfer rates are significantly lower than those available on a wired LAN. *IEEE Standard 802.11a* covers equipment operating at 5.7GHz with transfer rates of 54Mb. The current standard is *802.11n*, which supports data transfer rates of 150Mbps, and can be configured to achieve a rate of up to 600Mbps. Although these rates are a significant improvement over the 54Mbps rate available with *802.11g*, they are a far cry from the gigabit rates available from wired Ethernet. It operates in the 5GHz band, but can also be used in the 2.4GHz band if it will not interfere with other *802.11* systems such as Bluetooth®. Under the *802.11n* standard, WLANs are capable of indoor ranges up to about 295 feet and outdoor ranges approaching 900'.

In an *IEEE 802.11* wireless network, access to and use of the network media is controlled by a media access protocol (MAC) known as carrier sense multiple access with collision avoidance (CSMA/CA). This protocol performs the same functions as Ethernet does in an *IEEE 802.3* network.

Wireless LANs communicate on spread spectrum radio waves. Spread spectrum technology, which works only with digital signals, spreads the information over a wide range of the available frequency band to make the signal less susceptible to noise. Spread spectrum transmission can use either the frequency hopping or direct sequence method:

- *Frequency hopping spread spectrum (FHSS)* – In FHSS, a large number of channels within the frequency band are used. Part of the signal is sent over one channel, then the transmission hops to another channel to send more of the signal, and then to another channel, and so on. In the U.S., the FCC regulates the number of hops (75 minimum), the hopping pattern, and the hop rate (2.5 hops per second minimum). The hop rate may also be stated as a dwell time (time between hops).
- *Direct sequence spread spectrum (DSSS)* – DSSS spreads the signal over a wider band. A DSSS signal maps each bit of data into a pattern while the signal is in digital form, which is referred to as a chip or chipping code. The ratio of chips per bit is known as the spreading ratio. The FCC requires the spreading ratio to be greater than 10, while *IEEE 802.11* calls for a spreading ratio of 11.

Two popular *IEEE 802.11*-compliant networking standards are AirPort™ and BreezeNet™. AirPort networking products, jointly developed by Apple and Lucent Technologies, are widely used in schools, universities, and similar settings, for wireless networking between AirPort™ base stations and AirPort™-enabled computers. BreezeNet is one product line of *IEEE 802.11*-compliant devices that uses DSSS radio technology. This RF technology does not suffer from problems encountered with IR technologies because it does not depend on a clear line of sight for signal transmission. It can penetrate walls, allowing it to work well in small network setups where wiring is impractical, or where the system must quickly set up and take down a complete network, such as for sporting events.

On Site

Bluetooth®

Bluetooth® is a communication protocol that operates in the narrow band between 2.402 and 2.480GHz. This is known as the industrial, security, medical (ISM) band, and is also used for devices such as baby monitors, garage door openers, and some cordless phones. To avoid band conflict, Bluetooth® transmits a very low-power signal of about 1 milliwatt, so paired devices must be within a few feet of each other in order to connect. In addition, Bluetooth® avoids potential conflict by using a protocol known as spread-spectrum frequency hopping, in which the frequency is constantly changing in a random pattern within a specified range.

5.2.0 Wireless LAN Equipment

A wireless LAN contains three major elements: an access point, an antenna, and a network interface card (NIC). The access point acts as the hub and transmitter for the wireless LAN. It is connected by cable to the wired network (*Figure 20*) and interacts with the wireless LAN through one or more antennas. *Figure 21* shows an example of a wireless LAN access point.

Each device on the wireless LAN requires an NIC (*Figure 22*). The NIC is similar to Ethernet cards used in wired LANs, except that it contains an antenna and an RF or IR transceiver.

There are several types of antennas used with wireless LANs. The type of antenna depends upon the application. Parabolic antennas may be used in building-to-building applications (*Figure 23*). The tubular antenna is designed to be suspended from a ceiling. The flat antenna can be mounted on a column. The location for mounting antennas used in a wireless network is important. For line-of-sight applications, the line of sight between antennas must not be obstructed by buildings, trees, telephone poles, or other obstructions.

Stations containing a wireless NIC can be formed into an ad hoc wireless network without a central control or connection to a wired LAN.

The addition of a wireless print server (*Figure 24*) allows such a network to use a shared printer. Wireless networks have limited range, which can limit their use. However, range extenders can be used to obtain greater coverage.

5.3.0 Wireless Network Security

One of the risks attached to wireless networks is that the network can be detected and accessed by anyone within range and having a computer with a wireless adapter. Having obtained access, intruders can use your internet access to avoid having to pay for their own. If they have malicious intent, they can read your email, access files on your server, and use packet sniffers to capture information such as user names, passwords, and credit card numbers. The key to securing a wireless network lies in understanding how the network functions. The following are some common methods of securing a wireless network:

- A wireless access point comes with a service set identifier (SSID), which it then assigns to all devices that are connected to the network. The SSID is essentially the name of the network. The access point periodically broadcasts the SSID so that nearby wireless users can find it. The SSID is a generic ID assigned by the access point manufacturer. It can be changed in the

---

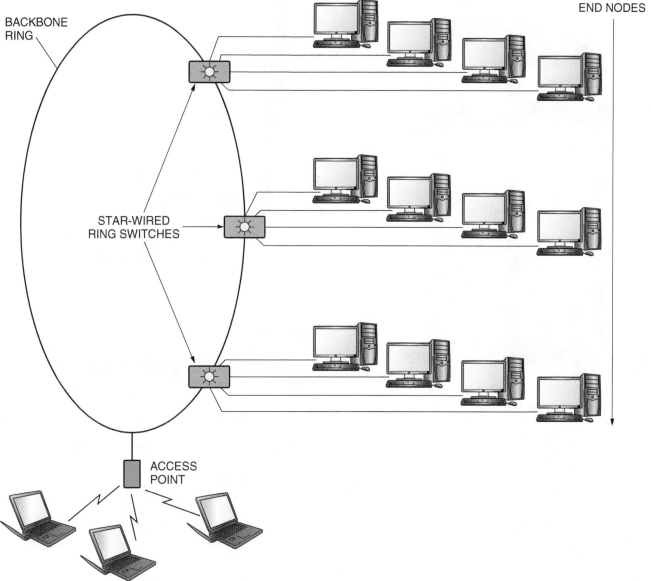

BACKBONE RING

STAR-WIRED RING SWITCHES

ACCESS POINT

303F20.EPS

Figure 20 Wireless LAN.

access point setup software, but a determined hacker can figure it out. Thus, changing the SSID is only the first line of defense in preventing unauthorized access.

- A wireless access point has a default user name and password assigned by the manufacturer. These values must be used to access the setup page for the access point. They should be changed and recorded during the setup process to prevent unauthorized access to the setup page.

- The encryption feature found on most access points should be activated when setting up the network. It encrypts the data being exchanged on the network so that only those with authorized access can read it. Access to the network can only obtained by entering the encryption

key, which can be either a phrase or randomly selected numbers and letters.

- One of the best ways to control network access is to install a firewall between the access point and the network devices.

6.0.0 SATELLITE COMMUNICATIONS

Satellites were first used for military communications. Then, in the early 1960s, other satellites began to be used for the transmission of television broadcasting to and from networks. Later, many satellites were used to broadcast television signals to cable systems and directly to individual subscribers. Today satellites are indispensable for the transmission and distribution of all types of communication, including radio/television network

Figure 21 Wireless access point.

Figure 22 Network interface card.

programming, telephone relay communication, and satellite phone service. Industrial applications include security monitoring, data monitoring, and surveillance. Many other satellites are used for global positioning/navigation, global mapping, weather monitoring, and scientific experiments or observations.

6.1.0 Satellite Communication System Overview

Earth stations transmit signals (uplink) to an orbiting satellite by means of a narrow radiated beam (*Figure 25*). The satellite receives this signal, amplifies it, shifts the frequency to a downlink frequency using a transponder, then retransmits the signal to other earth station(s) at the same time. This downlink signal covers an area known as the satellite footprint. There are three general types of uplink/downlink configurations in use that determine the footprint of the downlink:

- *Point-to-point* – This configuration is primarily for commercial communications carrier systems. The signal is transmitted from one single earth station to another single earth station.
- *Point-to-multipoint* – This configuration is the one primarily used for commercial network television or radio purposes. Various com-

mercial network signals are transmitted from specific spots to a satellite. The signals are then rebroadcast to broad areas on Earth for reception by commercial television or radio over-the-air broadcasters, commercial cable television (CATV) systems, private master antenna television (MATV) systems, and home direct broadcast service (DBS) or other television receive-only (TVRO) satellite antennas and receivers (*Figures 26* and *27*).

- *Multipoint-to-point* – This configuration is used for a very small aperture terminal (VSAT) where mobile or fixed-station point-of-sale ter-

On Site

Wi-Fi Hot Spots

Wi-fi is a term used to denote any *802.11* wireless network. Hot spots are open networks available to large numbers of users. They are commonly found in airports, libraries, and coffee shops. In some of these locations, it is a subscriber service available for a monthly or one-time fee. If you are sitting in an airport and want to check your email or download some information, you simply log on to the network, provide a credit card number, and use the service. Keep in mind that when you use such a service, you become part of an ad hoc network with all the others using the service at that location. You should have a firewall and disable file sharing on your computer. If you are going to access your company's network, it should be a virtual private network (VPN).

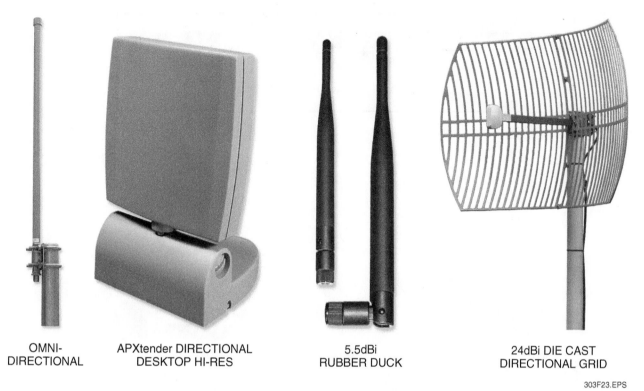

OMNI-DIRECTIONAL

APXtender DIRECTIONAL DESKTOP HI-RES

5.5dBi RUBBER DUCK

24dBi DIE CAST DIRECTIONAL GRID

303F23.EPS

Figure 23 Common wireless LAN antennas.

minals or monitoring equipment are linked to a home office via small- to medium-sized satellite antennas. The other major use is for mobile satellite-link telephone systems that allow users at any location on the planet to link to a communications carrier.

303F24.EPS

Figure 24 Print server.

When the earth stations are out of sight of a single satellite, the signal must be relayed between satellites, as shown in *Figure 28*.

6.2.0 Areas of Service

Satellite service can vary greatly depending on the downlink antenna size of the satellite. The coverage of a single satellite may incorporate a large portion of the hemisphere or a single metropolitan area. The greater the beam transmitted, the larger the service area will be. The larger the downlink antenna, the narrower the transmitted beam, and the smaller the service area will be to receive the signal. The following terms are used to identify different beams:

- *Spot* – Beams that cover a limited area and are used principally for point-to-point voice and

On Site

Microwave Ovens

Microwave ovens operate in the same frequency band as *IEEE 802.11* wireless LANs. However, this is a problem only at short distances. Wireless network devices must be kept more than 16.5' (5 meters) away from microwave ovens.

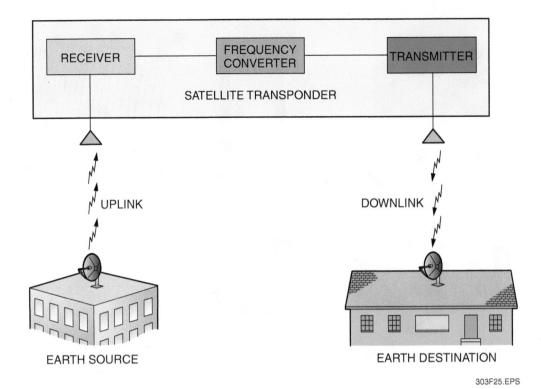

Figure 25 Overview of satellite uplink and downlink.

data communication. They are seldom used in television service.

- *National* – Beams that cover all or a significant portion of a single country. Most domestic US satellites are in this category. DirecTV®, Dish Network®, Wild Blue™, Hughesnet®, and Starband™ fall within this category.
- *Regional* – Beams that cover a group of countries, such as western Europe.
- *Global* – Beams that cover the entire area visible from the satellite. They are used for international communication.

6.3.0 Satellite Orbits

Most communications satellites use a geostationary (geosynchronous) orbit or a low-earth orbit (LEO). These orbits are nearly perfect circles. A few, including amateur radio satellites, use elongated, elliptical orbits.

6.3.1 Satellites in Geostationary Orbits

A geostationary (geosynchronous) orbit is the plane at which the property of centrifugal force, pushing an object outward into space, equals the

303F26.EPS

Figure 26 Typical cable TV TVRO antennas.

303F27.EPS

Figure 27 Satellite TV disks.

NCCER — *Contren® Learning Series* 33303-11

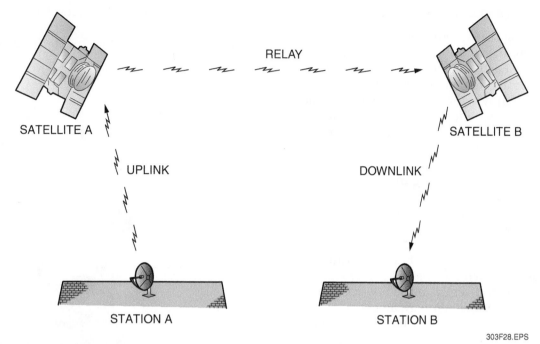

Figure 28 Signal path for long distance communication.

gravitational force (the force that pushes an object downward to the Earth). It is within this orbit that an object, moving in synchronism with the Earth's rotation, remains at a fixed elevated position and appears to be stationary. *Figure 29* illustrates a geostationary orbit. The geostationary orbit for Earth is a finite area about 22,247 miles directly above and in line with the equator. As a result, only a specific number of satellites may be

accommodated within this orbit. In order to maximize this number, the distance between satellites is critical and engineered to be minimal.

Established positions within the geosynchronous orbit are known as orbital slots. The arcs of slots, called orbitals, are defined by their longitudinal coordinates. Satellite orbitals for every country are allocated by international agreement. Within these orbitals, slots are assigned by the regulatory bodies of the country to various licensees. The United States regulatory authority is the Federal Communication Commission (FCC). The orbital arcs allocated to the United States are 62–103° and 120–146° west longitude for C-band satellites and 62–105° and 120–136° west longitude for Ku-band satellites.

On Site

IEEE Wireless Network Standards

There are actually a number of standards for wireless networking. Here are a few:

- *IEEE 802.11 – Working Group for Wireless LANs*
- *IEEE 802.11a – High Data Rate Extension (6/12/24Mbit/s, opt. 9/18/36/54 Mbit/s)*
- *IEEE 802.11b – High Data Rate Extension (5.5/11Mbit/s)*
- *IEEE 802.11e – MAC Enhancements for Quality of Service*
- *IEEE 802.11f – Recommended Practice for Inter Access Point Protocol*
- *IEEE 802.11g – Standard for Higher Rate (>20 Mbps) Extensions in the 2.4 GHz Band*
- *IEEE 802.11h – SMa, Spectrum Managed 802.11a*
- *IEEE802.11n – WLAN Enhancements for Higher Throughput*

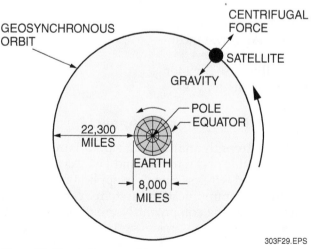

Figure 29 Geostationary orbit.

 Wireless Communication

6.3.2 *Satellites in Low-Earth Orbit*

Low-earth orbit (LEO) satellites are fleets of satellites operating in various low-altitude orbits over, or close to, the Earth's north and south geographic poles. These satellites are not stationary; they are continuously moving over the Earth's surface. The numbers of satellites in any orbit can vary; however, the more satellites in a particular orbit, the more reliable the communications handled by that group of satellites will be. There must be one or more satellites moving in the line-of-sight range of every point on the Earth covered by the orbit to allow continuous communications operation. A LEO system operates like a cellular telephone network, except that the transponders are moving in space instead of being mounted on fixed towers on Earth. The signals from the satellites are relayed directly to the nearest conventional communications network available.

The advantages of LEO satellites are three-fold. First, the distances at which the systems operate are very short compared to the 44,500-mile round-trip (about two seconds delay) that is required for a geostationary satellite link. This allows high-speed or voice communications with very little detectable delay between transmission and reception if the source and destination are within the range of one or two satellites. Second, very little power is required to communicate with the satellite. This makes the use of a mobile satellite phone (*Figure 30*) feasible for communications. Third, the low power requirements allow use in areas, such as deserts or over water, that are not accessible to standard cell phones or other terrestrial telephone systems.

7.0.0 TEST EQUIPMENT

This section gives an overview of some common items of test equipment used to test and monitor parameters in wireless systems:

- RF field strength analyzer
- RF analyzer/standing wave meter
- RF power meter
- Satellite signal meter
- Spectrum analyzer

7.1.0 RF Field Strength Analyzer

RF field strength analyzers (*Figure 31*) are used to measure wide and narrow band FM, AM, and single-sideband wireless signals in applications such as mobile telecommunications systems, cable TV, and satellite-receiving equipment. Normally, they are capable of measuring wireless signals via an antenna, but they can also be used to make direct-connection measurements. The typical RF field strength meter

contains a built-in digital frequency counter and RF spectrum analyzer. Most are capable of scanning 150 or more channels. With menu-driven mode se-

303F30.EPS

Figure 30 Typical mobile satellite phone.

303F31.EPS

Figure 31 RF field strength analyzer.

lection, they are capable of displaying frequency spectrums, single- or multi-channel frequency and level data via a bar graph presentation, and the exact signal frequency via a frequency counter. Most have the capability of outputting measured data to a computer for further analysis via an RS-232 output port.

7.2.0 RF Analyzer/Standing Wave Meter

An RF analyzer meter (*Figure 32*) is a portable, microprocessor-controlled instrument that transmits a low-power RF signal used to check and adjust antennas and related feedlines, as well as RF networks. It can be used to measure the standing wave ratio on an antenna line. Standing waves are voltage (or current) signals that are reflected back and forth on an RF transmission line. They occur as a result of impedance mismatches between the line and load (antenna). These reflected signals add to or subtract from the main (incident) signal on the line, causing areas of voltage peaks and voltage minimums (nulls). The ratio of the maximum value to the minimum value of the standing waves on the line is defined as the standing wave ratio (SWR) or voltage standing wave ratio (VSWR).

In a system where the transmission line impedance is perfectly matched to that of the load, the SWR is 1:1 because the voltage and current are in the same proportions everywhere along the line. In actual conditions, an SWR of 1:1 is never achieved and the SWR is always larger than 1:1. The value of the SWR is an important indicator of the performance of an antenna system. A high SWR value indicates a severe mismatch between the antenna and transmission line. For example, if an open or short exists in a transmission line, the SWR can be as high as 30:1 or greater. A high SWR can cause significant signal loss between an antenna and receiver. A high SWR on the transmission line between a transmitter and the antenna can severely damage the transmitter as a result of high power being reflected back into the transmitter. A high SWR can also cause arcing in waveguide-type transmission lines.

An RF analyzer can also be used to measure transmission line parameters such as loss, impedance, and electrical length. Some models also measure capacitance and inductance, R and X components of impedance, parallel R and X, and impedance at the far end of a feedline. Higher quality instruments can be used to check feedlines with impedances ranging from 25 to 450 ohms (Ω). Some can automatically determine if a load is inductive or reactive and display the value of the series coil or capacitor that should be added to eliminate series reactance and yield the lowest SWR.

7.3.0 RF Power Meter

RF power meters (*Figure 33*) are used in conjunction with the appropriate power sensor to measure the power of various RF signals found in wireless systems. Different model instruments cover different frequency ranges. A typical power meter may be capable of measuring power levels ranging from –70dBm to +45dBm and cover a frequency range of 10kHz to 100GHz. In newer instruments, all major functions are menu driven and/or selected by front panel touch-sensitive buttons. The power level of the measured RF signal expressed in dBm is typically displayed on a 4½ or 5½ digit LCD display.

303F32.EPS

Figure 32 RF analyzer/standing wave meter.

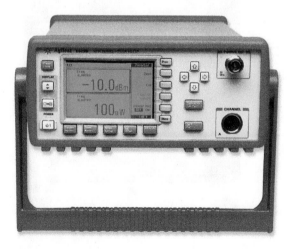

303F33.EPS

Figure 33 RF power meter.

7.4.0 Satellite Signal Meter

There are several types of portable, battery-operated satellite signal meters (*Figure 34*) that are used to monitor the strength of the IF signal output from the satellite antenna LNB (down converter) as an aid when aligning a satellite dish. As a minimum, the tester normally provides a direct readout of signal strength in dBµV, dBmV, and/or by a bar graph-type indication. It also has a built-in variable frequency audio tone generator driven by the received input signal. This feature allows for no-look aiming of the antenna. Correct alignment of the antenna is indicated when there is a maximum deflection of the meter and the highest audio pitch tone.

More complex satellite testers have other features to aid in the evaluation and optimization of satellite receiver installations. Typically, these features include the capability of measuring the DC voltage supplied to the LNB, the intermediate frequency (IF), the local oscillator frequency, and the carrier-to-noise (C/N) ratio. Most have the capability of powering the LNB in order to allow for independent evaluation of the antenna dish/LNB part of the system. Some units have factory-programmed worldwide channel listings and provide for multiple channels of user-programmed memory. Models are also available that allow the focusing of two satellite dishes by being able to observe meter swings on two separate meters (one for each dish) at the same time.

8.0.0 ANTENNA INSTALLATION

Antenna installations can be simple, involving a single antenna, or complex, involving multiple antennas mounted on a common tower (*Figure 35*). Depending on the size, weight, and construction, single antennas can be mounted in numerous ways. Larger ones are generally mounted on a tower, the roof of a building, or on a post in the ground. Smaller antennas are commonly installed on a pole, roof, or the side of a building. Regardless of the type, antennas must be installed in accordance with the manufacturer's instructions and local codes, using the appropriate mounting hardware and supports. Pole and bracket assemblies must be plumb. In line-of-sight applications, the line-of-sight between antennas, or between the antenna and a satellite, must not be blocked by buildings, trees, telephone poles, or other obstructions.

NEC Article 810 covers the installation and grounding requirements for communications antennas. Some guidelines are as follows:

- Antennas shall be securely supported, and lead-in wires shall be securely attached to the antenna.
- Antennas and lead-in wires shall not be attached to an electric service mast or to any poles that carry light and power lines having over 250V between conductors.
- Antennas and lead-in wires shall be kept away from and shall not cross over light and power conductors, so as to avoid accidental contact. If possible, they should not be run under open light and power conductors.
- On the outside of a building, lead-in conductors shall be at least 2' away from open light and power conductors if there is 250V or less between conductors, or at least 10' away if there is over 250V between conductors.
- Lead-in conductors shall be kept at least 6' away from a lightning rod system or else be bonded together in accordance with the NEC®.
- Lead-in wires shall not be run in electric boxes unless there is an effective, permanently in-

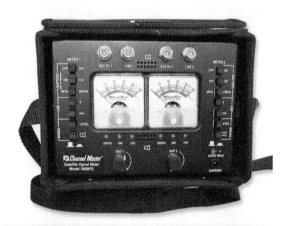

303F34.EPS

Figure 34 Satellite testers.

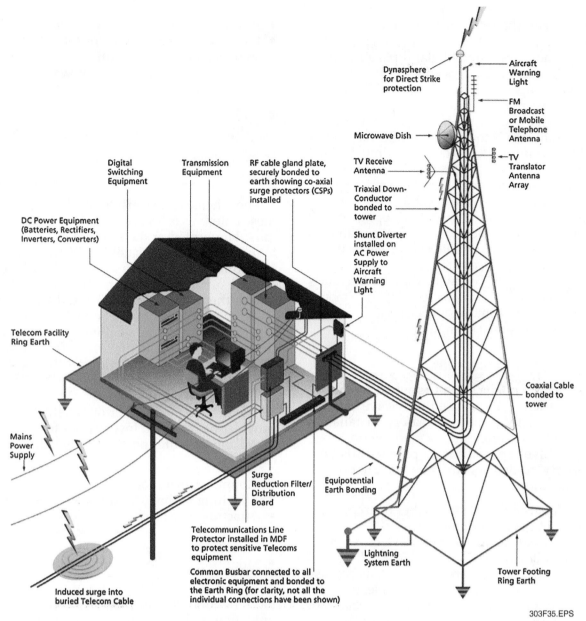

Figure 35 Telecommunications antenna installation overview.

stalled barrier to separate the light and power wires from the lead-in wires.

In the figure, you can see the typical grounding scheme for a group of tower-mounted antennas associated with a telecommunications facility. For these and simpler single-antenna installations, *NEC Section 810.21* gives the following guidelines:

- Grounding wires must be copper, aluminum, copper-clad steel, or similar corrosion-resistant material. They need not be insulated and shall not be smaller than No. 10 copper, No. 8 aluminum, or No. 17 copper-clad steel or bronze.
- Grounding wires can be run inside or outside of a building. They shall be run as straight as possible and they must be securely fastened in place.

- Grounding wires may be attached directly to a surface without the need for insulating supports. They shall be protected from damage or be large enough to compensate for lack of protection.
- The grounding conductor shall be connected to the nearest accessible location on the building or structure grounding electrode, the grounded interior metal water piping system, the power service accessible means external to enclosures, the metallic power service raceway, the service entrance enclosure, or the grounding electrode conductor.

To protect wireless equipment from lightning and induced electromagnetic disturbances, each coaxial or other type of antenna lead-in cable should be protected by a surge protector. The surge protector should be installed as close as possible to where the feeders enter the building. This is often done at a grounded gland plate installed in the wall where the cables enter the building.

For proper operation, satellite and other line-of-sight antennas must be properly aligned (focused) with the signal source. When the physical installation of the antenna is complete, the antenna must be aimed to specific angles to obtain optimum signal reception. Smaller antennas can easily be positioned by hand in order to focus them. Larger antennas may have a motor-driven mechanism to aid in alignment. Initially, a compass and topographical maps are used in conjunction with the manufacturer-supplied alignment data to coarsely focus the antenna so that it receives a signal as measured on a signal strength meter. Following this, the antenna is moved slowly in very small increments on its mount so as to obtain a maximum signal level indication as displayed on the signal strength meter.

8.1.0 Antenna Placement

There are two major concerns when installing antennas such as those used to receive satellite and broadcast TV. The first is safety. Installation of these antennas often means working on a roof, which involves special safety practices. Follow these safety guidelines any time you are doing a roof installation:

- Wear boots or shoes with rubber or crepe soles that are in good condition.
- Always wear fall protection devices, even on shallow-pitch roofs.
- Rain, frost, and snow are all dangerous because they make a roof slippery. If possible, wait until the roof is dry; otherwise, wear special roof shoes with skid-resistant cleats in addition to fall protection.
- Brush or sweep the roof periodically to remove any accumulated dirt or debris.
- Check and comply with any federal, local, and state code requirements when working on roofs.
- Be alert to hazards such as live power lines.
- Use common sense. Taking chances can lead to injury or death.

When working outdoors or in high-heat conditions for extended periods of time, take precautions to avoid heat exhaustion and exposure to the sun's ultraviolet rays. Preventive measures include the following:

- Wear a hard hat.
- Wear light clothing that is made of natural fibers.
- If possible, wear tinted glasses or goggles.
- Use a sun protection factor (SPF) 30 or higher sunblock on exposed skin.
- Drink adequate amounts of water to prevent dehydration, especially in arid climates.

> **NOTE**
> Always follow OSHA and company safety guidelines.

The other special consideration when installing these antennas is line of sight. That is, the antenna must be pointed at the satellite or broadcast antenna. In the case of satellite TV, there must be a clear line of sight to the satellite. This involves a careful survey of the property to find a location where the dish is not obstructed by buildings, trees, or other obstacles. The installation instructions provided by the satellite TV service specify the direction that the dish must face. In the case of broadcast TV, information on the locations of nearby transmitting antennas is available on the Internet at www.antennapoint.com.

The dish manufacturer or satellite TV service provider has detailed instructions for placing and installing an antenna. These instructions typically cover:

- Mounting on a roof
- Attaching to the sidewall of a frame structure
- Attaching to a brick wall
- Attaching to a concrete block wall
- Mounting on a pole installed on the ground
- Mounting on a deck or other wood surface

Always follow the antenna installation manual. The following are some general installation guidelines to keep in mind:

- If the antenna is being attached to a roof or sidewall, be sure it is securely attached to a truss (rafter) or wall stud with the lag screws prescribed. Sheathing such as OSB or plywood under the roofing or siding does not hold the antenna in a strong wind.
- In a roof installation, always use roof sealant in the openings you drill. Failure to do so is likely to result in a leak.
- If the antenna is being installed on a brick or concrete wall, drill holes and install the antenna mounting bracket using the prescribed expanding anchors.
- For a concrete block wall, use the prescribed hollow-wall anchors.
- If the antenna is to be mounted on a pole in the ground, guy wires may be needed to support it.

- Do not mount the antenna under an eave or overhang that might block reception.

9.0.0 NOISE/ELECTROMAGNETIC INTERFERENCE

Devices such as wireless phones, microwave ovens, commercial data links, and nearby antennas can radiate RF energy that interfere with the performance of wireless communication devices and networks. Electrical power lines, electric motors, power supplies and other sources provide electromagnetic interference (EMI). RFI and EMI can distort signals, causing static and audible noise in audio and snow in TV pictures.

Noise is defined as any unwanted electrical signals that are induced onto or superimposed on power or signal lines. Note that voltage and current transients/surges are commonly referred to by some as noise signals. Noise interference appears between the terminals of a circuit. Noise is classified in two ways: common-mode noise and normal-mode noise. Common-mode noise occurs between the line and ground or neutral and ground, but not between each line. The noise signals on each of the current-carrying conductors are in phase and equal in magnitude; thus, no voltage signal is generated between the conductors by the noise. Normal-mode noise, also called traverse-mode noise, occurs between the current-carrying conductors (line-to-line) or line-to-neutral. A voltage is generated between the ground and neutral lines because noise is only present on two of the conductors.

Electrical cables are vulnerable to receiving EMI noise from nearby sources. The transfer of noise can occur over one or more paths by radiation, conduction, and/or inductive and capacitive coupling, which is energy transfer between circuits or conductors caused by mutual capacitance between them. Note that optical fibers neither emit nor receive EMI. Improper bonding, shield-

ing, and grounding of cable shields and equipment can increase the susceptibility of a cable or device to EMI. A spectrum analyzer can be used to locate the source of RFI. While monitoring the affected device with the spectrum analyzer, turn off other devices and observe the resulting affect on the distorted signal.

In the United States, the Federal Communications Commission (FCC) is responsible for specifying EMI limits to prevent unacceptable levels of electromagnetic pollution (interference) being released into the environment. FCC regulations establish the maximum permissible emissions of electronic devices. For computing devices, the FCC separates its regulations for digital EMI into two categories: Class A computing devices (industrial and commercial) and Class B computing devices (residential). The FCC rules set limits on two kinds of emissions: conductive emissions and radio frequency (RF) emissions. Conductive emissions travel through the wires in the power cord. RF emissions radiate from devices into space.

Wireless communication takes many forms:

- Satellite TV and communication
- Infrared security, entertainment control systems, home automation, and computer interfaces
- Wireless microphone systems
- Assisted listening systems
- Cellular phone systems
- RF communication
- Wireless computer networks

In many cases, wireless systems interface with wired systems; wireless LANs and cellular telephone systems are good examples. As technology advances, an increase in the use of wireless systems can be anticipated. For that reason, it is essential for the EST to understand, and be able to work with, both wired and wireless systems.

Review Questions

1. Which of the following operates in the highest frequency range?

   a. FM radio
   b. Cellular phone
   c. Infrared system
   d. UHF TV

2. Which of the following bands represents the highest frequencies?

   a. L
   b. X
   c. Ku
   d. Ka

3. The frequency range for FM radio signals is _____.

   a. 87.5MHz–108MHz
   b. 30MHz–300MHz
   c. 530kHz–1,600kHz
   d. 800MHz–4.1GHz

4. Which of these components acts as the transducer in a radio transmitter?

   a. Local oscillator
   b. Microphone
   c. Mixer
   d. Power amplifier

5. The process in which low-capacity communications channels are combined into a single channel is know as _____.

   a. superheterodyning
   b. multiplexing
   c. modulation
   d. companding

6. The device that produces the carrier frequency in a transmitter is the _____.

   a. antenna
   b. decoder
   c. oscillator
   d. detector

7. Which type of antenna is made of two parallel wires with the ends connected?

   a. Folded dipole
   b. Parabolic
   c. Helical
   d. Phased array

8. The two types of IR systems are _____.

   a. direct and indirect
   b. indirect and diffused
   c. diffused and direct
   d. serial and parallel

9. In a wireless LAN, the device that acts as the direct interface between the wired network and the wireless network is the _____.

   a. NIC
   b. antenna
   c. receiver
   d. access point

10. A wireless network is identified by its _____.

    a. WAP
    b. SSID
    c. CDMA
    d. LAN

11. The network encryption key and the access point password are the same thing.

   a. True
   b. False

12. A point-to-multipoint satellite uplink/down-link configuration is used primarily for _____.

   a. commercial communications
   b. commercial TV and radio
   c. point-of-sale terminals
   d. wireless LANs

13. If a transmission line shows a VSWR of 30:1, it indicates that the _____.

   a. cable is open or shorted
   b. antenna and transmitter are perfectly matched
   c. antenna and transmitter are slightly mis-matched
   d. transmission has the correct impedance

14. Antenna alignment is done using a(n) _____.

   a. spectrum analyzer
   b. frequency counter
   c. RF analyzer/standing wave meter
   d. signal strength meter

15. Which of these items of test equipment would be used to locate the source of RFI?

   a. Spectrum analyzer
   b. RF analyzer/standing wave meter
   c. RF field strength analyzer
   d. Satellite tester

Trade Terms Introduced in This Module

Ad hoc: A Latin term meaning formed for a special purpose.

Companding: A method used for compressing and then expanding a wireless transmission. Wireless microphones compress the signal at the transmitter and expand it at the receiver.

Emitter: A semiconductor device that emits an infrared signal. It is used as a transmitting device in IR systems.

Micron: A unit of length equal to one-millionth of a meter.

Microwave: A radio frequency in the range of 0.3 to 300GHz. In alarm technology, an RF system using motion detectors that operate in the microwave frequency range.

Oscillator: An electronic device that produces a pure sine wave at a specified frequency or range of frequencies.

Packet sniffer: Also called packet analyzer. It is a hardware or software tool used to intercept network traffic.

RS-232: An interface standard for connecting serial devices. The standard supports a 25-pin or 9-pin D-type connector.

Transducer: A device that converts one form of energy to another. A microphone and a speaker are transducers used for opposite purposes.

Voltage standing wave ratio (VSWR): A measure of the impedance match between two RF components.

Additional Resources

This module presents thorough resources for task training. The following resource material is suggested for further study.

The Essential Guide to RF and Wireless. Upper Saddle River, NJ: Prentice Hall.

Handbook of Radio and Wireless Technology. New York, NY: McGraw-Hill.

Wireless Personal Communication Systems. Reading, MA: Addison-Wesley.

Figure Credits

CONTREN® LEARNING SERIES — USER UPDATE

NCCER makes every effort to keep its textbooks up-to-date and free of technical errors. We appreciate your help in this process. If you find an error, a typographical mistake, or an inaccuracy in NCCER's Contren® materials, please fill out this form (or a photocopy), or complete the online form at www.nccer.org/olf. Be sure to include the exact module number, page number, a detailed description, and your recommended correction. Your input will be brought to the attention of the Authoring Team. Thank you for your assistance.

Instructors – If you have an idea for improving this textbook, or have found that additional materials were necessary to teach this module effectively, please let us know so that we may present your suggestions to the Authoring Team.

NCCER Product Development and Revision
3600 NW 43rd Street, Building G, Gainesville, FL 32606

Fax: 352-334-0932
Email: curriculum@nccer.org
Online: www.nccer.org/olf

❏ Trainee Guide   ❏ AIG   ❏ Exam   ❏ PowerPoints   Other _____

Craft / Level: _____   Copyright Date: _____

Module Number / Title: _____

Section Number(s): _____

Description: _____

_____

_____

_____

_____

Recommended Correction: _____

_____

_____

_____

_____

Your Name: _____

Address: _____

_____

Email: _____   Phone: _____

Site Survey, Project Planning, and Documentation

33304-11

Trainees with successful module completions may be eligible for credentialing through NCCER's National Registry. To learn more, go to **www.nccer.org** or contact us at **1.888.622.3720**. Our website has information on the latest product releases and training, as well as online versions of our *Cornerstone* newsletter and Pearson's Contren® product catalog.

Your feedback is welcome. You may email your comments to **curriculum@nccer.org**, send general comments and inquiries to **info@nccer.org**, or use the User Update form at the back of this module.

Objectives

When you have completed this module, you will be able to do the following:

1. Describe the general procedure or steps involved when estimating a job for the purpose of submitting a bid.
2. Describe the general procedure or steps required to properly plan and complete a job once a contract for the job has been awarded.
3. Interpret contractual documents, working drawings, and specifications pertaining to a job to determine the requirements and scope of the work.
4. Perform a site survey in order to establish or confirm the installed locations of new and/or existing equipment and the routing of the related cabling.
5. Develop a schedule for completing a job or task from start to finish that efficiently accomplishes the work and is also compatible with the work performed by other trades.
6. Recognize and interpret the various types of forms and other documentation used when estimating and planning a project.

Performance Tasks

Under the supervision of the instructor, you should be able to do the following:

1. Interpret contract documents in order to determine the requirements for a selected job.
2. Perform a survey in order to accomplish the following:
   - Compare the working drawings for the site against the actual building structure to identify specific locations and the work to be performed there.
   - Confirm the installed locations of new and/or existing equipment and the routing of the related cabling.
   - Measure the routing and length of selected cable pathways and raceways to verify measurements shown on floor plans and/or estimate takeoff sheets.
3. Use task and labor hours data recorded on estimating forms and/or takeoff sheets for a selected job to develop a detailed schedule for accomplishing the job.

Trade Terms

Change order	Exclusions	Record drawings	Specifications
Contract	Punch list	Scope of work	Takeoff
Direct cost			

Contents

Topics to be presented in this module include:

Figures and Tables ————————————————

1.0.0 INTRODUCTION

There are three important factors that affect a company's ability to perform a professional and profitable installation. They are:

- Performing an effective site survey
- Proper and effective planning
- Acquiring and correctly interpreting the drawings, specifications, and other documents that define the project

This module begins by providing a brief overview of the process for estimating and bidding on a job. This overview is intended to provide an understanding of the scope of the estimating task and familiarize you with some of the terminology used. Understanding the estimating process and terminology is important so that you can communicate with the estimator after a contract has been awarded. Also, you will need to use some of the working papers generated during the estimating process to help you plan how to best do the job.

One of the most important aspects of job planning and estimating is to completely understand what is expected of you. Some projects simply require you to install and test systems in accordance with design concepts established by an architect or engineer. That is, they design it, and you build it. Other projects require you to develop the system concept and then implement the concept with systems and components. On these jobs, all of the work is performed under a single contract. This is a design-build approach, also known as design-construct or single responsibility.

2.0.0 THE JOB ESTIMATING AND BIDDING PROCESS

An estimate is a forecast of future costs. The accuracy and reliability of an estimate depend on the amount and quality of the information known at the time the estimate is made.

The person who does the estimating must have knowledge of the entire business operation, from the office accounting system to field production and bill collecting. The estimator needs to understand the installation crew's capabilities, the pricing practices of competitors, the customer's needs, and the number and availability of suppliers. For these reasons, estimating is normally done by managers, supervisors, or experienced technicians who have the required knowledge and skills. Depending on the size of your company, these individuals may work alone or as a team to develop an estimate.

Figure 1 shows a typical systematic approach used for estimating a job in preparation for submitting a bid. Depending on the job, some of the tasks may or may not be required.

2.1.0 Management Decision to Bid

A company wants to bid on those jobs that give it the best opportunity to be successful and make a profit. Not all jobs are desirable, and a company has limitations. Several factors are involved when management decides to bid a job:

- The initial decision to bid is often based on how well the company understands the products

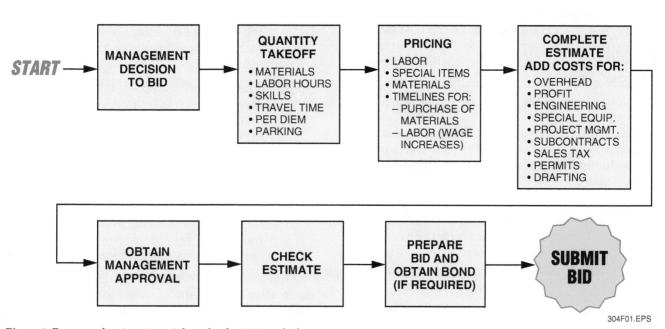

Figure 1 Process of estimating a job and submitting a bid.

304F01.EPS

 Site Survey, Project Planning, and Documentation

and processes involved and the extent to which it has the necessary equipment and skills.

- The time required to do the job must be compatible with the overall company schedule of prior commitments. The capacity of the schedule is determined by the willingness to add new employees, expand the management workload, and/or acquire new equipment.
- The job must be judged by the impact it will have on the working capital of the firm and also the company's ability to meet bonding and insurance requirements.
- The decision to bid may be affected by the desire to maintain a good relationship with existing customers or to build new customer relationships.
- The company may recognize an opportunity to obtain a project due to the small number of competitors.
- The company may be one of a few that possess the technical skills to perform a particular kind of job.
- The company may have an interest in opening new product lines to expand the business.

2.2.0 The Estimating Process

One of the most important elements of a cost estimate is the quantity takeoffs. These must be prepared carefully because they are the source for identifying and purchasing the equipment and materials to be used on the project. The takeoffs are also used to determine the types and amounts of labor and equipment that are needed. An error or oversight at this level can have serious cost and schedule impact later on.

Another important element of the cost estimating process is historical data. Most contractors keep detailed records of the costs on previous jobs. Such data could include the following:

- Prices of materials, labor costs, equipment, and other expenses
- Labor productivity rates for certain tasks
- Installation crew size and mixes that are most productive for certain types of work
- Past history with the customer or other contractors

The historical data may not be used directly in preparing a new estimate, but it can be very effective as a reality check of the estimate.

The following are the steps in the estimating process:

Step 1 Using the drawings and specifications, perform a takeoff of all the materials needed to complete the job. This information is placed on a quantity takeoff sheet, such as the worksheet shown in *Figure 2*.

Step 2 Establish the amount of time it takes for the designated crew size and crew composition to perform a specific amount of work. These are known as production figures. Most companies keep records of these figures from past jobs. Also, commercially published manuals are available that contain this information, but these should only be used if company figures are not available.

Step 3 Determine the cost of equipment to be used on the project. If the contractor does not own this equipment, it may have to be rented. Although cost data may be available from past projects, it is usually better to contact suppliers to determine the current cost.

Step 4 Take the total material quantities from the worksheet and place them on a summary sheet (*Figure 3*). Material prices are obtained from suppliers or, if material is stored by the company, by using current inventory costs.

Step 5 The estimator multiplies the labor hours by the hourly cost of labor to determine the labor costs for each project element. If subcontractors are used instead of employees, quotes for specific tasks should be obtained from subcontractors.

Step 6 The total direct cost of material, labor, and equipment is then found and placed on the summary sheet. One can also find the unit cost (the total cost divided by the total number of units of material to be put into place).

Step 7 The costs of material, labor, and equipment for each activity are then placed on a recapitulation sheet (*Figure 4*), which serves as a summary for the entire project.

Step 8 Costs for special items, such as travel expenses, licensing fees, bonds, subcontractor services, taxes, engineering and design, supervision, and other requirements are added to the direct costs to come up with the total job cost.

This estimating process described may also be done after a job is under contract in order to determine materials and labor for additional work or as a result of a change order. Change orders are described later in this module.

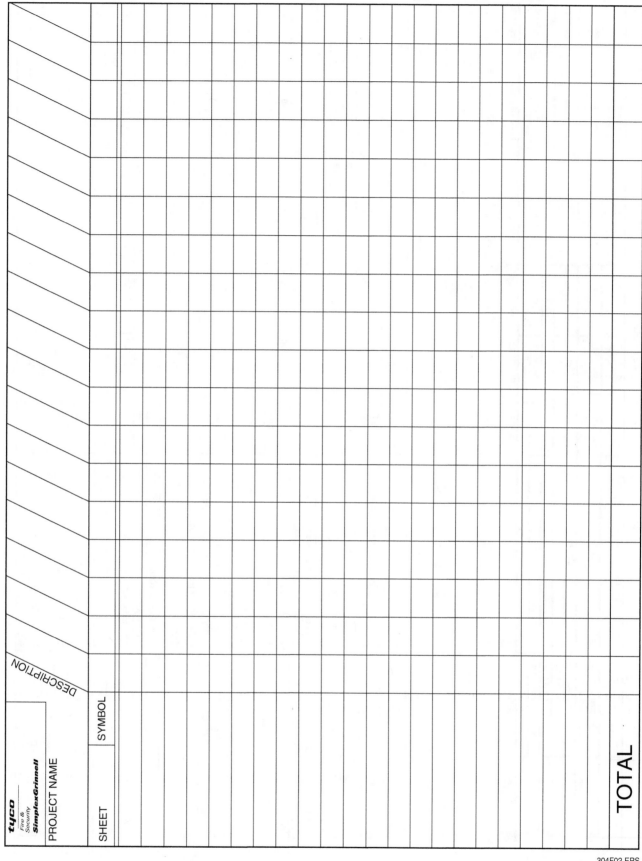

Figure 2 Quantity takeoff worksheet.

304F02.EPS

 Site Survey, Project Planning, and Documentation

Sequoia auditorium-rev 2.xls - Estimate

Note: Anything with Blue Text Can be Changed

PROJECT:		Rev. 02/19/00
SYSTEM: Sound System		
SALES: Dick Fyten		

		Tech Rt ->	$80.00	Per Hour	$80.00	Accum.Hrs.	189.00
Material Mark Up %	50.00%	Hardware % 2.00%	Install Rt ->	$60.00	Per Hour	$60.00	Accum.Hrs. -
Sub Contract Mark Up %	20.00%	Engr Rt ->	$90.00	Per Hour	$90.00	Accum.Hrs. -
Preferred Margin	33.33%	Frgt % ->	3.00%	Of Material Cost (Enter as 0.00)		
2ND OR 3RD Shift		Warnty % ->	5.00%	Of Material Cost (0.00)		
1 1/2 OR 2 (1.5 or 2)		Engr % ->	3.00%	Of Total Cost (Enter as 0.00)		
Subcontract Margin %	16.67%	True Sell Margin	2%	Gross Margin 33.33%		PM Hours
Project Management %		Sell price (inc. Bond)	$68,053.72	Sell Price Margin Excl. Sub & Bond 33.33%		

QTY.	VENDOR	PART NO.	DESCRIPTION	MATERIAL UNIT	TECH HRS UNIT	SYS ENGR UNIT	INSTALL UNIT	MATERIAL NET	SYS ENGR NET	TECH HRS NET	INSTALL NET
			"ENTER HERE AND BELOW"								
			COMPLETE SYSTEM WITH GTI LABOR								
			CLUSTER								
3	Electro Voice	QRx-153/75	EV three way flyable loudspeakers	1,099.00				3,297.00			
3	Electro Voice	CPS-3	EV power amplifier (FR loudspeaker)	949.00				2,847.00			
2	Electro Voice	QRx-118S	EV single 18" subwoofer	649.00				1,298.00			
1	Electro Voice	P1201	EV power amplifier (subs)	759.00				759.00			
1	Electro Voice	M118S	Sub module for 118S	79.00				79.00			
1	Electro Voice	DX38	EV system controller	799.00				799.00			
1	ASHLEY	GQX3101	Ashly dual 1/3 octave house EQ	519.00				519.00			
1000	WPW	25227B	12 GA wire for cluster installation	0.20	40.00			200.00		40.00	
1	GTI		Labor to terminate								
1			Mounting hardware	1,000.00				1,000.00			
			STAGE MONITORS								
4	Electro Voice	QRX-112-75	EV 12" two way floor monitor	712.00				2,848.00			
2	Electro Voice	CPS2	2 Channel Power amplifier	549.00				1,098.00			
2	ASHLEY	GQX3102	Ashly dual 1/3 octave house EQ	709.00				1,418.00			
300	WPW	25226B	14 GA wire for monitor installation	0.14				42.00			
1	GTI		Labor to install		4.00					4.00	
			RACK ELECTRONICS								
1	JVC	XV-SAM65GD	DVD Player	162.00				162.00			
1	GTI		Labor to install		1.00					1.00	
			RACKS								
1	Middle Atlantic	SR-40-28	Equipment Rack with 40 Rack Units	625.00				625.00			
1	Middle Atlantic	RSH-4	Rack Shelf For DVD	63.33				63.33			
1	Middle Atlantic	USC 6/12	Sequencer	279.00				279.00			
1	Middle Atlantic	PD1415 CNS	15 Amp 12 outlet corded outlet strip	61.50				61.50			
1	Middle Atlantic	MPR	Sequencing Raceway	21.00				21.00			

304F03.EPS

Figure 3 Sample summary sheet.

XYZ TELESYSTEMS, INC.
20202 Hopeful Drive
Redding, CA 96002

Proposal: **204-1000**

Date: 5/30/2010

Jane Doe
1805 Sequoia Street
Redding, CA 96001

RE: Sequoia Auditorium

Dear Jane Doe,

We are pleased to provide the following proposal:

QTY	Vendor	Part Number	Description	Unit Price	Ext. Price
			COMPLETE SYSTEM WITH GTI LABOR		
			CLUSTER		
3	Electro Voice	QRx-153/75	EV three way flyable loudspeakers	$1,866.43	$5,599.30
3	Electro Voice	CPS-3	EV power amplifier (FR loudspeaker)	$1,611.69	$4,835.06
2	Electro Voice	QRx-118S	EV single 18" subwoofer	$1,102.20	$2,204.39
1	Electro Voice	P1201	EV power amplifier (subs)	$1,289.01	$1,289.01
1	Electro Voice	M118S	Sub module for 118S	$134.17	$134.17
1	Electro Voice	DX38	EV system controller	$1,356.94	$1,356.94
1	ASHLY	GQX3101	Ashly dual 1/3 octave house EQ	$881.42	$881.42
1000	WPW	25227B	12 GA wire for cluster installation	$0.34	$339.66
1	GTI		Labor to terminate	$3,200.00	$3,200.00
1			Mounting hardware	$1,698.30	$1,698.30
			STAGE MONITORS		
4	Electro Voice	QRX-112-75	EV 12" two way floor monitor	$1,209.19	$4,836.76
2	Electro Voice	CPS2	2 Channel Power amplifier	$932.37	$1,864.73
2	ASHLY	GQX3102	Ashly dual 1/3 octave house EQ	$1,204.09	$2,408.19
300	WPW	25226B	14 GA wire for monitor installation	$0.24	$71.33
1	GTI		Labor to install	$320.00	$320.00
			RACK ELECTRONICS		
1	JVC	XV-SAM65GD	DVD Player	$275.12	$275.12
1	GTI		Labor to install	$80.00	$80.00
			RACKS		
1	Middle Atlantic	SR-40-28	Equipment Rack with 40 Rack Units	$1,061.44	$1,061.44
1	Middle Atlantic	RSH-4	Rack Shelf For DVD	$107.55	$107.55
1	Middle Atlantic	USC 6/12	Sequencer	$473.83	$473.83
1	Middle Atlantic	PD1415 CNS	15 Amp 12 outlet corded outlet strip	$104.45	$104.45
1	Middle Atlantic	MPR	Sequencing Raceway	$35.66	$35.66

304F04.EPS

Figure 4 Sample recapitulation sheet.

2.3.0 Completing the Estimate

Completing the estimate consists of properly organizing the pricing information and adding the appropriate costs for overhead and profit. Overhead consists of all the other direct costs of doing a job besides the labor, material, and special items. Included are such things as test equipment, ladders, cable pullers, and miscellaneous materials. Overhead also includes all the indirect costs of being in business. Included are salaries for office personnel, rent, utilities, advertising, accounting fees, building depreciation, fire insurance, and so on. Finally, to all other costs, a markup or profit is added. Profit is the company's reward for taking a risk and is a return on its investment.

2.4.0 Management Approval

Once the estimate has been completed, the takeoff is reviewed to make sure all major variables affecting the cost of the job have been taken into consideration. Also, the overall schedule is checked to see if the job can fit into the expected workload along with the company's other job commitments. The price, expected profit, competition for the job, and other bidding variables are weighed by management to determine the appropriate bidding strategy. The decision may also be influenced by the company's desire to open a new market or product line.

2.5.0 Preparing and Submitting the Bid

Before sending the formal bid to a customer, the estimate is rechecked for mistakes and to preclude the possibility of making a large cost error. All the calculations are normally double-checked by a finance person or another qualified individual.

The formal bid is prepared on the company's bid form or on one supplied by the customer. Some customers require a bond to assure them of the contractor's ability to perform the job as specified for the bid price or to assure them that sub-

contractors and suppliers are properly paid. Once the formal bid is prepared and any bond procured, the bid is submitted to the customer by the date specified. Normally, additional information on references, past performance, safety record, and guarantees are included along with the bid to bring into focus all of the other important considerations that may influence the customer to select your company over the other bidders.

3.0.0 REVIEW OF JOB REQUIREMENTS

The first thing the supervisor should do is discuss the job with the estimator. This is necessary in order to find out the specific job requirements and to understand the expected job performance standards. As applicable, the estimator should provide the supervisor with copies of all construction drawings and specifications for the job. Any clarifications arrived at by the estimator with the customer should also be discussed. It is important that this information be accurately passed on from the supervisor to the installation team crew members in order to eliminate disputes and unneeded work at the job site. The estimator should also point out any special safety hazards, environmental hazards, building code restrictions, and any other important information that applies to the job.

> **NOTE**
>
> It is not unusual for bidders to recommend alternatives to the equipment and materials listed in the project specs and drawings. That is, the bidder may provide a compliant bid, but offer a lower cost approach or suggest upgraded technology as an alternative. It is important that these changes, if accepted, are reflected in the project documentation, or are provided to the planners and installers before the project gets under way.

3.1.0 Construction Drawings

Because of their importance to the tasks of job planning and installation, the content and sequence of a typical set of construction (working) drawings (*Figure 5*) are briefly reviewed here.

- *Title sheets* – Title sheets are placed at the beginning of a set of or at the beginning of a major section of drawings. They provide an index to the other drawings, list of abbreviations used on the drawings and their meanings, list of symbols used on the drawings and their meanings, and various other project data, such as the project location, size of land parcel, and building size. When making an estimate or perform-

---

ing work related to drawings, it is important to get a complete set of drawings and specifications, including the title sheets, so that you can better understand the abbreviations and symbols used.

- *Architectural drawings* – Architectural drawings include line, form, material, finish, arrangement, and other features of the project, typically in the following order: site plan (plot plan), floor plan, exterior elevations, interior elevations, sections, details, and schedules.
- *Structural drawings* – Structural drawings may be included as part of the architectural drawings or can be a separate section within a drawing set. Structural drawings provide the detail necessary to produce the structural support components of the building, such as walls, floors, ceilings, and loadbearing structures. They also show important information pertaining to equipment and cabling installation, such as the locations of equipment rooms, entrance facilities, telecommunications closets, spaces, and pathways.
- *Plumbing plans* – Plumbing plans show the size and location of water and gas systems if not included in the mechanical section.
- *Mechanical plans* – Mechanical plans show temperature control and ventilation equipment, including ducts, louvers, and registers.
- *Electrical plans* – Electrical plans show all electrical equipment, conduit, and outlets. They often show the details for telecommunications and other low-voltage system installations. They may include plot or site plans showing the entry points for all electrical and communications services. They typically include power and communications riser diagrams and floor plans that show the location of all outlets, fixtures, panels, and backboards.
- *Fire protection plans* – Fire protection plans show the location and details of suppression and alarm features. These plans may be detailed or shown in a general outline format.

Normally, the dimensions for a building or details of structures within a building are shown in the working drawings. When they are not, it becomes necessary to use an architectural ruler, tape, or wheel to measure the dimensions of a room or the length of a pipe or cable run on a scaled drawing. It is also useful to verify dimensions with the drawn object to see if the scale is correct or to see if the stated dimension is correct. If discrepancies are found, it may require further verification with the architect or engineer. When it becomes necessary for you to scale a drawing (by measuring the length of an object on the drawing and converting it to the actual length using the given scale), you should always proceed as follows:

- Determine the scale used for the particular view. The scale should be indicated directly below the drawing, or the legend may indicate the scale used on the whole sheet. Many different scales are used in a set of working drawings; different scales may even be used on the same sheet.
- Verify the stated scale by measuring a line for which the length is written on the drawing. For example, if a scale is given as ¼" = 1', a dimension written on the drawing as 4' should measure 1 inch long on the drawing.

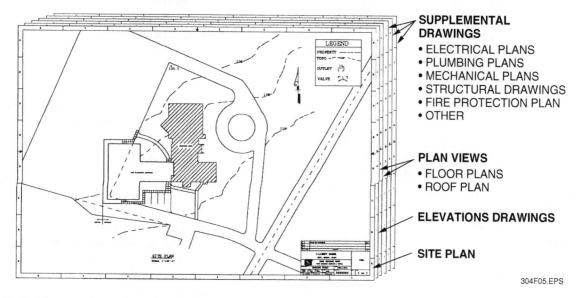

Figure 5 Typical format of a working drawing set.

3.1.1 Shop Drawings

Shop drawings are plans that are made by the company to provide greater detail than what is shown on construction (contract) drawings. These drawings provide details showing sizing, how cables will be run, panel and device locations, termination details, coordination with other trades, and other details to guide you or another EST to do the job according to the specifications and codes. The authority having jurisdiction (AHJ) and the architect may also require shop drawings.

3.1.2 As-Built Drawings

Most owners demand proper documentation in the form of as-built drawings, or record drawings, especially for existing buildings that undergo continual occupancy changes and remodeling. As-built drawings are drawings that show all changes made from the original job plan to the finished state, including change orders and field changes. They show the details of how something was actually built or how field conditions were found. The as-built configuration is typically marked on the architect's original plan (usually in another color, such as red) along with appropriate explanatory notes. These marked-up drawings then become part of the permanent drawing file for a building. As-built drawings are often used as the basis for planning future retrofit construction jobs.

As-built drawings are frequently neglected or improperly prepared. For this reason, a drawing set used as the basis for retrofit construction may not reflect the site as it actually exists, and the drawings should be carefully checked for accuracy during the site survey.

Record drawings are not as detailed. They show only major changes and modifications to construction or shop drawings. Unless all changes to the original plans are included, final drawings should be marked *Record Drawings*.

3.2.0 Specifications

Specifications are written statements provided by architectural and engineering firms to the general contractor and, subsequently, to the subcontractors. Specifications define the quantity and quality of work to be done and the materials to be used. They are very important to the architect and owner because they guarantee compliance by the contractors to the standards set for the job. Specifications consist of various elements that may differ somewhat for particular construction jobs. Specifications have several important purposes:

- Clarifying information that cannot be shown on the drawings
- Identifying the scope of work for each trade, work standards, types of materials to be used, and the responsibility of various parties to the contract
- Providing information on details of construction
- Serving as a guide for contractors bidding on the job
- Serving as a standard of quality for materials and workmanship
- Serving as a guide for compliance with building codes and zoning ordinances
- Serving as the basis of agreement between the owner, architect, and contractors in settling disputes

Two types of information are contained in a set of specifications: special and general conditions, and technical aspects of construction.

3.2.1 Special and General Conditions

Special and general conditions cover the nontechnical aspects of the contractual arrangements. Special conditions cover topics such as safety and temporary construction. General conditions cover the following points of information:

- Contract terms
- Responsibilities for examining the construction site
- Types and limits of insurance
- Permits and payments of fees
- Use and installation of utilities
- Supervision of construction
- Other pertinent items

Misunderstandings often occur in the general conditions section of the construction contract. Therefore, these conditions are usually much more explicit on large, complicated construction projects.

3.2.2 Technical Aspects

The technical aspects section of specifications classifies the work to be done into major categories and subsequently identifies the standards that apply to each part. The categories are usually organized in the order in which the work will be performed; for example, site work comes before carpentry or electrical. The technical sections include information on materials that are specified by standard numbers, such as NFPA standards, and by standard national testing organizations, such as the American Society of Testing Materials International (ASTM) and Underwriters Laboratories (UL).

Most specifications written for large construction jobs in North America are organized in a format developed by the Construction Specifications Institute (CSI) and called the *Uniform Construction Index*. In this format, all construction is divided into divisions. Each division is then broken down into sections and subsections.

3.3.0 Scope of Work

A scope of work statement describes the overall requirements for the job and is typically a less formal document than the specifications. On large jobs, the scope of work will be supplemented by detailed specifications. On smaller jobs, it might be only the definition of what is to be done, thus relying on the contractor's expertise to lay out and install the system.

Typically, the scope of work describes the following in general terms:

• Work to be performed by the contractor
• Work to be performed by others
• Standards to use
• Sections of the specification that apply
• Identification, labeling, and documentation systems to be used
• Testing and acceptance methods to be used
• When and how the installation is to be turned over to the customer
• Terms and conditions

In responding to the scope of work, which is often part of a request for proposal (RFP) or request for quote (RFQ), the bidder will include such items as the following:

• A detailed description of the work being bid by the contractor

---

On Site

Conflicts

Which document rules if there is a conflict between the construction drawings and the project specifications? Because they are the more detailed of the two documents, the specifications typically take precedence over the drawings unless the specifications themselves say otherwise. This makes a compelling argument for reading the specs.

• Prices for equipment, materials, and labor
• Conditions of the bid and assumptions on which the bid is based
• Alternate equipment being proposed

An example of a contractor's bid letter is provided in *Appendix A*.

3.4.0 Exclusions

Contracts, specifications, and other contractual documents define the work to be performed. In some cases, the prime contractor or owner state in the contract that a certain portion of the work is to be excluded or performed by another company. This is known as excluded work. More often, however, the bidder chooses not to perform certain parts of the work because it is outside their experience or capability. Exclusions are most commonly found in the scope of work statement the company sends when it bids on a job. Always make sure you fully understand all such exclusions.

4.0.0 JOB PLANNING AFTER THE CONTRACT AWARD

After the acceptance of the bid by the customer and the awarding of the contract, the estimated cost for the job minus the company's profit and fixed overhead becomes the working budget. This is the amount within which all tasks must be accomplished and all materials purchased. From this point on, the ability of the company to successfully accomplish the job and make a profit rests to a great extent on your efficiency and productivity and on that of others assigned to do the work. An important step toward being successful is to have a systematic approach or plan for doing the job. For long-term jobs, an overall plan can be divided into smaller plans that apply to the specific work to be performed on a daily, weekly, or monthly basis. Regardless of job size or duration, the plan for accomplishing the work should include the following elements (*Figure 6*):

Step 1 Review the job requirements.

Step 2 Perform a site survey.

Step 3 Schedule the work.

Step 4 Schedule needed equipment and materials.

Step 5 Assemble the crew.

Step 6 Install all equipment, wiring, and accessories, and then check out the system.

Step 7 Perform quality control checks during and after.

 Site Survey, Project Planning, and Documentation

Step 8 Perform a walk-through with the owner or owner's representative.

Step 9 Perform any required punch list work.

Step 10 Perform any required code inspections.

Step 11 Clean up the site (daily and final).

Step 12 Support the commissioning process as required.

Step 13 Provide user training.

5.0.0 NEW CONSTRUCTION SITE SURVEY, PLANNING, AND DOCUMENTATION

Once the drawing set and specifications have been reviewed, and their requirements are fully understood, one or more site surveys are performed. This section focuses on the site survey for new construction. Retrofit construction is covered later in the module. A site survey involves a meeting and a walk around the job site with all the key people (stakeholders) involved with the project. As appropriate, this group can include the lead installer, project manager, client/end user, contractor, architect/designer, and others representing the various construction trade subcontractors. In addition to observing conditions at the site, the site survey allows you to do the following:

- Introduce yourself and other key installation members to the customer and other contractors.
- Explain the work to be performed for the customer to the other contractors.
- Get copies of the general contractor's and other contractors' installation schedules, and check them against the original schedule and your own. Make sure to document any changes that could affect your work and the ability to complete the job on time.
- Verify the drawing set against the actual physical location/structure.
- Determine the progress of other contractors' work, such as HVAC, electrical, and plumbing.

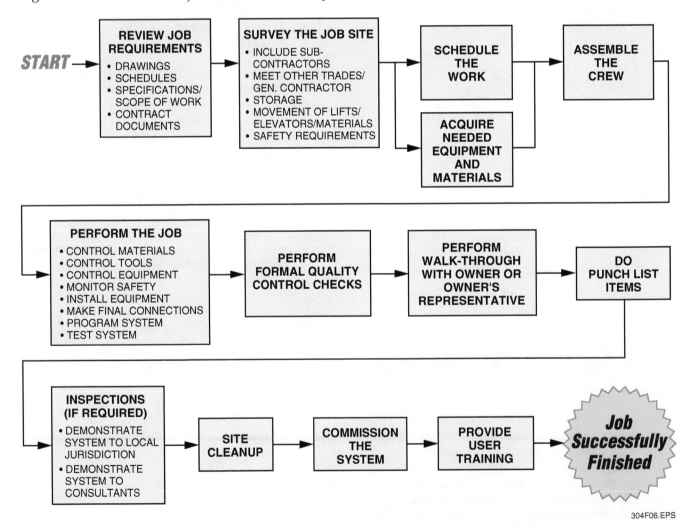

Figure 6 Process for planning and completing a job.

More than one survey may be needed, and you must be prepared to visit the site as each phase of the job becomes ready for installation. For example, the locations of pathways should be investigated and confirmed when concrete forms are being placed, not after the concrete is poured.

To perform a site survey, be sure to bring the tools or equipment needed to perform any work and/or required inspections. These items can include a hard hat and other personal protective equipment, ladders, a flashlight, measuring wheel, 50-foot tape or laser measuring device, handheld tape recorder, still camera and/or video camera, sketch pad, architect's rule, and other appropriate items.

During the site survey, you and the other stakeholders walk through the entire project area. While doing so, compare the working drawings against the actual building structure to identify specific locations and the work to be performed there. The walk-through provides you with a chance to see equipment, raceways, pathways, and structures that are already in place. It also provides a forum where all stakeholders can mutually agree upon the specific locations for installing or mounting equipment, cutting holes, and installing raceways. These locations should then be marked both in your drawing set and physically at the site. Ideally, photos or videos of these areas should be taken for reference in the future. If necessary, sketches should be drawn of any unusual installation conditions not shown on the working drawings. Again, photos of these situations can prove to be useful later in the project.

A checklist should be used during the site survey to check that all questions and items of concern identified during the estimating phase of the project have been answered. If new concerns or problems arise during the survey, mutually agreed upon solutions can often be made to correct the problem while still at the site. *Figure 7* shows an example of a checklist for a proposed telecommunications installation.

As applicable, during a site survey you should perform the following tasks:

- Determine the physical location, size, and type of construction for all equipment rooms, closets, and other enclosures.
- Determine the location of system grounding points and prime power inputs for your equipment.
- Determine the exact measurements of locations where equipment is to be installed in order to make sure there is adequate space to install and maintain the equipment or device.
- Determine the routing for various cable runs, and verify the lengths of cable pathways and raceways. Verify the required quantity of each type of cable being used.
- Check cable paths for obstacles, such as concrete or cinder-block walls, paneled walls, drywall ceilings without access, and vaulted or atrium ceilings with skylights
- Confirm that all walls, partitions, half-walls, and other structures are located as shown on the drawings.
- Determine the building access points and routes for moving large items of equipment and cable reels into the building and to their final points of installation or use.
- As applicable, determine the locations for all terminals, workstations, speakers, monitors, control panels, motion detectors, cameras, and other equipment.
- Determine the status and acceptability of the work performed by related trades.
- Determine the availability of a secure location for receiving and storing tools and materials.

After the walk-through has been completed, the entire job should be summarized and checked with all the stakeholders to make sure nothing is missing and that all functionality for the systems being installed has been verified. All information gathered during the site survey should be placed into the project file. This information will become

Long Lead Items

An experienced planner knows that some items of material or equipment may need to be ordered well in advance. It is a good idea during the estimating process to check availability of any questionable items. The key? Don't assume anything! You never know when labor disputes, material shortages, or high demand might affect availability of materials or equipment you need for a job. Long lead items also need to be considered before you commit to a project schedule. If a key component or material is not available when you need it, the entire schedule could be thrown off. If the contract has a penalty or incentive clause related to schedule performance, such a delay could cost the company money.

useful later, especially if new team members are assigned to the installation after it starts.

During a site survey, alternative methods or procedures may be identified. If there is agreement among the stakeholders and the authority having jurisdiction (AHJ) to implement any recommendations, then the appropriate documentation should be executed. This documentation may be in the form of a change order or a letter signed by all parties.

CHECKLIST FOR SITE SURVEY NEW CONSTRUCTION				
ITEM	DESCRIPTION OF OPERATION	DATE CHECKED	Y	N
1	Is the general contractor responsible for construction and finish of the closets and pathways?			
2	Is there an electrical contractor on the project and are they responsible for the pathways?			
3	Is the electrical contractor responsible for the telecommunications grounding and bonding within the building?			
4	Are the closets completed and ready for use by the telecommunications vendor? If not, when will they be ready?			
5	Are the pathways completed and ready for use by the telecommunications vendor? If not, when will they be ready?			
6	Is the grounding and bonding system installed and ready for use by the telecommunications vendor? If not, when will it be ready?			
7	Is there space on the job site available for storage and staging of the materials and tools? Will it be secured?			
8	Will the space be under the control of the telecommunications vendor? If not, you must work out responsibility for loss or damage.			
9	Is the space inside or outside?			
10	Does the general contractor conduct construction progress meetings? When? Can the telecommunications vendor attend?			
11	Does the general contractor have a posted safety plan?			
12	What is the schedule for inspections by local code authorities?			
13	Is the building equipped with suspended ceilings? Are they used to handle environmental air?			
14	Where can the telecommunications vendor set up field operations?			
15	Is there a way to get large cable reels and other heavy materials to the top floor of the building?			
16	Are lifts required on the job site?			
17	Is the building a hardhat area?			
18	When will the telecommunications vendor be allowed access to all spaces in the building requiring telecommunications work?			
19	When will the walls receive final finishes?			
20	Are elevators and loading docks available and appropriate for the job?			
21	Is rigging required?			
22	Does any equipment require floor loading modifications?			
23	Verify position of sprinklers and water pipes.			
24	Are any alarm systems activated?			

304F07.EPS

Figure 7 Example of a site survey checklist.

NCCER – Contren® Learning Series 33304-11

6.0.0 SCHEDULING THE WORK

No matter how large or small, all jobs can be divided into manageable activities (tasks). These tasks can be performed in a particular sequence and within a specific amount of time in order to accomplish the overall job. For larger jobs, the estimator may have already generated a preliminary schedule for doing the job. If such a schedule exists, it can be used as the basis for making an updated or revised schedule. Otherwise, the schedule will have to be generated from scratch.

The first phase of schedule development involves identifying all the different tasks and determining the time that each takes to complete. This information can normally be found on the estimator's takeoff sheets or estimating forms. Recorded on the forms is the estimated labor in hours or units required to perform the different tasks involved in the project (*Figure 8*). These labor hours/units are one of the factors used to determine the time interval (hours, days, weeks) required to do each of the tasks. The other factor is how many people are assigned to do the task. For example, if 40 hours are required to install a local area network, and two people are assigned to do it, then it will take 20 hours to complete the task. Note that the labor hours/units marked on the takeoff sheets represent the job productivity (performance standards) expected by the estimator and management to do the job. As such, they are also the basis for determining the job's price and working budget. For this important reason, any differences of opinion between the supervisor and estimator that may exist about the labor hours allocated for a particular task should always be clarified before proceeding with the scheduling.

After all the tasks and their durations have been determined, all job constraints must be identified and taken into consideration. Typically, these are factors that prevent one task from starting until another task is finished. For example, cable runs cannot be installed until the cable trays are in place. A major area of constraints involves the relationships of the tasks being performed by you and those performed by the other trades. To avoid unnecessary delays and/or other problems, it is extremely important that all scheduling of your work be closely coordinated with that of the other trades. If the job is at a building under construction, you should obtain a current copy of the general contractor's schedule to use as a reference. It covers all the trades working on the project and shows their specific time frames for accomplishing work on the project. On a large project, it may take several meetings with the other trades to work out the details for dovetailing your work

schedule with theirs. Other constraints that can impact the job's schedule are the availability of special materials or resources, such as installers and leased equipment. Other factors for which time may need to be allocated are:

- Preparation of shop drawings
- The permitting process
- Third-party quality inspections
- In-process acceptance testing of equipment

Once all the factors are identified, the actual schedule can be produced. With the exception of small jobs, this is typically done on a computer using one of several commercially available project scheduling software packages. The project schedule normally begins with the award of contract and ends on acceptance of the job by the customer. Depending on the software being used, the schedule can take many forms. Two examples of schedules for the installation of a telecommunications system are shown in *Figures 9* and *10*.

After work on the job begins and throughout its progress, actual performance should be continuously compared with the schedule to make sure that the work goes as planned. Tracking performance is not always easy because there is usually a reluctance to get involved with the detailed paperwork at the production end. The actual method used to track job performance varies among different companies. Regardless of the method, accurate record keeping must be done to support it. Accuracy depends on the supervisor's and workers' attitudes, type of records kept, detail required, forms used, analysis of the totals, general enthusiasm of the record keeper, and the extent to which the data is used. No matter how bothersome, it is necessary that job performance be recorded and tracked to make sure that the job is being done within the allocated time and cost budget. Recording actual job performance data on current jobs is also important because the data is often used as the basis for establishing job performance standards or production rates on future jobs.

7.0.0 ACQUIRING THE NEEDED MATERIALS/EQUIPMENT

While the schedule is being prepared, the materials and equipment required can be ordered or otherwise accumulated for use on the job. Because the estimator has previously recorded the types of materials needed on the job estimating forms, they can be used as the basis for ordering the materials (*Figure 11*). Their use also allows you to verify any assumptions the estimator made about the availability, price, and common order quantities

PROJECT NUMBER: XXXXXXXXXXXXXXXX
PROJECT NAME: ANYWHERE ELEMENTARY SCHOOL:
ADDRESS: ANYWHERE, USA

ITEM	TASK DESCRIPTION	UNITS/QUANTITY	LABOR RATE	TOTAL PRICE
1	Installing horizontal wires (2 per run)	9	0	$0.00
2	Installing horizontal wires (4 per run)	47	0	$0.00
3	Installing faceplates and jacks	208	0	$0.00
4	Installing relay racks	2	0	$0.00
5	Installing patch panels	6	0	$0.00
6	Terminating wires at patch panels	208	0	$0.00
7	Certifying Category 5 wires	208	0	$0.00
8	Installing surface raceway	56	0	$0.00
9	Installing surface-mount blocks	56	0	$0.00
10	Installing backbone fiber-optic (F/O) cables	300	0	$0.00
11	Installing F/O cables	4	0	$0.00
12	Terminating F/O cables	12	0	$0.00
13	Testing F/O cables	6	0	$0.00
14	Installing fire-/smoke-rated partition penetrations	59	0	$0.00
15	Mounting hubs on relay racks	4	0	$0.00
16	Installing backbone conduit from main building to portables	120	0	$0.00
17	Installing horizontal conduit between backbone conduit and portables	48	0	$0.00
18				
19				
20				
Total labor				**$0.00**

Total cost			**$0.00**
Materials markup		50%	**$0.00**
Labor markup		50%	**$0.00**
State sales tax on materials		6%	**$0.00**
Total price to customer			**$0.00**

304F08.EPS

Figure 8 Example of a labor estimate.

of the materials during the material pricing stage. Make sure that the quantities of materials like cable, alarm boxes, and conduit recorded on various forms are first summarized to find the total quantity of like items needed before you place an order. Ordering by larger quantities normally allows the materials to be purchased at lower prices from the distributors. When preparing purchase orders for the various materials, make sure to ask that a copy of the material safety data sheet (MSDS) be supplied for any items containing hazardous chemicals or materials, such as batteries and firestop compound, so that they can be made readily available to workers at the job site.

If a job involves the use of special equipment or materials, it is important that the orders for these items be placed as early as possible. Written assurances must be received from the supplier or material manufacturer that the items will be available when you need them. The same should be done if the purchase or leasing of special equipment is required. Failure to cover these important aspects may result in unnecessary job delays and possible cost overruns.

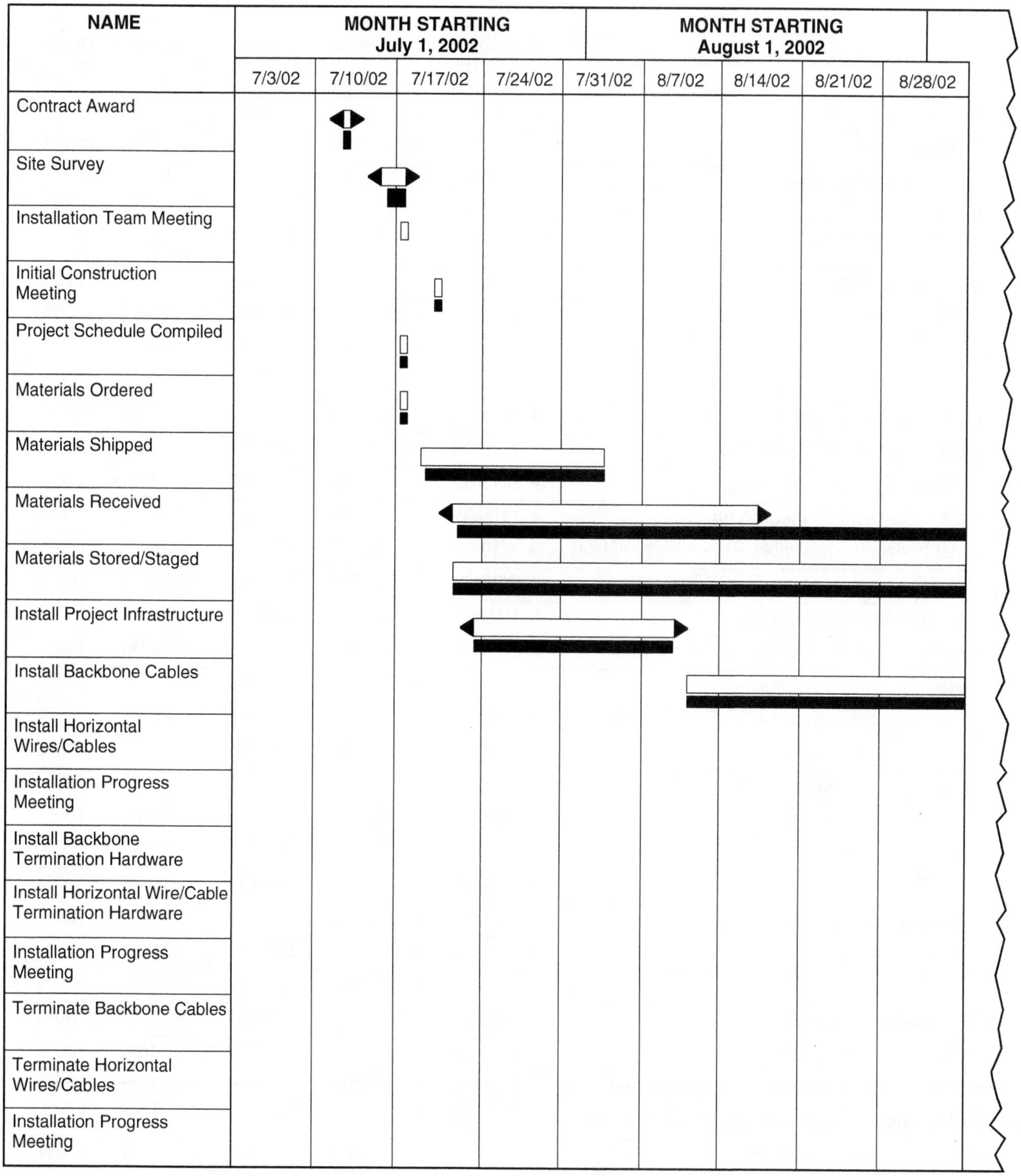

Figure 9 Example of a graphic-type project schedule (Gantt chart).

304F09.EPS

Name	Start Constraint	Finish Constraint	Actual Start	Actual Finish	Percent Done
Contract awarded	✶ 7/12/95	✶ 7/12/95	✶ 7/12/95	✶ 7/12/95	100%
Site survey	✶ 7/14/95	✶ 7/17/95	✶ 7/15/95	✶ 7/17/95	100%
Installation team meeting	✶ 7/17/95	✶ 7/17/95	✶ 7/17/95	✶ 7/17/95	100%
Initial construction meeting	✶ 7/20/95	✶ 7/20/95	7/20/95	7/20/95	100%
Project schedule compiled	✶ 7/17/95	✶ 7/17/95	7/17/95	7/17/95	100%
Materials ordered	✶ 7/17/95	✶ 7/17/95	7/17/95	7/17/95	100%
Materials shipped	✶ 7/18/95	✶ 8/4/95	✶ 7/18/95	✶ 8/4/95	100%
Materials received	✶ 7/20/95	✶ 8/19/95	✶ 7/20/95	✶ 9/25/96	100%
Materials stored/staged	✶ 7/20/95	✶ 10/20/95	7/20/95	10/20/95	100%
Install project infrastructure	✶ 7/21/95	✶ 8/10/95	✶ 7/21/95	✶ 8/10/95	100%
Install backbone cables	✶ 8/11/95	✶ 9/8/96	8/11/95	9/8/96	100%
Install horizontal wires/cables	✶ 9/11/95	✶ 12/22/96	9/11/95	12/22/96	100%
Installation progress meeting	✶ 9/11/95	✶ 9/11/95	9/11/95	9/11/95	100%
Install backbone connecting hardware	✶ 1/2/96	✶ 1/18/96	1/2/96	1/18/96	100%
Install horizontal wire/cable connecting hardware	✶ 1/19/96	✶ 2/26/96	1/19/96	2/26/96	100%
Installation progress meeting	✶ 2/26/96	✶ 2/26/96	2/26/96	2/26/96	100%
Terminate backbone cables	✶ 1/19/96	✶ 2/9/96	1/19/96	2/9/96	100%
Terminate horizontal wires/cables	✶ 2/27/96	✶ 3/29/96	2/27/96	3/29/96	100%
Installation progress meeting	✶ 3/29/96	✶ 3/29/96	3/29/96	3/29/96	100%
Label all facilities as per ANSI/TIA/EIA-606	✶ 3/22/96	✶ 4/5/96	3/22/96	4/5/96	100%
Test backbone cables	✶ 4/5/96	✶ 4/12/96	4/5/96	4/12/96	100%
Test horizontal wires/cables	✶ 4/15/96	✶ 5/3/96	4/15/96	5/3/96	100%
Compile all test results	✶ 5/6/96	✶ 5/8/96	5/6/96	5/8/96	100%
Installation progress meeting	✶ 5/3/96	✶ 5/3/96	5/3/96	5/3/96	100%
Punch list	✶ 5/9/96	✶ 5/10/96	5/9/96	5/10/96	100%
Correct all items on punch list	✶ 5/16/96	✶ 5/17/96	✶ 5/16/96	✶ 5/17/96	100%
Final punch list	✶ 5/16/96	✶ 5/16/96	5/16/96	5/16/96	100%
Customer punch list	✶ 5/17/96	✶ 5/17/96	5/17/96	5/17/96	100%
Customer acceptance	✶ 5/20/96	✶ 5/20/96	5/20/96	5/20/96	100%
Prepare "as-built" package	✶ 5/20/96	✶ 5/24/96	5/20/96	5/24/96	100%
Provide all project documents to customer	✶ 5/27/96	✶ 5/27/96	5/27/96	5/27/96	100%
Return surplus materials to distributor for storage	✶ 5/20/96	✶ 5/22/96	5/20/96	5/22/96	100%
Complete billing to customer	✶ 5/28/96	✶ 5/28/96	5/28/96	5/28/96	100%
Review billing with customer	✶ 5/29/96	✶ 5/29/96	5/29/96	5/29/96	100%
Receive final payment	✶ 5/30/96	✶ 5/30/96	5/30/96	5/30/96	100%
Close project	✶ 5/31/96	✶ 5/31/96	5/31/96	5/31/96	100%
Clear punch list	✶ 5/10/96	✶ 5/13/96	5/10/96	✶ 5/13/96	100%
Clear customer punch list	✶ 5/17/96		5/17/96	✶ 5/20/96	100%

✶ indicates completion

304F10.EPS

Figure 10 Example of a text-type project schedule.

Only authorized personnel can order materials and equipment. Materials and equipment are ordered using a document called a purchase order (PO). A PO identifies the quantity and quality of materials being ordered, along with delivery place and date, insurance requirements for protection during transit, and other information. The person responsible for receiving the material should have a copy of the purchase order to be sure that what is received is what was ordered. The delivery receipt and PO are compared to make sure all materials have been received.

The responsible person should take the time to inspect the shipment to be sure it is not damaged. If damage is observed, it should be noted on the delivery receipt so that the supplier can reorder the damaged material at no additional cost.

8.0.0 ASSIGNING THE INSTALLATION CREW

After completing the site survey and scheduling the job, the members of the installation crew can be selected and assembled. The size of the crew can vary over the duration of the job, depending on the specific work that is scheduled and the number of people allocated to perform the work. It is important that the members chosen for the crew have the skills needed to accomplish the job. Those with special skills may only be needed at specific points during the process.

Before the start of the installation, the project manager, supervisor, or lead technician should hold a meeting with the entire crew. At this meeting, the lead person should define the responsibilities for everyone involved. This allows everyone to become associated with one another and familiar with all aspects of the job. It also provides an opportunity for crew members to ask questions.

Ideally, lists of required materials, equipment, and special tools for each of the major tasks shown on the job schedule should be made and given to the supervisor and lead technicians. Make sure such lists also include any fall protection or other special safety equipment. Throughout the job, these lists can serve as checklists for the crew members to use while loading material and equipment into trucks at the shop before leaving for the job site. At the job site, the checklists can be used to make sure that all materials and equipment are available before starting a task. Failure to do this can result in missing materials or equipment and possible job delays.

9.0.0 COMPLETING THE INSTALLATION

It is important to emphasize here that while accomplishing the job, the lead installer and crew members must be aware of the installation schedule and should plan for upcoming events. This is necessary in order to avoid job delays and prevent possible problems and mistakes. Planning ahead pertains to the current task and all tasks scheduled to be performed in the near future. Planning ahead includes coordination with the following:

- The customer and general contractor in relation to schedules, drawing updates, building access, storage, and required meetings
- Other trades in relation to work area arrangements, schedules, and safety issues
- Vendors/suppliers in relation to the availability and timely delivery of needed project materials and equipment
- Local, state, and other agencies as necessary to obtain the necessary permits and to schedule required tests and inspections

The installation practices for low-voltage equipment and systems are regulated by several codes and standards. All work performed during the installation must comply with the prevailing codes and standards.

Job controls provide a means of measuring performance against and correcting any apparent deviations from the way a job was initially planned. The effectiveness of job control methods is measured by whether the job is done within the allocated time and budget. In addition to crew efficiency and performance, control of materials, tools, equipment, and safety are important elements of job control.

Bill of Materials

Job Name: **Anytown Office Building RFC #175**   Job # : 5002

Date: 09/08/10   CO. # :

	Quantity	Model #	Description	Unit Cost	Extended
1)	4	757-5A-RS70	25 Candela Speaker Strobe, Red, Wall Mount	122.82	$491.29
2)	0			0.00	$0.00
3)	0			0.00	$0.00
4)	250	3/4 Conduit	Conduit	0.00	$0.00
5)	250	990	Fire Wire Strobe	0.15	$36.29
6)	250	991	Fire Wire Speaker	0.15	$36.29
7)	100	Greenfield	Greenfield	0.40	$40.32
8)	12	Connectors	Connectors	1.21	$14.52
9)	25	3/4" Coupling	Coupling	1.21	$30.24
10)	10	3/4" Connectors	Coupling	1.21	$12.10
11)	6	4x4x Connectors	Box	1.21	$7.26
12)	1	Fire Patch	Fire Patching Compound	19.35	$19.35
13)	5	Blank Plates	4x4 blank plate	2.02	$10.08
14)	0	0	0	0.00	$0.00
15)	0	0	0	0.00	$0.00
16)	0	0	0	0.00	$0.00
17)	0	0	0	0.00	$0.00
18)	0	0	0	0.00	$0.00
19)	0	0	0	0.00	$0.00
20)	0	0	0	0.00	$0.00

Sub-Total:		$697.74
Sales Tax:	9.00%	$62.80
Freight:	7%	$48.84
Escalation:	10.00%	$69.77
Material Total:		**$879.15**

304F11.EPS

Figure 11 Example of a material list.

10.0.0 INCORPORATING QUALITY CONTROL/ACCEPTANCE TESTS

Regardless of the size of the job, inspections and acceptance tests must be incorporated to verify the quality of the work. These procedures are normally performed at various points throughout the job.

Quality cannot be controlled unless all work is done to meet a standard. As an EST, you must be vigilant about meeting quality standards and have a clear idea of the standard of quality required for every component of the work. To be professional, you must consistently set high standards of qual-

ity. Many of these quality standards, such as how to properly prepare a cable termination or ground equipment, are learned through your formal apprenticeship and other training programs and job experience. If you are responsible for the work of other ESTs, you need to communicate the standards of quality that you require of them.

Quality standards are defined by a job's contract and specifications. Typically, these include electrical and other inspections as the job progresses, followed by system acceptance testing at the end of the job. Quality control inspections and the monitoring of system acceptance testing are usually performed by designated inspectors. Such third-party inspections can lead to the success of a project and reduce needless delays, costs, general disruptions, and tension between owners, contractors, and material suppliers.

As you work to meet the job requirements, it is wise to document that you have met these requirements. The documentation can help you make a case in the event of a dispute. More importantly, the recording of the inspection, check, or measurement helps to ensure that no quality control inspection or acceptance test is overlooked and that no work goes unperformed. Informal records can be kept in a daily diary. However, most large projects require formal documentation. This is normally done using preprinted forms. Preprinted forms are used because they are clearly organized, with entries for specific kinds of information.

In the case of fire alarm systems, *National Fire Alarm and Signaling Code (NFPA 72)* requires that permanent maintenance, inspection, and testing records be filled out and retained until the next test and for one year thereafter. It also describes the types of information that must be included on inspection and testing forms. An example of a preprinted inspection and testing form typically used by security and fire alarm contractors is shown in *Appendix B*.

11.0.0 COMPLETING THE PUNCH LIST

As a result of quality inspections, acceptance testing, or for other reasons, rework or repairs may be necessary to satisfy the requirements of the contract and obtain customer acceptance of the job. These items are usually written up on a document called a punch list. Any rework shown on a punch list that is a result of poor workmanship by the contractor is normally within the scope of the contract or specifications and must be done within the cost budget allocated for the job.

A punch list can require rework to repair equipment or materials that have been damaged by individuals other than those employed by the contractor or for latent damage. Latent damage is damage that happens because of conditions beyond the control of the low-voltage contractor, such as settling of the building and water damage not present at the time of initial installation. Rework resulting from damage caused by others or latent damage is usually outside the scope of the contract. This rework should only be done after a price for doing the additional work has been accepted, and a change order to perform the work has been received from the owner or other contracting entity.

12.0.0 COMPLETING THE JOB

Cleanup of the work site must always be done at the completion of the job. This is extremely important, particularly if dealing with any hazardous material residues. The basic rule is to leave the job site cleaner and better looking than you found it. The extra effort put into site cleanup can also result in repeat business or referrals from the customer.

Commissioning is an important part of the project closeout. At the conclusion of the project, a qualified person or agency, commonly referred to as the commissioning authority, supervises the acceptance of the system. The purpose of this activity is to verify the following:

- The system design meets the functional requirements of the owner.
- All systems are properly installed in accordance with the project drawings and specifications.
- All the equipment/systems and software are operating properly.
- All documentation is accurate and complete.

Another closeout activity involves instructing the customer's employees in the proper operation of all new equipment and software. The customer should also be given all the documentation specified in the contract, such as operation and maintenance manuals for the equipment, copies of inspection and testing reports, and certificates of completion.

13.0.0 RETROFITTING INSTALLATIONS

Retrofit is defined as installing new systems and equipment in existing structures. In contrast, system modification involves replacing or adding new equipment to systems already in place in an existing structure. Generally, installing a retrofit system is more complicated and labor intensive than an installation in new construction.

13.1.0 Planning Retrofit Installations

The overall process for planning, scheduling, and accomplishing a retrofit installation is basically the same as that described for new construction. The process includes the following:

- Working with the customer to determine the system specifications
- Obtaining, reviewing, and verifying existing as-built drawings
- Performing a site survey
- Developing an installation plan
- Determining the safety concerns and requirements for the project
- Installing in compliance with codes and standards
- Performing a changeover from the old system to the new system
- Testing the installation
- Providing final documentation
- Site cleanup, including removal of old equipment and cabling

The focus of the remainder of this section is on some common conditions or situations unique to retrofit installations.

13.2.0 Performing the Site Survey

In order to perform a site survey at an existing site and structure, it is important to coordinate with the customer. Make arrangements with the owner to enter the property and to have access to all areas. If necessary, obtain proper security clearances, and inform the customer's department managers and supervisory personnel that a survey will be made. Close scheduling is important, especially if surveying locations such as schools, hospitals, nursing homes, apartments, or dormitories, so that your survey activities do not disturb the occupants or interfere with the normal operation of the facility.

If the job will involve doing work outside the building, such as trenching for pipelines or communication lines, begin the survey with a complete walk-around and walk-through of the entire property. This will allow you to get a general idea of the type and use of the occupancy and the location of special equipment and features contained within and outside the building. An existing plot plan, if available, or an accurate sketch drawn by you should clearly show the following types of information:

- Property lines and the relative location of the building or buildings
- Locations of pertinent public utilities, such as electric, phone, water, and fire hydrants
- Location and identification of all adjacent buildings and exposures

It is also important to coordinate with the customer to determine what existing equipment is to be reused. If some equipment or cabling is to be reused, determine if it conforms to the requirements for the new system. For example, non-plenum-rated cables installed prior to the requirement for using plenum-rated cables may have to be removed. Many localities require that such cables be replaced. Always check with local inspectors to determine the local requirements.

If the customer has as-built drawings, obtain a copy. As mentioned earlier, as-built drawings are frequently neglected or improperly prepared. For this reason, a drawing set used as the basis for retrofit construction may not reflect the actual site. You should carefully check the drawings for accuracy during the site survey. If drawings are not available, you must make accurate sketches drawn to scale (typically $\frac{1}{8}$" = 1') during the site survey for the areas involved in the installation. The content of these sketches can vary greatly, depending on your particular application and the complexity of the installation. Typically, your sketches should indicate the location, size, elevation, and type of material for all major supporting structural items such as walls, shafts, beams, trusses, girders, joists, and other floor and roof framing components. You should locate accurately and draw to scale all details of suspended ceilings and ceiling obstructions, such as lights, intercom speakers, and air diffusers. In addition, there are other types of information you will commonly note and record during a site survey for retrofit construction:

- Determine the size and available space in existing equipment rooms, telecommunications closets, and work areas. This information is important in order to determine if the new equipment, equipment racks, and mounting boards will fit into these areas. Determine if it is necessary to build new closet areas or to expand the existing ones.

- Determine the location and routing of existing pathways. Note if any of these pathways are vacant, have usable space, or will have available space after existing cables have been removed. Also, note if the existing pathways are capable of supporting the added weight of more cables. Determine if any new pathways are required.
- Determine if the existing power and grounding systems comply with the prevailing electrical code. Also, determine if the existing building power and grounding systems can support the new system requirements, or if modifications must be made in these areas to comply with the requirements.
- Determine the accessibility for running cables or pipes through walls, floors, and ceilings.
- Determine if the use of additional equipment or temporary cabling is required in order to install the new system without interrupting the customer's operation.
- Determine if any safety hazards exist.
- Determine if firestopping materials need to be used.

In order to minimize any disruption to the customer's normal business operations, the retrofit system often must be installed while the existing system remains operational. Also, it is common for the installation to be performed during off-peak business hours of the day or night so as not to conflict with the customer's business operation. Once the new system is in place and tested, a transfer (changeover) from the old system to the new one can take place. This changeover must be planned very carefully and coordinated with the customer. A poorly planned changeover can cause unnecessary and costly interruptions in your schedule and in the customer's operation. Depending on the extent of the retrofit, the changeover may occur all at one time or by phasing in selected portions of the system over time.

Following the changeover, all unused equipment and cabling from the old system should be removed. However, if it is impractical to remove unused items, a tag or label should be placed on each component that is no longer in use. The tag should state when the equipment was abandoned (month/day/year) and should be signed by the contractor.

Removal of abandoned cable is an important part of a retrofit. It should be done in a manner that does not disturb or damage other cables. Abandoned cables left in the building take up valuable pathway space and place an unnecessary strain on the supporting hardware. Also, if a fire occurs, unused cables can provide additional fuel and a pathway that may contribute to the spread of fire.

14.0.0 ADDITIONAL DOCUMENTATION

Nearly every step of the construction process is controlled by codes, standards, specifications, drawings, and other documents. Many codes and specifications were described earlier in this and other modules. The purpose of this section is to introduce some other documents and contractual terms that should be understood by the on-site supervisor and installation crew members, such as:

- Addenda
- Liens
- Stop-work orders
- Requests for information (RFIs)
- Change orders
- Project logs
- Certificates of completion
- Operator and maintenance manuals
- Activation/deactivation reports
- Wiring certifications

14.1.0 Addenda

Addenda are contractual documents issued prior to the bid date and award of the contract. They describe changes that have been made to the drawings and/or specifications. Addenda are usually incorporated as part of the original set of specifications.

14.2.0 Liens

Liens are claims that one person has upon the property of another as security for a debt. Common-law liens were created as a result of someone

On Site

Abandoned Cable

The *NEC®* requires the removal of abandoned data and communication cable. For example, if a network is being upgraded, old Cat 3 cable may need to be replaced with cable capable of handling more bandwidth. If so, the old Cat 3 cable will have to be removed. The cost of removing and disposing of such cable can be an important factor in a bid.

holding property or performing work on the personal property of another. Until the person holding the property was paid, that person could keep the item and had the right at some point to sell it for the value of services provided. In the construction industry, when contractors don't pay their material suppliers, the suppliers have the right under state lien laws to place a lien on the completed work of the contractor until such time that the contractor pays the suppliers.

If the contractor does not pay the bill, the supplier has the right to file a claim against the property on which the lien has been placed. This is just one example of how a lien works.

The filing of liens must follow a strict timetable and form established by each state and incorporated jurisdiction. Liens often involve an owner's real property as the valuable property. It is suggested that contractors filing liens consult a qualified attorney and also consider the consequences that filing liens might have on business relationships.

14.3.0 Stop-Work Orders

A stop-work order is a legal ban on any further work on a project or part of a project. It is usually issued by the owner or the owner's representative. Once issued, the order prevents any payment to the contractor for any work done during the time the stop order is in effect.

14.4.0 Requests for Information

A request for information (RFI) is used to clarify conflicting information found on plans, specifications, and the job site. If a discrepancy is noted on the working drawings, a conflict arises when shop drawings are prepared, or a conflict is seen on the site, an RFI may be issued to the architect, engineer, or contractor. There is a hierarchy that is usually followed. For example, should you notice a discrepancy on the plans, you should notify your supervisor. The supervisor generates the RFI. The information should be as specific as possible and have the time and date on it. The supervisor passes the RFI to the superintendent or project manager, who then passes it to the general contractor. The general contractor then relays the RFI to the architect or engineer. A sample RFI form is shown in *Figure 12*.

14.5.0 Change Orders

A change order (*Figure 13*) is a supplement to the job contract. It is prepared whenever additions, deletions, a change in arrangement of work, qual-

ity of materials, or job completion time is made. Sometimes changes are required because the owner has changed a certain detail in the building or because a structural beam has interfered with the installation of a piece of equipment, causing the equipment to be relocated. The on-site supervisor and crew members may or may not be involved in the generation and approval of a change order. However, they are always involved in carrying out the required work changes.

14.6.0 Project Logs

One very important document that should be kept by the supervisor on the job is a project log, or diary. This log should be started on the first day at the job site and stop when the job is totally completed. All important activities and job milestones that happen on the job each day should be recorded. Contractual milestones that are missed or otherwise affected by the work of other contractors should be logged and accompanied by detailed notes.

The primary purpose for the log is to inform upper management of what happened or is happening on your job. It also provides a historical record that is useful for many purposes. The information placed in the project log might include the following:

- Client name and address
- Principal contact
- Description of the work performed
- Contractual milestones accomplished or missed
- Equipment and material installed
- Equipment used
- Personnel used
- Special requirements
- Time schedule
- Problems encountered and how they were solved
- What was learned
- Toolbox (safety) meetings

The log should be kept in a bound book, not a loose-leaf cover. The log is an admissible legal document if claims or conflicts arise. In addition to noting conflicts that arise, simple statements noting normal job progress in the current working area should be made.

14.7.0 Certificates of Completion

A certificate of completion is a document that summarizes the installation of a system and the associated acceptance testing. It is provided to the property owner and to the AHJ. An example of

XYZ, Inc
General Contractors
123 Main Street
Bigtown, USA 10001
(111) 444-5555

PROJECT:

TO:

R.F.I.

Request for Information

XYZ Project # _____

Date: _____

R.F.I. # _____

RE:

**Specification
Reference:** _____

**Drawing
Reference:** _____

SUBJECT:

REQUIRED:

**Date Information
is Required:** _____

**XYZ, Inc
By:** _____

REPLY:

**Distribution: Superintendent
        Field File**

By: _____

Date: _____

304F12.EPS

Figure 12 Request for information form.

Company Name
Company Address

Job Change Order
Project Name
Project Location
Project Number

Change Order No. _____

Initiated by: _____ Date: _____

Details of request or nature of change:

Labor $ _____ Materials $ _____ TOTAL $ _____

_____ _____
Verified by (Signature)   Date

The undersigned hereby accepts the prices quoted on this Job Change Order and agrees to pay _____ the amount stipulated at the TOTAL above upon satisfactory delivery of the goods and services described above.
Acceptance also authorizes the materials to be ordered and/or the work to be performed.

_____
Authorized signature
_____
Title
_____
Date

Date work complete: _____ Verified by: _____

Figure 13 Change order form.

a certificate of completion typical of that used by security and fire alarm contractors is shown in *Appendix C*.

14.8.0 Operation and Maintenance Manuals

Many projects have a contractual requirement to supply one or more copies of an operation and maintenance manual for major item of equipment. These documents are normally turned over to the customer or designated person/organization at the end of the job. Operation and maintenance (O & M) manuals provide the customer and any technician servicing the equipment in the future with an invaluable source of information. The release of the O & M manuals is usually accompanied by a training session for building operations personnel, which is an essential part of the job to ensure that the customer knows how to operate and maintain the system. O & M manuals typically contain the following types of information for each piece of equipment:

- Principles of operation
- Functional description of the circuitry
- Illustrations showing the location and function of each control and indicator
- Step-by-step procedures for the proper use of equipment
- Safety considerations and precautions that should be observed when operating equipment
- Schematics, wiring diagrams, mechanical layouts, and parts lists for the specific unit
- Maintenance schedules and actions
- Troubleshooting recommendations for common problems

14.9.0 Activation/Deactivation Reports

The activation/deactivation report is primarily used during retrofit construction. During a changeover or tie-in to existing systems, it may be necessary to take existing alarms off line and take systems out of service. Before doing so, you should notify the AHJ, alarm company, and building management that the system will be out of service for a specific period of time. You should obtain the signature of the building manager, which acknowledges notification that the system is out of service, and obtain another signature when the system is reactivated.

14.10.0 Wiring Certification Diagrams and Lists

A wiring certification is used to check and record information about each wire in an installation or piece of equipment. It can take many forms and have various kinds of information, depending on the requirements that apply to various segments of the industry. Typical forms are a point-to-point wiring diagram or wiring list. Typically, each wire is checked for proper labeling, correct point-to-point terminations, its no-load resistance, and its resistance to ground. This information is important because some of the wires run in an initial installation may be reserved for future additions to the system. Also, the wiring certification provides a historical record for use in the future for comparison purposes to check the wires after an electrical disturbance, such as a lightning strike, power surge, or the failure of an individual piece of equipment. The wiring certification diagram or wiring list normally becomes part of the as-built drawings that are given to the owner at the time of the final check and acceptance testing.

On Site

Project Log

Among the items that should be recorded in the daily log or diary are the following:

- A record of project-related telephone calls and the nature of the call
- A record of any significant conversations and events in which commitments/agreements were made with other parties to the project
- Unexpected conditions that could affect cost or schedule

SUMMARY

The cost estimating process is a critical part of every project. People who participate in this process must be able to interpret project drawings and specifications so that they can accurately determine the amount and types of equipment and materials required, the numbers and types of personnel needed, and the time it will take to perform the work. The job supervisor or lead technician holds a key position. This person is given the responsibility and authority to get the job done within a given time and with a specified amount of resources. Because of the importance of this position, it is essential that the person selected have knowledge and experience in the functions of organizing, planning and scheduling, as well as directing and controlling labor, materials, tools, and equipment. This person must make several types of important work decisions. These decisions relate to job planning and accomplishment. The supervisor who plans ahead can be sure that the correct materials, equipment, and personnel are available and ready.

Your knowledge and skills are enhanced by your understanding of the elements and terminology involved in the job-planning process. In addition, skill in the use of contract documents and specifications increases your ability to plan ahead and meet contractual commitments.

1. When making a job estimate, the equipment and materials required for the job are recorded on a _____.
   a. summary sheet
   b. takeoff sheet
   c. recapitulation sheet
   d. survey checklist

2. When scaling a pipe run shown on a construction drawing, you measure 5¼". What is the actual length of the pipe run if the drawing scale is ⅛" = 1'?
   a. 21'
   b. 27'
   c. 35'
   d. 42'

3. If you are bidding on a project and decide not to bid on a portion of the work, it is referred to as a(n) _____.
   a. change order
   b. addendum
   c. exclusion
   d. RFI

4. When planning a job after the contract is awarded, the first step is to _____.
   a. review the job requirements
   b. perform a site survey
   c. schedule the work
   d. assemble the crew

5. A site survey is performed only once on any project.
   a. True
   b. False

6. Assume you are developing a schedule for a job based on an 8-hour day and a 40-hour work week. Also assume that the estimator's takeoff sheets show that a certain task requires 128 hours to complete. If two employees are doing the work, how many working days should your schedule show to complete the task?
   a. 8 days
   b. 10½ days
   c. 13 days
   d. 16 days

7. Materials are ordered from a supplier using a form called a _____.
   a. material list (ML)
   b. purchase order (PO)
   c. material safety data sheet (MSDS)
   d. material invoice

8. Proof that the quality control inspections and/or acceptance tests for a specific job have been accomplished is usually documented _____.
   a. on the working drawings
   b. in the contract
   c. in the addenda
   d. on preprinted inspection forms

9. Rework or repair work identified on a punch list is normally accomplished _____.
   a. before the quality inspections
   b. before completing the estimate
   c. after the project closeout
   d. after the acceptance testing

10. Contractual documents that describe changes that have been made to drawings and/or specifications prior to the award of the contract are called _____.
    a. addenda
    b. exclusions
    c. change orders
    d. project logs

Trade Terms Introduced in This Module

Change order: The formal document that modifies the original agreement covering work to be performed. This includes, but is not limited to, work added or deleted, changes to location of services, different materials, and the time allotted for the project. It can also include damages to your work caused by others or by natural forces.

Contract: The legal framework of an agreement between at least two parties in which one agrees to perform some task for the other in return for a specified payment.

Direct cost: The cost for labor wages, fringe benefits, and wage-related taxes and insurance for all workers and supervisors directly engaged in the work; also, the cost for all materials, engineering and design, permits and fees, and tools and equipment used in an installation to accomplish the finished job.

Exclusions: A designated section or statements in contractual documents that identify work, items, and other entities exempted from or not covered by the requirements of the contract or scope of work.

Punch list: A list of repairs or other work that must be done to satisfy the requirements of the contract and obtain customer acceptance of the job.

Record drawings: Drawings that show major changes and modifications made from the original job plan to the finished state.

Scope of work: A written document that defines the quantity and quality of work to be done and the materials to be used in the construction of a building or other structure. Its content and format are generally of a less formal nature than those used in the specifications.

Specifications: A formal written document that defines the quantity and quality of work to be done and the materials to be used in the construction of a building or other structure. It also provides additional information not contained on a related set of working drawings.

Takeoff: The process of surveying, measuring, itemizing, and counting all materials and equipment needed for a construction job as indicated by the drawings or job site survey.

CONTRACTOR'S BID LETTER

11/4/10

Smith and Smith, A.I.A.
123 Rainbow Drive
P.O. Box 1234
Anytown, NY
13288

Dear Mr. Smith,

Based on our discussions, I have attached our proposal to install an integrated sound system at your facility, Anytown Church, Anytown, NY. The attached proposal includes the project scope, project implementation, system costs, exclusions and warranty.

Briefly, ABC Systems, Inc. is a 30-year-old New City company with a focus on delivering the highest quality integrated systems.

Our engineering, integration, and installation skills have earned ABC national recognition in our industry for customer satisfaction. Our goal is to provide the highest quality systems in a prompt, efficient and cost-effective manner.

Please let me know if you have any additional questions regarding this proposal.

We look forward to working with you on this project.

Sincerely,

Jack Ryan, SET
Installation Manager

304A01.EPS

ABC Systems

<u>**Proposal For:**</u>

Smith and Smith, A.I.A.
123 Rainbow Drive
P.O. Box 1234
Anytown, NY
13288

<u>**Project:**</u>

Anytown Church

<u>Installation of sound system</u>

- **Overview:** New Sound System as per plans and specifications. ABC will provide all specified equipment or equal as approved

- **Equipment:** Describe what equipment will be used.

Quantity	Manufacturer	Part No.	Description
1	Lowell	L250-36	Rack
1	HAS		Mini Rolltop
1	FSR	SPC-20	Power Distribution
2	ETA	PD11LVP	Power Conditioners
2	AKG	D880	Handheld Mic
1	AKG	D880S	Handheld Mic
1	AKG	GN50E	Podium Mic
1	AKG	CK80	Capsule
2	ATSO	MS-12E	Stand
2	ATSO	PB21XE	Boom
6	Whirlwind	MKQ25-BLK	Cable
3	Shure	ULXS14/93	Laviler Package
3	Shure/Count	B3W4FF05BSL	Mic
3	Shure	ULXS14/30	Headset Package
3	Country	E60W5lsl	Earset Microphone
1	Shure	UA220	Antenna / Splitter
1	Shure	UA844	Active Antenna
2	BSS	Ar-133	Direct Box
2	Whirlwind	MKQ25-BLK	Cable
1	Tascam	CD-A500	Player
1	Middle	Rack Kit	Rack Kit for CD Player
1	Allen	WZ12:12DX	Mixer

-1-

304A02.EPS

1	Shure	SCM810	Automixer
1	Shure	P4800	Processor
1	Crown	CT600	Amp
1	Crown	CT8200	Amp
1	Crown	CT600	Amp
1	Crown	Ct600	Amp
12	EAW	LS432	Speaker
2	JBL	Control 25	Speaker
1	Listen	LT-800-0072-1-14-3	Transmitter
4	Listen	LR-300-072-0-a-c	Receiver
1	Listen	LA-122	Remote Antenna
4	Listen	LA-161	Ear Bud Element
1	Listen	LA-326	Rack Kit

- **Engineering, Installation, Testing and Training:** ABC will provide engineering services, as required, and will install all equipment and cable for the system, with provisions as listed in this proposal. The system will be thoroughly tested for proper operation and an informal training session will be conducted. Smith and Smith will designate those needing to attend the training.

Project Implementation

Project Engineering, including:
- System review and evaluation for minimum customer impact.
- Drawings, as necessary, to support installation of cable and components.
- Submittals to include materials list, manufacturers' data sheets and installation information.

Project Coordination, including:
- Prepare project schedule and schedule of values
- Procurement and staging of materials
- Assure all pre-installation tasks are completed

Installation Services, including:
- Wire and Back Boxes
- Install system components
- Termination of control equipment
- Site clean-up
- Programming, as necessary
- Testing and certification

Customer Training, including:
- Provide operation and maintenance manuals
- Provide as-built documentation, if required
- Review system operation with designated staff
- Provide ABC support and service contact information

- 2 -

304A03.EPS

Project Costs

The cost for the project as described is $47,608.00

Cost includes all equipment, materials, project implementation, applicable taxes,and ground freight as required to provide a complete and operating system.

Exclusions

The following is not included in our scope of work:

- 120 VAC electrical circuits for control equipment
- Conduit, raceway and standard electrical boxes
- Unseen conditions, structural work or asbestos removal
- Core drilling and/or building penetrations
- Patching, painting, millwork or finishes
- Permits and fees (unless specifically provided for elsewhere in this proposal)
- Overtime labor or accelerated schedule
- ABC will recommend and/or coordinate any of the above on customer request

Customer Responsibilities

It is the responsibility of owner to:

- Designate representative(s) empowered to act as agent for owner. This will insure accurate communications throughout the project.
- Provide facility access as required by ABC.
- Prepare the facility prior to the start of work.
- Verify that any owner-furnished or existing equipment that may be provided for the project is in good working order and covered under separate warranty or service contract.
- Provide hardware/software modifications to systems furnished by others.

Warranty and Support

- ABC warrants the system to be free from defects for a period of one (1) year from the date of acceptance or first beneficial use, whichever occurs first. Warranty service will be performed in a timely manner during ABC's normal business hours.

- Warranty is not extended for abuse, negligence, vandalism, accidents, misuse of equipment, power surges, lightning or other acts of God or nature.

- Additional services including extended/expanded warranty coverage, maintenance, training, telephone support, inspection,and monitoring agreements are available from ABC.

- 3 -

304A04.EPS

Payment Terms

20% (mobilization costs) due upon proposal acceptance. Remainder due 30 days from date of invoice, with approved credit.

Proposed by:

ABC Systems, Inc.

_____   _____   _____

               Signature   *Title*   *Date*

Accepted by:

Anytown Church

_____   _____   _____

               Signature   *Title*   *Date*

304A05.EPS

INSPECTION AND TESTING FORM

INSPECTION AND TESTING FORM

SERVICE ORGANIZATION	PROPERTY NAME (User)
Name:	Name:
Addr.:	Addr.:
City, State	City, State
Representative:	Owner Contact:
License #:	Telephone:
Telephone:	

MONITORING ENTITY	APPROVING AGENCY
Contact:	Contact:
Telephone:	Telephone:
Monitoring Account Ref. No.:	

TYPE TRANSMISION		SERVICE	
	McCulloh		Weekly
	Multiplex		Monthly
	Digital		Quarterly
	Reverse Priority		Semi-Annually
	Radio Frequency		Annually
	Other (specify)		Other (specify)

PANEL MANUFACTURER:_____ Model #:_____
Circuit Styles: _____ Number of Circuits:_____
Software Rev.: _____ Last date system had any service performed: _____
Last date that any software or configuration was revised: _____

ALARM INITIATING DEVICES AND CIRCUIT INFORMATION

QTY	CIRCUIT STYLE		SPECIAL TESTING INFORMATION
		MANUAL STATIONS	
		ION DETECTORS	
		PHOTO DETECTORS	
		DUCT DETECTORS	
		HEAT DETECTORS	
		WATERFLOW SWITCHES	
		SUPERVISORY SWITCHES	
		OTHER:	

ALARM INDICATING DEVICES AND CIRCUIT INFORMATION

QTY	CIRCUIT STYLE		SPECIAL TESTING INFOMATION				
		BELLS					
		HORNS					
		CHIMES					
		STROBES					
		SPEAKERS					
		OTHER:					
	TOTAL NUMBER OF ALARM INDICATING CIRCUITS			YES		NO	ARE CIRCUITS SUPERVISED?

304A06.EPS

SUPERVISORY SIGNAL INITIATING DEVICES AND CIRCUIT INFORMATION

QTY	CIRCUIT STYLE		COMMENTS
		Building Temp.	
		Site Water Temp.	
		Site Water Level	
		Fire Pump Power	
		Fire Pump Running	
		Fire Pump Auto Position	
		Fire Pump or Pump Controller Trouble	
		Generator Running	
		Generator in Auto Position	
		Generator or Generator Controller TBL	
		Switch Transfer	
		Other:	

SIGNALING LINE CIRCUITS See NFPA 72 table 3-6.1 QTY: _____ Style(s): _____

SYSTEM POWER SUPPLIES

    a) **Primary** (Main): Nominal Voltage _____, Amps _____

    Overcurrent Protection: Type _____, Amps _____

    Location (Panel Number) _____ Circuit Number _____

    Disconnecting (Means) Location: _____

    b) **Secondary** (Standby):

    _____Storage Battery: Amp-hr rating _____

    Calculated capacity to operate system, in hours: _____24 _____60

    _____ Engine-driven generator dedicated to fire alarm system:

    Location of fuel storage: _____

    Type Battery ___Dry Cell ___Nickel Cadmium ___ Sealed Lead-Acid ___Lead-Acid

    ___Other (specify): _____

    c) **Emergency or Standby system used as a backup to primary power supply, instead of using a secondary power supply:**

    ____ Emergency system described in NFPA 70, Article 700

    ____ Legally required standby described in NFPA 70, Article 701

    ____ Optional standby system described in NFPA 70, Article 702, which also meets the performance requirements of Article 700 or 701.

PRIOR TO ANY AND ALL TESTING

NOTIFICATION MADE:	YES	NO	WHO	TIME
Monitoring Entity				
Building Occupants				
Building Management				
Other (specify)				
AHJ (Notified) of any impairments				

304A07.EPS

SYSTEM TEST AND INSPECTIONS

TYPE	VISUAL	FUNCTIONAL	COMMENTS
Control Panel			
Interface Eq.			
Lamps/LEDs			
Fuses			
Primary Power Supply			
Disconnect Switches			
Ground Fault Monitoring			

SECONDARY POWER

TYPE	VISUAL	FUNCTIONAL	COMMENTS
Battery Condition			
Load Voltage			
Discharge Test			
Charge Test			
Specific Gravity			

Transient Suppressors			
Remote Annunciators			
Notification Appliances			
Audible			
Visual			
Speakers			
Voice Clarity			

INITIATING AND SUPERVISORY DEVICE TEST AND INSPECTIONS

LOC. & S/N	DEVICE TYPE	VISUAL CHECK	FUNCTIONAL TEST	FACTORY TEST	FACTORY SETTING	MEAS. SETTING	PASS	FAIL
Comments:								

304A08.EPS

EMERGENCY COMMUNICATIONS EQUIPMENT

	VISUAL	FUNCTIONAL	COMMENTS
Phone Set			
Phone Jacks			
Off-Hook Indicator			
Amplifier(s)			
Tone Generator(s)			
Call in Signal			
System Performance			

INTERFACE EQUIPMENT	VISUAL	DEVICE OPERATION	SIMULATED OPERATION
SPECIAL HAZARD SYSTEMS			

SPECIAL PROCEDURES:

_____

_____

_____

_____

_____

_____

_____

_____

COMMENTS:

_____

_____

_____

_____

ON/OFF PREMISES MONITORING	YES	NO	TIME	COMMENTS
Alarm Signal				
Alarm Restoral				
Trouble Signal				
Trouble Restoral				
Supervisory Signal				
Supervisory Restoral				

304A09.EPS

NOTIFICATIONS THAT TESTING IS COMPLETE

NOTIFICATION MADE:	YES	NO	WHO	TIME
Monitoring Entity				
Building Occupants				
Building Management				
Other (specify)				
AHJ (Notified) of any impairments				

The following did not operate correctly: _____

_____

_____

_____

_____

_____

_____

_____

Who was notified of the above indicated operation problem:

_____AHJ   Name: _____

_____Owner   Name: _____

_____Building Engineer   Name: _____

_____Service Company Supervisor   Name: _____

_____ Other   Name: _____

                   Title: _____

SYSTEM RESTORED TO NORMAL OPERATION:

Date _____ Time: _____ By: _____

Signature: _____

This testing was performed in accordance with applicable NFPA Standards.

Print Name of Inspector: _____ Date: _____

Signature of Inspector: _____ Time: _____

Name of Owner or Representative: _____ Date: _____

Signature of Owner or Rep.: _____ Time: _____

304A10.EPS

CERTIFICATE OF COMPLETION

Certificate of Completion

Name of Protected Property:	
Address:	
Rep. of Protected Property (name & phone):	
Authority Having Jurisdiction:	
AHJ Address & Phone Number:	

1) Type(s) of System or Service:

_____Local - NFPA 72 chapter 3   If alarm is transmitted to location(s) off Premise, list where received:

_____

_____**Emergency Voice/Alarm Service** - NFPA 72 chapter 3

    _____Qty of voice/alarm channels Single _____   Multiple:_____

    _____Qty of speakers installed   _____Quantity of speaker zones:

    _____Qty of telephones or telephone jacks included in system

Auxiliary - NFPA 72 chapter 4   Indicate type of connection:

    _____ Local energy   _____Shunt   _____Parallel telephone

    Location for receipt of signals: _____

    Telephone number for receipt of signals: _____

Remote Station - NFPA 72 chapter 4

    Alarm: _____   Supervisory: _____

Proprietary - NFPA 72 chapter 4 If alarms are retransmitted to public fire service communications center or others, indicate location and telephone number of the organization receiving alarm:

    _____

    Indicate how alarm is retransmitted: _____

Central Station -NFPA 72 chapter 4

    The Prime Contractor: _____

    Central Station Location: _____

    Means of transmission of signals from the protected premises to the central station:

    _____ McCulloh   _____ Multiplex   _____ One-way Radio

    _____ Digital Alarm Communicator   _____ Two-way Radio   _____ Others

Means of transmission of alarms to the public fire service communications center:
1)
2)
System Location:

	Organization Name/Phone	Representative Name/Phone
Installer		
Supplier		
Service Organization		

Location of Record (As-Built) Drawings:
Location of Owner's Manuals:
Location of Test Reports:
A contract, dated _____ , for test and inspection in accordance with NFPA standard(s) No.(s) _____ , dated _____ is in effect.

2) Certification of System Installation (Fill out after installation is complete and wiring checked for opens, shorts, ground faults, and improper branching but prior to conducting operational acceptance test.)

304A11.EPS

This system has been installed in accordance with the NFPA standards as listed below, was inspected by_____ on
_____, includes the device listed below and has been in service since _____.

        _____ NFPA 72 chapters 1 3 4 5 6 7 (circle all that apply)

        _____ NFPA 70, *National Electrical Code*, Article 760

        _____ Manufacturer's instructions

        _____ Other (specify) _____

Signed: _____ Print Name:_____

Organization: _____ Date: _____

3) Certification of System Operation All operational features and functions of this system were tested by
_____ on _____ and found to be operating properly in accordance with:

        _____ NFPA 70, *National Electrical Code*, Article 760

        _____ Manufacturer's instructions

        _____ Other (specify) _____

Signed: _____ Print Name: _____

Organization: _____ Date: _____

4) Alarm Initiating Devices and Circuits (Fill in all blanks to indicate quantity of devices. Enter -0- for none.)

MANUAL

    a) ____Manual Stations _____Noncoded, Activating _____Transmitters _____Coded

    b) ____Combination Manual Fire Alarm and Guard's Tour Coded Station

AUTOMATIC   Complete: _____ Partial: _____

    a) ___Smoke Detectors ____Ion ____Photo

    b)___Duct Detectors ____Ion ____Photo

    c)___Heat Detectors ____Fixed ____Rate-of-Rise ____Fixed Tempt Rate-of-Rise ____Rate-Compensating

    d)___Sprinkler Water Flow Switches: ____Noncoded, Activating ____Transmitters ____Coded

    e)___Other(s): _____

5) Supervisory Signal Initiating Devices and Circuits (Fill in all blanks to indicate quantity of devices. Enter -0- for none.)

GUARD'S TOUR

    a)____Coded Stations

    b)____Noncoded Stations Activating _____Transmitters

    c)____Compulsory Guard Tour Systems Comprised of _____ Transmitter Stations & ____ Intermediate Stations

           Note: Combination devices recorded under 4(b) and 5(a).

SPRINKLER SYSTEM

    a)____Coded Valve Supervisory Signaling Attachments

    b)____Building Temperature Points

    c)____Site Water Temperature Points

    d)____Site Water Supply Level Points

ELECTRIC FIRE PUMP

    e)____Fire Pump Power

    f)____Fire Pump Running

    g)____Phase Reversal

ENGINE-DRIVEN FIRE PUMP

    h)____Selector in Auto Position

    i)____Engine or Control Panel Trouble

    j)____Fire Pump Running

304A12.EPS

ENGINE-DRIVEN GENERATOR

    k)____ Selector in Auto Position

    l)____ Control Panel Trouble

    m)____ Transfer Switches

    n)____ Engine Running

OTHER SUPERVISORY FUNCTION(s)

_____

_____

_____

_____

6) Alarm Notification Appliances and Circuits Qty. of indicating appliance circuits connected to the system: _____

Types and quantities of alarm indicating appliances installed:

    a)____ Bells ____ Inches

    ____ Speakers

    b)____ Horns

    c)____ Chimes

    d)____ Other: _____

    e)____ Visual Signals   Type: _____ ____ with audible ____ without audible

    f)____ Local Annunciator Location: _____

7) Signaling Line Circuits (See NFPA 72 table 3-6.1)

    _____ Quantity   _____ Style

8) System Power Supplies

    a) PRIMARY (Main)   Nominal Voltage: _____   Current Rating: _____ amps

    Overcurrent Protection: Type _____   Current Rating: _____ amps

    Location: _____ circuit number: _____

    b) SECONDARY (Standby)

    ____ Storage Battery: Amp-Hour Rating _____

        Calculated capacity to drive system, in hours: ____ 24 minimum   ____ 60 minimum

    ____ Engine-driven generator dedicated to fire alarm system:

        Location of fuel storage: _____

    c) EMERGENCY OR STANDBY SYSTEM   used as backup to Primary Power Supply, instead of a Secondary Power Supply:

    ____ Emergency System described in NFPA 70, Article 700

    ____ Legally Required Standby System described in NFPA 70, Article 701

    ____ Optional Standby System described in NFPA 70, Article 702, which also meets the

        performance requirements of Article 700 or 701

9) System Software

    a) Operating System Software Revision Level(s): _____

    b) Application Software Revision Level(s): _____

    c) Revision Completed by: Print Name: _____

                           Print Firm's Name: _____

10) Comments

_____ (signed) for Central Station or Alarm Service Company

_____ (title) _____ (date)

304A13.EPS

Frequency of routine test and inspections, if other than in accordance with the referenced NFPA standard(s):

_____
_____
_____
_____
_____
_____

System deviations from the referenced NFPA standards(s) are:

_____
_____
_____
_____
_____

_____(signed) for Central Station or Alarm Service Company
_____ (title) _____ (date)

Upon completion of the system(s) satisfactory test(s) witnessed (if required by the authority having jurisdiction):

AHJ special comments:
_____
_____
_____
_____
_____
_____
_____
_____

_____(signed) representative of the authority having jurisdiction
_____ (title) _____ (date)

Notes:

I (we), the customer (end user or representative), have:	(check all that apply)
	Received a copy of the owner's manual(s) and been provided with any keys/codes needed to access any user controls
	Been shown how to use all customer functions and controls
	Been advised how to contact the monitoring station and fire department prior to testing
	Been advised who to contact for repair and maintenance service
	Been advised who to contact for new employee training relating to use of this fire alarm system
	Been advised who to contact to help you (the customer) with a false alarm reduction plan
	Been advised of required testing frequency
	Received an exact copy of this certificate of completion

_____ (signed) _____ (Print Name)
_____ (title) _____ (Date)
_____ (witness) _____ (Print Witness' Name)

304A14.EPS

Additional Resources

This module presents thorough resources for task training. The following resource material is suggested for further study.

Project Supervision. Gainesville, FL: The National Center for Construction Education and Research.

Mike's Basic Guide to Cabling Computers and Telephones. Prairie Wind Communications.

Information Transport Systems Installation Methods Manual. Tampa, FL: BICSI.

Figure Credits

CONTREN® LEARNING SERIES — USER UPDATE

NCCER makes every effort to keep its textbooks up-to-date and free of technical errors. We appreciate your help in this process. If you find an error, a typographical mistake, or an inaccuracy in NCCER's Contren® materials, please fill out this form (or a photocopy), or complete the online form at www.nccer.org/olf. Be sure to include the exact module number, page number, a detailed description, and your recommended correction. Your input will be brought to the attention of the Authoring Team. Thank you for your assistance.

Instructors – If you have an idea for improving this textbook, or have found that additional materials were necessary to teach this module effectively, please let us know so that we may present your suggestions to the Authoring Team.

NCCER Product Development and Revision
3600 NW 43rd Street, Building G, Gainesville, FL 32606

Fax: 352-334-0932
Email: curriculum@nccer.org
Online: www.nccer.org/olf

❏ Trainee Guide ❏ AIG ❏ Exam ❏ PowerPoints Other _____

Craft / Level: _____ Copyright Date: _____

Module Number / Title: _____

Section Number(s): _____

Description: _____

_____

_____

_____

Recommended Correction: _____

_____

_____

_____

Your Name: _____

Address: _____

_____

Email: _____ Phone: _____

Fundamentals of Crew Leadership

National Center for Construction Education and Research

President: Don Whyte
Director of Product Development: Daniele Stacey
Fundamentals of Crew Leadership Project Manager: Patty Bird
Production Manager: Tim Davis
Quality Assurance Coordinator: Debie Ness
Editor: Chris Wilson
Desktop Publishing Coordinator: James McKay
Production Assistant: Laura Wright

Editorial and production services provided by Topaz Publications, Liverpool, NY
Lead Writer/Project Manager: Tom Burke
Desktop Publisher: Joanne Hart
Art Director: Megan Paye
Permissions Editors: Andrea LaBarge, Alison Richmond

10 9 8 7 6 5 4 3 2 1

V.1 1/11

Prentice Hall
is an imprint of

www.pearsonhighered.com

ISBN 13: 978-0-13-610652-4

FOREWORD

Work gets done most efficiently if workers are divided into crews with a common purpose. When a crew is formed to tackle a particular job, one person is appointed the leader. This person is usually an experienced craftworker who has demonstrated leadership qualities. To become an effective leader, it helps if you have natural leadership qualities, but there are specific job skills that you must learn in order to do the job well.

This module will teach you the skills you need to be an effective leader, including the ability to communicate effectively; provide direction to your crew; and effectively plan and schedule the work of your crew.

As a crew member, you weren't required to think much about project cost. However, as a crew leader, you need to understand how to manage materials, equipment, and labor in order to work in a cost-effective manner. You will also begin to view safety from a different perspective. The crew leader takes on the responsibility for the safety of the crew, making sure that workers follow company safety polices and have the latest information on job safety issues.

As a crew leader, you become part of the chain of command in your company, the link between your crew and those who supervise and manage projects. As such, you need to know how the company is organized and how you fit into the organization. You will also focus more on company policies than a crew member, because it is up to you to enforce them within your crew. You will represent your team at daily project briefings and then communicate relevant information to your crew. This means learning how to be an effective listener and an effective communicator.

Whether you are currently a crew leader wanting to learn more about the requirements, or a crew member preparing to move up the ladder, this module will help you reach your goal.

This program consists of an Annotated Instructor's Guide (AIG) and a Participant's Manual. The AIG contains a breakdown of the information provided in the Participant's Manual as well as the actual text that the participant will use. The Participant's Manual contains the material that the participant will study, along with self-check exercises and activities, to help in evaluating whether the participant has mastered the knowledge needed to become an effective crew leader.

For the participant to gain the most from this program, it is recommended that the material be presented in a formal classroom setting, using a trained and experienced instructor. If the student is so motivated, he or she can study the material on a self-learning basis by using the material in both the Participant's Manual and the AIG. Recognition through the National Registry is available for the participants provided the program is delivered through an Accredited Sponsor by a Master Trainer or ICTP instructor. More details on this program can be received by contacting the National Center for Construction Education and Research at www.nccer.org.

Participants in this program should note that some examples provided to reinforce the material may not apply to the participant's exact work, although the process will. Every company has its own mode of operation. Therefore, some topics may not apply to every participant's company. Such topics have been included because they are important considerations for prospective crew leaders throughout the industries supported by NCCER.

A Note to NCCER Instructors and Trainees

If you are training through an Accredited NCCER Sponsor company, note that you may be eligible for dual credentials upon completion. When submitting Form 200, indicate completion of the two module numbers that apply to *Fundamentals of Crew Leadership* – 46101-11 (from NCCER's Contren® Management Series) or the applicable module in a craft training program, such as Module 26413 from NCCER's Electrical curriculum. Transcripts will be issued accordingly.

Contents

Topics to be presented in this module include:

Contents (continued)

Figures and Tables

Acknowledgments

This curriculum was revised as a result of the farsightedness
and leadership of the following sponsors:

ABC South Texas Chapter
HB Training & Consulting
Turner Industries Group, LLC

University of Georgia
Vision Quest Academy
Willmar Electric Service

This curriculum would not exist were it not for the dedication and unselfish energy of
those volunteers who served on the Authoring Team. A sincere thanks is extended to the following:

John Ambrosia
Harold (Hal) Heintz
Mark Hornbuckle
Jonathan Liston

Jay Tornquist
Wayne Tyson
Antonio "Tony" Vazquez

NCCER Partners

American Fire Sprinkler Association

Associated Builders and Contractors, Inc.

Associated General Contractors of America

Association for Career and Technical Education

Association for Skilled and Technical Sciences

Carolinas AGC, Inc.

Carolinas Electrical Contractors Association

Center for the Improvement of Construction Management and Processes

Construction Industry Institute

Construction Users Roundtable

Design Build Institute of America

Merit Contractors Association of Canada

Metal Building Manufacturers Association

NACE International

National Association of Manufacturers

National Association of Minority Contractors

National Association of Women in Construction

National Insulation Association

National Ready Mixed Concrete Association

National Technical Honor Society

National Utility Contractors Association

NAWIC Education Foundation

North American Technician Excellence

Painting & Decorating Contractors of America

Portland Cement Association

SkillsUSA

Steel Erectors Association of America

U.S. Army Corps of Engineers

University of Florida, M.E. Rinker School of Building Construction

Women Construction Owners & Executives, USA

Objectives

Upon completion of this section, you should be able to:

1. Describe the opportunities in the construction and power industries.
2. Describe how workers' values change over time.
3. Explain the importance of training and safety for the leaders in the construction and power industries.
4. Describe how new technologies are beneficial to the construction and power industries.
5. Identify the gender and minority issues associated with a changing workforce.
6. Describe what employers can do to prevent workplace discrimination.
7. Differentiate between formal and informal organizations.
8. Describe the difference between authority, responsibility, and accountability.
9. Explain the purpose of job descriptions and what they should include.
10. Distinguish between company policies and procedures.

1.0.0 INDUSTRY TODAY

Today's managers, supervisors, and crew leaders face challenges different from those of previous generations of leaders. To be a leader in industry today, it is essential to be well prepared. Today's crew leaders must understand how to use various types of new technology. In addition, they must have the knowledge and skills needed to manage, train, and communicate with a culturally diverse workforce whose attitudes toward work may differ from those of earlier generations and cultures. These needs are driven by changes in the workforce itself and in the work environment, and include the following:

- A shrinking workforce
- The growth of technology
- Changes in employee attitudes and values
- The emphasis on bringing women and minorities into the workforce
- The growing number of foreign-born workers
- Increased emphasis on workplace health and safety
- Greater focus on education and training

1.1.0 The Need for Training

Effective craft training programs are necessary if the industry is to meet the forecasted worker demands. Many of the skilled, knowledgeable craftworkers, crew leaders, and managers—the so-called baby boomers—have reached retirement age. In 2010, these workers who were born between 1946 and 1964, represented 38 percent of the workforce. Their departure leaves a huge vacuum across the industry spectrum. The Department of Labor (DOL) concludes that the best way for industry to reduce shortages of skilled workers is to create more education and training opportunities. The DOL suggests that companies and community groups form partnerships and create apprenticeship programs. Such programs could provide younger workers, including women and minorities, with the opportunity to develop job skills by giving them hands-on experience.

When training workers, it is important to understand that people learn in different ways. Some people learn by doing, some people learn by watching or reading, and others need step-by-step instructions as they are shown the process. Most people learn best by a combination of styles. It is important to understand what kind of a learner you are teaching, because if you learn one way, you tend to teach the way you learn. Have you ever tried to teach somebody and failed, and then another person successfully teaches the same thing in a different way? A person who acts as a mentor or trainer needs to be able to determine what kind of learner they are addressing and teach according to those needs.

The need for training is not limited to craftworkers. There must be supervisory training to ensure there are qualified leaders in the industry to supervise the craftworkers.

1.1.1 Motivation

As a supervisor or crew leader, it is important to understand what motivates your crew. Money is often thought to be a good motivator. Although that may be true to some extent, it has been proven to be a temporary solution. Once a person has reached a level of financial security, other factors come into play. Studies show that many people tend to be motivated by environment and conditions. For those people, a great workplace may provide more satisfaction than pay. If you give someone a raise, they tend to work harder for a period of time. Then the satisfaction dissipates and they may want another raise. People are often motivated by feeling a sense of accomplishment. That is why setting and working toward recognizable goals tends to make employees more productive. A person with a feeling of involvement or a sense of achievement is likely to be better motivated and help to motivate others.

1.1.2 Understanding Workers

Many older workers grew up in an environment in which they were taught to work hard and stay with the job until retirement. They expected to stay with a company for a long time, and companies were structured to create a family-type environment.

Times have changed. Younger workers have grown up in a highly mobile society and are used to rapid rewards. This generation of workers can sometimes be perceived as lazy and unmotivated, but in reality, they simply have a different perspective. For such workers, it may be better to give them small projects or break up large projects into smaller pieces so that they feel repetitively rewarded, thus enhancing their perception of success.

- *Goal setting* – Set short-term and long-term goals, including tasks to be done and expected time frames. Help the trainees understand that things can happen to offset the short-term goals. This is one reason to set long-term goals as well. Don't set them up for failure, as this leads to frustration, and frustration can lead to reduced productivity.
- *Feedback* – Timely feedback is important. For example, telling someone they did a good job last year, or criticizing them for a job they did a month ago, is meaningless. Simple recognition isn't always enough. Some type of reward should accompany positive feedback, even if it is simply recognizing the employee in a public way. Constructive feedback should be given in private and be accompanied by some positive action, such as providing one-on-one training to correct a problem.

1.1.3 Craft Training

Craft training is often informal, taking place on the job site, outside of a traditional training classroom. According to the American Society for Training and Development (ASTD), craft training is generally handled through on-the-job instruction by a qualified co-worker or conducted by a supervisor.

The Society of Human Resources Management (SHRM) offers the following tips to supervisors in charge of training their employees:

- *Help crew members establish career goals.* Once the goals are established, the training required to meet the goals can be readily identified.
- *Determine what kind of training to give.* Training can be on the job under the supervision of a co-worker. It can be one-on-one with the supervisor. It can involve cross-training to teach a new trade or skill, or it can involve delegating new or additional responsibilities.
- *Determine the trainee's preferred method of learning.* Some people learn best by watching, others from verbal instructions, and others by doing. When training more than one person at a time, try to use all three methods.

Communication is a critical component of training employees. The SHRM advises that supervisors do the following when training their employees:

- *Explain the task, why it needs to be done, and how it should be done.* Confirm that the trainees understand these three areas by asking questions. Allow them to ask questions as well.
- *Demonstrate the task.* Break the task down into manageable parts and cover one part at a time.
- *Ask trainees to do the task while you observe them.* Try not to interrupt them while they are doing the task unless they are doing something that is unsafe and potentially harmful.
- *Give the trainees feedback.* Be specific about what they did and mention any areas where they need to improve.

1.1.4 Supervisory Training

Given the need for skilled craftworkers and qualified supervisory personnel, it seems logical that companies would offer training to their employees through in-house classes, or by subsidizing outside training programs. While some contractors have their own in-house training programs or participate in training offered by associations and other organizations, many contractors do not offer training at all.

There are a number of reasons that companies do not develop or provide training programs, including the following:

- Lack of money to train
- Lack of time to train
- Lack of knowledge about the benefits of training programs
- High rate of employee turnover
- Workforce too small

- Past training involvement was ineffective
- The company hires only trained workers
- Lack of interest from workers
- Lack of company interest in training

For craftworkers to move up into supervisory and managerial positions, it will be necessary for them to continue their education and training. Those who are willing to acquire and develop new skills have the best chance of finding stable employment. It is therefore critical that employees take advantage of training opportunities, and that companies employ training as part of their business culture.

Your company has recognized the need for training. Your participation in a leadership training program such as this will begin to fill the gap between craft and supervisory training.

1.2.0 Impact of Technology

Many industries, including the construction industry, have made the move to technology as a means of remaining competitive. Benefits include increased productivity and speed, improved quality of documents, greater access to common data, and better financial controls and communication. As technology becomes a greater part of supervision, crew leaders need to be able to use it properly. One important concern with electronic communication is to keep it brief, factual, and legal. Because the receiver has no visual or auditory clues as to the sender's intent, the sender can be easily misunderstood. In other words, it is more difficult to tell if someone is just joking via e-mail because you can't see their face or hear the tone of their voice.

Cellular telephones, voicemail, and handheld communication devices have made it easy to keep in touch. They are particularly useful communication sources for contractors or crew leaders who are on a job site, away from their offices, or constantly on the go.

Cellular telephones allow the users to receive incoming calls as well as make outgoing calls. Unless the owner is out of the cellular provider's service area, the cell phone may be used any time to answer calls, make calls, and send and receive voicemail or email. Always check the company's policy with regard to the use of cell phones on the job.

Handheld communication devices known as smart phones allow supervisors to plan their calendars, schedule meetings, manage projects, and access their email from remote locations. These computers are small enough to fit in the palm of the hand, yet powerful enough to hold years of information from various projects. Information can be transmitted electronically to others on the project team or transferred to a computer.

2.0.0 GENDER AND CULTURAL ISSUES

During the past several years, the construction industry in the United States has experienced a shift in worker expectations and diversity. These two issues are converging at a rapid pace. At no time has there been such a generational merge in the workforce, ranging from The Silent Generation (1925–1945), Baby Boomers (1946–1964), Gen X (1965–1979), and the Millennials, also known as Generation Y (1980–2000).

This trend, combined with industry diversity initiatives, has created a climate in which companies recognize the need to embrace a diverse workforce that crosses generational, gender, and ethnic boundaries. To do this effectively, they are using their own resources, as well as relying on associations with the government and trade organizations. All current research indicates that industry will be more dependent on the critical skills of a diverse workforce—a workforce that is both culturally and ethnically fused. Across the United States, construction and other industries are aggressively seeking to bring new workers into their ranks, including women and racial and ethnic minorities. Diversity is no longer solely driven by social and political issues, but by consumers who need hospitals, malls, bridges, power plants, refineries, and many other commercial and residential structures.

Some issues relating to a diverse workforce will need to be addressed on the job site. These issues include different communication styles of men and women, language barriers associated with cultural differences, sexual harassment, and gender or racial discrimination.

2.1.0 Communication Styles of Men and Women

As more and more women move into construction, it becomes increasingly important that communication barriers between men and women are broken down and that differences in behaviors are understood so that men and women can work together more effectively. The Jamestown, New York Area Labor Management Committee (JALMC) offers the following explanations and tips:

- *Women tend to ask more questions than men do.* Men are more likely to proceed with a job and figure it out as they go along, while women are more likely to ask questions first.
- *Men tend to offer solutions before empathy; women tend to do the opposite.* Both men and women should say what they want up front, whether it's the solution to a problem, or simply a sympathetic ear. That way, both genders will feel understood and supported.
- *Women are more likely to ask for help when they need it.* Women are generally more pragmatic when it comes to completing a task. If they need help, they will ask for it. Men are more likely to attempt to complete a task by themselves, even when assistance is needed.
- *Men tend to communicate more competitively, and women tend to communicate more cooperatively.* Both parties need to hear one another out without interruption.

This does not mean that one method is more effective than the other. It simply means that men and women use different approaches to achieve the same result.

2.2.0 Language Barriers

Language barriers are a real workplace challenge for crew leaders. Millions of workers speak languages other than English. Spanish is commonly spoken in the United States. As the makeup of the immigrant population continues to change, the number of non-English speakers will rise dramatically, and the languages being spoken will also change. Bilingual job sites are increasingly common.

Companies have the following options to overcome this challenge:

- Offer English classes either at the work site or through school districts and community colleges.
- Offer incentives for workers to learn English.

As the workforce becomes more diverse, communicating with people for whom English is a second language will be even more critical. The following tips will help when communicating across language barriers:

- Be patient. Give workers time to process the information in a way that they can comprehend.
- Avoid humor. Humor is easily misunderstood and may be misinterpreted as a joke at the worker's expense.
- Don't assume that people are unintelligent simply because they don't understand what you are saying.
- Speak slowly and clearly, and avoid the tendency to raise your voice.
- Use face-to-face communication whenever possible. Over-the-phone communication is often more difficult when a language barrier is involved.
- Use pictures or drawings to get your point across.
- If a worker speaks English poorly but understands reasonably well, ask the worker to demonstrate his or her understanding through other means.

2.3.0 Cultural Differences

As workers from a multitude of backgrounds and cultures are brought together, there are bound to be differences and conflicts in the workplace.

To overcome cultural conflicts, the SHRM suggests the following approach to resolving cultural conflicts between individuals:

- *Define the problem from both points of view.* How does each person involved view the conflict? What does each person think is wrong? This involves moving beyond traditional thought processes to consider alternate ways of thinking.
- *Uncover cultural interpretations.* What assumptions are being made based on cultural programming? By doing this, the supervisor may realize what motivated an employee to act in a particular manner.

- *Create cultural synergy.* Devise a solution that works for both parties involved. The purpose is to recognize and respect other's cultural values, and work out mutually acceptable alternatives.

2.4.0 Sexual Harassment

In today's business world, men and women are working side-by-side in careers of all kinds. As women make the transition into traditionally male industries, such as construction, the likelihood of sexual harassment increases. Sexual harassment is defined as unwelcome behavior of a sexual nature that makes someone feel uncomfortable in the workplace by focusing attention on their gender instead of on their professional qualifications. Sexual harassment can range from telling an offensive joke or hanging a poster of a swimsuit-clad man or woman, to making sexual comments or physical advances.

Historically, sexual harassment was thought to be an act performed by men of power within an organization against women in subordinate positions. However, the number of sexual harassment cases over the years, have shown that this is no longer the case.

Sexual harassment can occur in a variety of circumstances, including but not limited to the following:

- The victim as well as the harasser may be a woman or a man. The victim does not have to be of the opposite sex.
- The harasser can be the victim's supervisor, an agent of the employer, a supervisor in another area, a co-worker, or a non-employee.
- The victim does not have to be the person harassed, but could be anyone affected by the offensive conduct.
- Unlawful sexual harassment may occur without economic injury to or discharge of the victim.
- The harasser's conduct must be unwelcome.

The Equal Employment Opportunity Commission (EEOC) enforces sexual harassment laws within industries. When investigating allegations of sexual harassment, the EEOC looks at the whole record, including the circumstances and the context in which the alleged incidents occurred. A decision on the allegations is made from the facts on a case-by-case basis. A crew leader who is aware of sexual harassment and does nothing to stop it can be held responsible. The crew leader therefore should not only take action to stop sexual harassment, but should serve as a good example for the rest of the crew.

Did you know?

Some companies employ what is known as sensitivity training in cases where individuals or groups have trouble adapting to a multicultural, multi-gender workforce. Sensitivity training is a psychological technique using group discussion, role playing, and other methods to allow participants to develop an awareness of themselves and how they interact with others.

Prevention is the best tool to eliminate sexual harassment in the workplace. The EEOC encourages employers to take steps to prevent sexual harassment from occurring. Employers should clearly communicate to employees that sexual harassment will not be tolerated. They do so by developing a policy on sexual harassment, establishing an effective complaint or grievance process, and taking immediate and appropriate action when an employee complains.

Both swearing and off-color remarks and jokes are not only offensive to co-workers, but also tarnish a worker's image. Crew leaders need to emphasize that abrasive or crude behavior may affect opportunities for advancement. If disciplinary action becomes necessary, it should be covered by company policy. A typical approach is a three-step process in which the perpetrator is first given a verbal reprimand. In the event of further violations, a written reprimand and warning are given. Dismissal typically accompanies subsequent violations.

2.5.0 Gender and Minority Discrimination

More attention is being placed on fair recruitment, equal pay for equal work, and promotions for women and minorities in the workplace. Consequently, many business practices, including the way employees are treated, the organization's hiring and promotional practices, and the way people are compensated, are being analyzed for equity.

Once a male-dominated industry, construction companies are moving away from this image and are actively recruiting and training women, younger workers, people from other cultures, and workers with disabilities. This means that organizations hire the best person for the job, without regard for race, sex, religion, age, etc.

To prevent discrimination cases, employers must have valid job-related criteria for hiring, compensation, and promotion. These measures must be used consistently for every applicant interview, employee performance appraisal, and hiring or promotion decision. Therefore, all workers responsible for recruitment, selection, and supervision of employees, and evaluating job performance, must be trained on how to use the job-related criteria legally and effectively.

3.0.0 BUSINESS ORGANIZATIONS

An organization is the relationship among the people within the company or project. The crew leader needs to be aware of two types of organizations. These are formal organizations and informal organizations.

A formal organization exists when the activities of the people within the work group are directed toward achieving a common goal. An example of a formal organization is a work crew consisting of four carpenters and two laborers led by a crew leader, all working together toward a common goal.

A formal organization is typically documented on an organizational chart, which outlines all the positions that make up an organization and shows how those positions are related. Some organizational charts even depict the people within each position and the person to whom they report, as well as the people that the person supervises. *Figures 1* and 2 show examples of organization charts for fictitious companies. Note that each of these positions represents an opportunity for advancement in the construction industry that a crew leader can eventually achieve.

An informal organization allows for communication among its members so they can perform as a group. It also establishes patterns of behavior that help them to work as a group, such as agreeing to use a specific training program.

An example of an informal organization is a trade association such as Associated Builders and Contractors (ABC), Associated General Contractors (AGC), and the National Association of Women in Construction (NAWIC). Those, along with the thousands of other trade associations in the U.S., provide a forum in which members with common concerns can share information, work on issues, and develop standards for their industry.

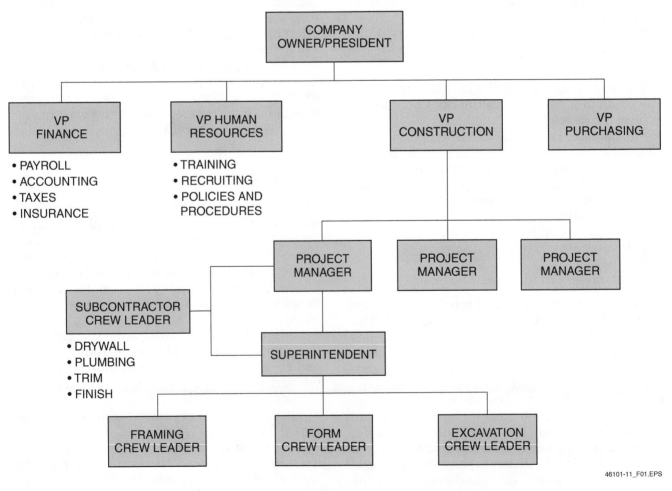

46101-11_F01.EPS

Figure 1 Sample organization chart for a construction company.

NCCER – *Contren® Learning Series* 46101-11

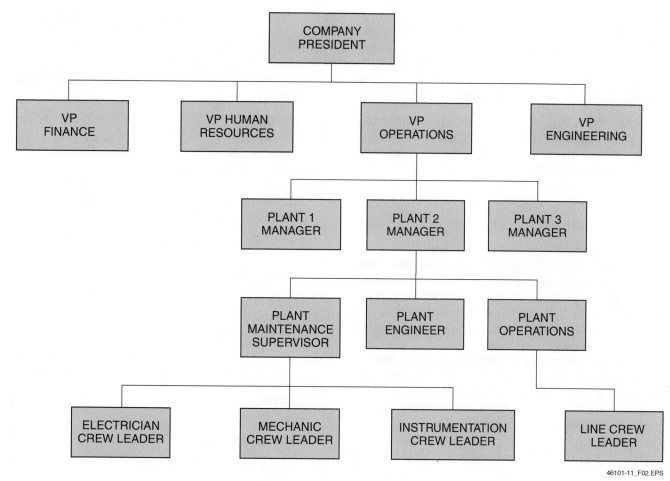

Figure 2 Sample organization chart for an industrial company.

Both types of organizations establish the foundation for how communication flows. The formal structure is the means used to delegate authority and responsibility and to exchange information. The informal structure is used to exchange information.

Members in an organization perform best when each member:

- Knows the job and how it will be done
- Communicates effectively with others in the group
- Understands his or her role in the group
- Recognizes who has the authority and responsibility

3.1.0 Division of Responsibility

The conduct of a business involves certain functions. In a small organization, responsibilities may be divided between one or two people. However, in a larger organization with many different and complex activities, responsibilities may be grouped into similar activity groups, and the responsibility for each group assigned to department managers. In either case, the following major departments exist in most companies:

- *Executive* – This office represents top management. It is responsible for the success of the company through short-range and long-range planning.
- *Human Resources* – This office is responsible for recruiting and screening prospective employees; managing employee benefits programs; advising management on pay and benefits; and developing and enforcing procedures related to hiring practices.
- *Accounting* – This office is responsible for all record keeping and financial transactions, including payroll, taxes, insurance, and audits.
- *Contract Administration* – This office prepares and executes contractual documents with owners, subcontractors, and suppliers.
- *Purchasing* – This office obtains material prices and then issues purchase orders. The purchasing office also obtains rental and leasing rates on equipment and tools.

- *Estimating*: This office is responsible for recording the quantity of material on the jobs, the takeoff, pricing labor and material, analyzing subcontractor bids, and bidding on projects.
- *Operations*: This office plans, controls, and supervises all project-related activities.

Other divisions of responsibility a company may create involve architectural and engineering design functions. These divisions usually become separate departments.

3.2.0 Authority, Responsibility, and Accountability

As an organization grows, the manager must ask others to perform many duties so that the manager can concentrate on management tasks. Managers typically assign (delegate) activities to their subordinates. When delegating activities, the crew leader assigns others the responsibility to perform the designated tasks.

Responsibility means obligation, so once the responsibility is delegated, the person to whom it is assigned is obligated to perform the duties.

Along with responsibility comes authority. *Authority* is the power to act or make decisions in carrying out an assignment. The type and amount of authority a supervisor or worker has depends on the company for which he or she works. Authority and responsibility must be balanced so employees can carry out their tasks. In addition, delegation of sufficient authority is needed to make an employee accountable to the crew leader for the results.

Accountability is the act of holding an employee responsible for completing the assigned activities. Even though authority and responsibility may be delegated to crew members, the final responsibility always rests with the crew leader.

3.3.0 Job Descriptions

Many companies furnish each employee with a written job description that explains the job in detail. Job descriptions set a standard for the employee. They make judging performance easier, clarify the tasks each person should handle, and simplify the training of new employees.

Each new employee should understand all the duties and responsibilities of the job after reviewing the job description. Thus, the time it takes for the employee to make the transition from being a new and uninformed employee to a more experienced member of a crew is shortened.

A job description need not be long, but it should be detailed enough to ensure there is no misun-

derstanding of the duties and responsibilities of the position. The job description should contain all the information necessary to evaluate the employee's performance.

A job description should contain, at minimum, the following:

- Job title
- General description of the position
- Minimum qualifications for the job
- Specific duties and responsibilities
- The supervisor to whom the position reports
- Other requirements, such as qualifications, certifications, and licenses

A sample job description is shown in *Figure 3*.

3.4.0 Policies and Procedures

Most companies have formal policies and procedures established to help the crew leader carry out his or her duties. A *policy* is a general state-

---

Position:
Crew Leader

General Summary:
First line of supervision on a construction crew installing concrete formwork.

Reports To:
Job Superintendent

Physical and Mental Responsibilities:
- Ability to stand for long periods
- Ability to solve basic math and geometry problems

Duties and Responsibilities:
- Oversee crew
- Provide instruction and training in construction tasks as needed
- Make sure proper materials and tools are on the site to accomplish tasks
- Keep project on schedule
- Enforce safety policies and procedures

Knowledge, Skills, and Experience Required:
- Extensive travel throughout the Eastern United States, home base in Atlanta
- Ability to operate a backhoe and trencher
- Valid commercial driver's license with no DUI violations
- Ability to work under deadlines with the knowledge and ability to foresee problem areas and develop a plan of action to solve the situation

46101-11_F03.EPS

Figure 3 Example of a job description.

---

ment establishing guidelines for a specific activity. Examples include policies on vacations, breaks, workplace safety, and checking out tools.

Procedures are the ways that policies are carried out. For example, a procedure written to implement a policy on workplace safety would include guidelines for reporting accidents and general safety procedures that all employees are expected to follow.

A crew leader must be familiar with the company policies and procedures, especially with regard to safety practices. When OSHA inspectors visit a job site, they often question employees and crew leaders about the company policies related to safety. If they are investigating an accident, they will want to verify that the responsible crew leader knew the applicable company policy and followed it.

Review Questions

1. The construction industry should provide training for craftworkers and supervisors _____.
   a. to ensure that there are enough future workers
   b. to avoid discrimination lawsuits
   c. in order to update the skills of older workers who are retiring at a later age than they previously did
   d. even though younger workers are now less likely to seek jobs in other areas than they were 10 years ago

2. Companies traditionally offer craftworker training _____.
   a. that a supervisor leads in a classroom setting
   b. that a craftworker leads in a classroom setting
   c. in a hands-on setting, where craftworkers learn from a co-worker or supervisor
   d. on a self-study basis to allow craftworkers to proceed at their own pace

3. One way to provide effective training is to _____.
   a. avoid giving negative feedback until trainees are more experienced in doing the task
   b. tailor the training to the career goals and needs of trainees
   c. choose one training method and use it for all trainees
   d. encourage trainees to listen, saving their questions for the end of the session

4. One way to prevent sexual harassment in the workplace is to _____.
   a. require employee training in which the potentially offensive subject of stereotypes is carefully avoided
   b. develop a consistent policy with appropriate consequences for engaging in sexual harassment
   c. communicate to workers that the victim of sexual harassment is the one who is being directly harassed, not those affected in a more indirect way
   d. educate workers to recognize sexual harassment for what it is—unwelcome conduct by the opposite sex

5. Employers can minimize all types of workplace discrimination by hiring based on a consistent list of job-related requirements.

   a. True
   b. False

6. Members tend to function best within an organization when they _____.

   a. are allowed to select their own style of clothing for each project
   b. understand their role
   c. do not disagree with the statements of other workers or supervisors
   d. are able to work without supervision

7. A formal organization is defined as a group of individuals who work independently, but share the same goal

   a. True
   b. False

8. A formal organization uses an organizational chart to _____.

   a. depict all companies with which it conducts business
   b. show all customers with which it conducts business
   c. track projects between departments
   d. show the relationships among the existing positions in the company

9. Which of the following is a function typically performed by the operations department of a company?

   a. Purchase materials
   b. Plan projects
   c. Prepare payrolls
   d. Recruiting and screening new hires

10. The company department that manages employee benefits and personnel recruiting is _____.

    a. Engineering
    b. Human Resources
    c. Purchasing
    d. Contract Administration

11. The power to make decisions and act on them in carrying out an assignment is _____.

    a. delegating
    b. responsibility
    c. decisiveness
    d. authority

12. Accountability is defined as _____.

    a. the power to act or make decisions in carrying out assignments
    b. giving an employee a particular task to perform
    c. the act of an employee responsible for the completion and results of a particular duty
    d. having the power to promote someone

13. A good job description should include _____.

    a. a complete organization chart
    b. any information needed to judge job performance
    c. the company dress code
    d. the company's sexual harassment policy

14. The purpose of a policy is to _____.

    a. establish company guidelines regarding a particular activity
    b. specify what tools and equipment are required for a job
    c. list all information necessary to judge an employee's performance
    d. inform employees about the future plans of the company

15. One example of a procedure would be the rules for taking time off.

    a. True
    b. False

Objectives

Upon completion of this section, you should be able to:

1. Describe the role of a crew leader.
2. List the characteristics of effective leaders.
3. Be able to discuss the importance of ethics in a supervisor's role.
4. Identify the three styles of leadership.
5. Describe the forms of communication.
6. Describe the four parts of verbal communication.
7. Describe the importance of active listening.
8. Explain how to overcome the barriers to communication.
9. List ways that leaders can motivate their employees.
10. Explain the importance of delegating and implementing policies and procedures.
11. Distinguish between problem solving and decision making.

1.0.0 INTRODUCTION TO LEADERSHIP

For the purpose of this program, it is important to define some of the positions that will be discussed. The term *craftworker* refers to a person who performs the work of his or her trade(s). The crew leader is a person who supervises one or more craftworkers on a crew. A superintendent is essentially an on-site supervisor who is responsible for one or more crew leaders or front-line supervisors. Finally, a project manager or general superintendent may be responsible for managing one or more projects. This training will concentrate primarily on the role of the crew leader.

Craftworkers and crew leaders differ in that the crew leader manages the activities that the skilled craftworkers on the crews actually perform. In order to manage a crew of craftworkers, a crew leader must have first-hand knowledge and experience in the activities being performed. In addition, he or she must be able to act directly in organizing and directing the activities of the various crew members.

This section explains the importance of developing effective leadership skills as a new crew leader. Effective ways to communicate with all levels of employees and co-workers, build teams, motivate crew members, make decisions, and resolve problems are covered in depth.

2.0.0 THE SHIFT IN WORK ACTIVITIES

The crew leader is generally selected and promoted from a work crew. The selection will often be based on that person's ability to accomplish tasks, to get along with others, to meet schedules, and to stay within the budget. The crew leader must lead the team to work safely and provide a quality product.

Making the transition from a craftworker to a crew leader can be difficult, especially when the new crew leader is in charge of supervising a group of peers. Crew leaders are no longer responsible for their work alone; rather, they are accountable for the work of an entire crew of people with varying skill levels and abilities, a multitude of personalities and work styles, and different cultural and educational backgrounds. Crew leaders must learn to put their personal relationships aside and work for the common goals of the team.

New crew leaders are often placed in charge of workers who were formerly their friends and peers on a crew. This situation can create some conflicts. For example, some of the crew may try to take advantage of the friendship by seeking special favors. They may also want to be privy to information that should be held closely. These problems can be overcome by working with the crew to set mutual performance goals and by freely communicating with them within permitted limits. Use their knowledge and strengths along with your own so that they feel like they are key players on the team.

As an employee moves from a craftworker position to the role of a crew leader, he or she will find that more hours will be spent supervising the work of others than actually performing the technical skill for which he or she has been trained. *Figure 4* represents the percentage of time craftworkers, crew leaders, superintendents, and project managers spend on technical and supervisory work as their management responsibilities increase.

The success of the new crew leader is directly related to the ability to make the transition from crew member into a leadership role.

3.0.0 BECOMING A LEADER

A crew leader must have leadership skills to be successful. Therefore, one of the primary goals of a person who wants to become a crew leader should be to develop strong leadership skills and learn to use them effectively.

There are many ways to define a leader. One straightforward definition is a person who influences other people in the achievement of a goal.

Fundamentals of Crew Leadership

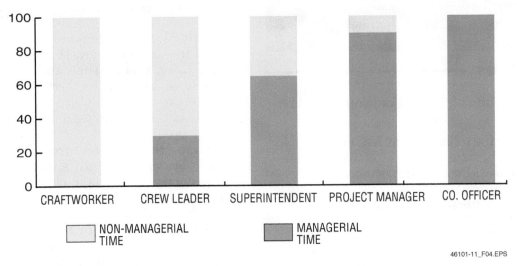

Figure 4 Percentage of time spent on technical and supervisory work.

Some people may have inherited leadership qualities or may have developed traits that motivate others to follow and perform. Research shows that people who possess such talents are likely to succeed as leaders.

3.1.0 Characteristics of Leaders

Leadership traits are similar to the skills that a crew leader needs in order to be effective. Although the characteristics of leadership are many, there are some definite commonalities among effective leaders.

First and foremost, effective leaders lead by example. In other words, they work and live by the standards that they establish for their crew members or followers, making sure they set a positive example.

Effective leaders also tend to have a high level of drive and determination, as well as a stick-to-it attitude. When faced with obstacles, effective leaders don't get discouraged; instead, they identify the potential problems, make plans to overcome them, and work toward achieving the intended goal. In the event of failure, effective leaders learn from their mistakes and apply that knowledge to future situations. They also learn from their successes.

Effective leaders are typically effective communicators who clearly express the goals of a project to their crew members. Accomplishing this may require that the leader overcome issues such as language barriers, gender bias, or differences in personalities to ensure that each member of the crew understands the established goals of the project.

Effective leaders have the ability to motivate their crew members to work to their full potential and become effective members of the team. Crew leaders try to develop crew member skills and encourage them to improve and learn as a means to contribute more to the team effort. Effective leaders strive for excellence from themselves and their team, so they work hard to provide the skills and leadership necessary to do so.

In addition, effective leaders must possess organizational skills. They know what needs to be accomplished, and they use their resources to make it happen. Because they cannot do it alone, leaders enlist the help of their team members to share in the workload. Effective leaders delegate work to their crew members, and they implement company policies and procedures to ensure that the work is completed safely, effectively, and efficiently.

Finally, effective leaders have the authority and self-confidence that allows them to make decisions and solve problems. In order to accomplish their goals, leaders must be able to calculate risks, absorb and interpret information, assess courses of action, make decisions, and assume the responsibility for those decisions.

3.1.1 Leadership Traits

There are many other traits of effective leaders. Some other major characteristics of leaders include the following:

- Ability to plan and organize
- Loyalty to their company and crew
- Ability to motivate
- Fairness

- Enthusiasm
- Willingness to learn from others
- Ability to teach others
- Initiative
- Ability to advocate an idea
- Good communication skills

3.1.2 *Expected Leadership Behavior*

Followers have expectations of their leaders. They look to their leaders to:

- Lead by example
- Suggest and direct
- Plan and organize the work
- Communicate effectively
- Make decisions and assume responsibility
- Have the necessary technical knowledge
- Be a loyal member of the group
- Abide by company policies and procedures

3.2.0 Functions of a Leader

The functions of a leader will vary with the environment, the group being led, and the tasks to be performed. However, there are certain functions common to all situations that the leader will be called upon to perform. Some of the major functions are:

- Organize, plan, staff, direct, and control work
- Empower group members to make decisions and take responsibility for their work
- Maintain a cohesive group by resolving tensions and differences among its members and between the group and those outside the group
- Ensure that all group members understand and abide by company policies and procedures
- Accept responsibility for the successes and failures of the group's performance
- Represent the group
- Be sensitive to the differences of a diverse workforce

3.3.0 Leadership Styles

There are three main styles of leadership. At one extreme is the autocratic or commander style of leadership, where the crew leader makes all of the decisions independently, without seeking the opinions of crew members. At the other extreme is the hands-off or facilitator style, where the crew leader empowers the employees to make decisions. In the middle is the democratic or collaborative style, where the crew leader seeks crew member opinions and makes the appropriate decisions based on their input.

The following are some characteristics of each of the three leadership styles:

Commander types:

- Expect crew members to work without questioning procedures
- Seldom seek advice from crew members
- Insist on solving problems alone
- Seldom permit crew members to assist each other
- Praise and criticize on a personal basis
- Have no sincere interest in creatively improving methods of operation or production

Partner types:

- Discuss problems with their crew members
- Listen to suggestions from crew members
- Explain and instruct
- Give crew members a feeling of accomplishment by commending them when they do a job well
- Are friendly and available to discuss personal and job-related problems

Facilitator types:

- Believe no supervision is best
- Rarely give orders
- Worry about whether they are liked by their crew members

Effective leadership takes many forms. The correct style for a particular situation or operation depends on the nature of the crew as well as the work it has to accomplish. For example, if the crew does not have enough experience for the job ahead, then a commander style may be appropriate. The autocratic style of leadership is also effective when jobs involve repetitive operations that require little decision-making.

However, if a worker's attitude is an issue, a partner style may be appropriate. In this case, providing the missing motivational factors may increase performance and result in the improvement of the worker's attitude. The democratic style of leadership is also used when the work is of a creative nature, because brainstorming and exchanging ideas with such crew members can be beneficial.

The facilitator style is effective with an experienced crew on a well-defined project.

The company must give a crew leader sufficient authority to do the job. This authority must be commensurate with responsibility, and it must be made known to crew members when they are hired so that they understand who is in charge.

A crew leader must have an expert knowledge of the activities to be supervised in order to be effective. This is important because the crew members need to know that they have someone to turn to when they have a question or a problem, when they need some guidance, or when modifications or changes are warranted by the job.

Respect is probably the most useful element of authority. Respect usually derives from being fair to employees, by listening to their complaints and suggestions, and by using incentives and rewards appropriately to motivate crew members. In addition, crew leaders who have a positive attitude and a favorable personality tend to gain the respect of their crew members as well as their peers. Along with respect comes a positive attitude from the crew members.

3.4.0 Ethics in Leadership

The crew leader should practice the highest standards of ethical conduct. Every day the crew leader has to make decisions that may have ethical implications. When an unethical decision is made, it not only hurts the crew leader, but also other workers, peers, and the company for which he or she works.

There are three basic types of ethics:

1. Business or legal
2. Professional or balanced
3. Situational

Business, or legal, ethics concerns adhering to all laws and regulations related to the issue.

Professional, or balanced, ethics relates to carrying out all activities in such a manner as to be honest and fair to everyone concerned.

Situational ethics pertains to specific activities or events that may initially appear to be a gray area. For example, you may ask yourself, "How will I feel about myself if my actions are published in the newspaper or if I have to justify my actions to my family, friends, and colleagues?"

The crew leader will often be put into a situation where he or she will need to assess the ethical consequences of an impending decision. For instance, should a crew leader continue to keep one of his or her crew working who has broken into a cold sweat due to overheated working conditions just because the superintendent says the activity is behind schedule? Or should a crew leader, who is the only one aware that the reinforcing steel placed by his or her crew was done incorrectly, correct the situation before the concrete is placed in the form? If a crew leader is ever asked to carry

through on an unethical decision, it is up to him or her to inform the superintendent of the unethical nature of the issue, and if still requested to follow through, refuse to act.

4.0.0 COMMUNICATION

Successful crew leaders learn to communicate effectively with people at all levels of the organization. In doing so, they develop an understanding of human behavior and acquire communication skills that enable them to understand and influence others.

There are many definitions of communication. Communication is the act of accurately and effectively conveying or transmitting facts, feelings, and opinions to another person. Simply stated, communication is the method of exchanging information and ideas.

Just as there are many definitions of communication, it also comes in many forms, including verbal, nonverbal, and written. Each of these forms of communication are discussed in this section.

4.1.0 Verbal Communication

Verbal communication refers to the spoken words exchanged between two or more people. Verbal communication consists of four distinct parts:

1. Sender
2. Message
3. Receiver
4. Feedback

Figure 5 depicts the relationship of these four parts within the communication process. In verbal communication, the focus is on feedback, which is used to verify that the sender's message was received as intended.

Did you know?

Research shows that the typical supervisor spends about 80 percent of his or her day communicating through writing, speaking, listening, or using body language. Of that time, studies suggest that approximately 20 percent of communication is written, and 80 percent involves speaking or listening.

Figure 5 Communication process.

4.1.1 *The Sender*

The sender is the person who creates the message to be communicated. In verbal communication, the sender actually says the message aloud to the person(s) for whom it is intended.

The sender must be sure to speak in a clear and concise manner that can be easily understood by others. This is not an easy task; it takes practice. Some basic speaking tips are:

- Avoid talking with anything in your mouth (food, gum, etc.).
- Avoid swearing and acronyms.
- Don't speak too quickly or too slowly. In extreme cases, people tend to focus on the rate of speech rather than what is being communicated.
- Pronounce words carefully to prevent misunderstandings.
- Speak with enthusiasm. Avoid speaking in a harsh voice or in a monotone.

4.1.2 *The Message*

The message is what the sender is attempting to communicate to the audience. A message can be a set of directions, an opinion, or a feeling. Whatever its function, a message is an idea or fact that the sender wants the audience to know.

Before speaking, determine what must be communicated, then take the time to organize what to say, ensuring that the message is logical and complete. Taking the time to clarify your thoughts prevents rambling, not getting the message across effectively, or confusing the audience. It also permits the sender to get to the point quickly.

In delivering the message, the sender should assess the audience. It is important not to talk down to them. Remember that everyone, whether in a senior or junior position, deserves respect and courtesy. Therefore, the sender should use words and phrases that the audience can understand and avoid technical language or slang. In addition, the sender should use short sentences, which gives the audience time to understand and digest one point or fact at a time.

4.1.3 *The Receiver*

The receiver is the person to whom the message is communicated. For the communication process to be successful, it is important that the receiver understands the message as the sender intended. Therefore, the receiver must listen to what is being said.

There are many barriers to effective listening, particularly on a busy construction job site. Some of these obstacles include the following:

- Noise, visitors, cell phones, or other distractions
- Preoccupation, being under pressure, or daydreaming
- Reacting emotionally to what is being communicated
- Thinking about how to respond instead of listening
- Giving an answer before the message is complete
- Personal biases to the sender's communication style
- Finishing the sender's sentence

Some tips for overcoming these barriers are:

- Take steps to minimize or remove distractions; learn to tune out your surroundings
- Listen for key points
- Take notes
- Try not to take things personally
- Allow yourself time to process your thoughts before responding
- Let the sender communicate the message without interruption
- Be aware of your personal biases, and try to stay open-minded

There are many ways for a receiver to show that he or she is actively listening to what is being said. This can even be accomplished without saying a word. Examples include maintaining eye contact, nodding your head, and taking notes. It may also be accomplished through feedback.

4.1.4 Feedback

Feedback refers to the communication that occurs after the message has been sent by the sender and received by the receiver. It involves the receiver responding to the message.

Feedback is a very important part of the communication process because it allows the receiver to communicate how he or she interpreted the message. It also allows the sender to ensure that the message was understood as intended. In other words, feedback is a checkpoint to make sure the receiver and sender are on the same page.

The receiver can use the opportunity of providing feedback to paraphrase back what was heard. When paraphrasing what you heard, it is best to use your own words. That way, you can show the sender that you interpreted the message correctly and could explain it to others if needed.

In addition, the receiver can clarify the meaning of the message and request additional information when providing feedback. This is generally accomplished by asking questions.

One opportunity to provide feedback is in the performance of crew evaluations. Many companies have formal evaluation forms that are used on a yearly basis to evaluate workers for pay increases. These evaluations should not come as a once-a-year surprise. An effective crew leader provides constant performance feedback, which is ultimately reflected in the annual performance evaluation. It is also important to stress the importance of self-evaluation with your crew.

4.2.0 Nonverbal Communication

Unlike verbal or written communication, nonverbal communication does not involve the spoken or written word. Rather, non-verbal communication refers to things that you can actually see when communicating with others. Examples include facial expressions, body movements, hand gestures, and eye contact.

Nonverbal communication can provide an external signal of an individual's inner emotions. It occurs simultaneously with verbal communication; often, the sender of the nonverbal communication is not even aware of it.

Because it can be physically observed, nonverbal communication is just as important as the words used in conveying the message. Often, people are influenced more by nonverbal signals than by spoken words. Therefore, it is important to be conscious of nonverbal cues because you don't want the receiver to interpret your message incorrectly based on your posture or an expression on your face. After all, these things may have nothing to do with the communication exchange; instead, they may be carrying over from something else going on in your day.

4.3.0 Written or Visual Communication

Some communication will have to be written or visual. Written or visual communication refers to communication that is documented on paper or transmitted electronically using words or visuals.

Many messages on a job have to be communicated in text form. Examples include weekly reports, requests for changes, purchase orders, and correspondence on a specific subject. These items are written because they must be recorded for business and historical purposes. In addition, some communication on the job will have to be visual. Items that are difficult to explain verbally or by the written word can best be explained through diagrams or graphics. Examples include the plans or drawings used on a job.

When writing or creating a visual message, it is best to assess the reader or the audience before beginning. The reader must be able to read the message and understand the content; otherwise, the communication process will be unsuccessful. Therefore, the writer should consider the actual meaning of words or diagrams and how others might interpret them. In addition, the writer should make sure that all handwriting is legible if the message is being handwritten.

Here are some basic tips for writing:

- Avoid emotion-packed words or phrases.
- Be positive whenever possible.
- Avoid using technical language or jargon.
- Stick to the facts.
- Provide an adequate level of detail.
- Present the information in a logical manner.
- Avoid making judgments unless asked to do so.
- Proofread your work; check for spelling and grammatical errors.
- Make sure that the document is legible.
- Avoid using acronyms.
- Make sure the purpose of the message is clearly stated.
- Be prepared to provide a verbal or visual explanation, if needed.

Here are some basic tips for creating visuals:

- Provide an adequate level of detail.
- Ensure that the diagram is large enough to be seen.
- Avoid creating complex visuals; simplicity is better.
- Present the information in a logical order.
- Be prepared to provide a written or verbal explanation of the visual, if needed.

4.4.0 Communication Issues

It is important to note that each person communicates a little differently; that is what makes us unique as individuals. As the diversity of the workforce changes, communication will become even more challenging because the audience may include individuals from different ethnic groups, cultural backgrounds, educational levels, and economic status groups. Therefore, it is necessary to assess the audience in order to determine how to communicate effectively with each individual.

The key to effective communication is to acknowledge that people are different and to be able to adjust the communication style to meet the needs of the audience or the person on the receiving end of your message. This involves relaying the message in the simplest way possible, avoiding the use of words that people may find confusing. Be aware of how you use technical language, slang, jargon, and words that have multiple meanings. Present the information in a clear, concise manner. Avoid rambling and always speak clearly, using good grammar.

In addition, be prepared to communicate the message in multiple ways or adjust your level of detail or terminology to ensure that everyone understands the meaning as intended. For instance, a visual person who cannot comprehend directions in a verbal or written form may need a map. It may be necessary to overcome language barriers on the job site by using graphics or visual aids to relay the message.

Figure 6 shows how to tailor the message to the audience.

VERBAL INSTRUCTIONS Experienced Crew	VERBAL INSTRUCTIONS Newer Crew	WRITTEN INSTRUCTIONS	DIAGRAM/MAP
"Please drive to the supply shop to pick up our order."	"Please drive to the supply shop. Turn right here and left at Route 1. It's at 75th Street and Route 1. Tell them the company name and that you're there to pick up our order."	1. Turn right at exit. 2. Drive 2 miles to Route 1. Turn LEFT. 3. Drive 1 mile (pass the tire shop) to 75th Street. 4. Look for supply store on right. . . .	

Different people learn in different ways. Be sure to communicate so you can be understood.

Figure 6 Tailor your message.

46101-11_F06.EPS

Read the following verbal conversations, and identify any problems:

Conversation I:

Judy: Hey, Roger…

Roger: What's up?

Judy: Has the site been prepared for the job trailer yet?

Roger: Job trailer?

Judy: The job trailer—it's coming in today. What time will the job site be prepared?

Roger: The trailer will be here about 1:00 PM.

Judy: The job site! What time will the job site be prepared?

_____

_____

_____

_____

_____

_____

Conversation II:

John: Hey, Mike, I need your help.

Mike: What is it?

John: You and Joey go over and help Al's crew finish laying out the site.

Mike: Why me? I can't work with Joey. He can't understand a word I say.

John: Al's crew needs some help, and you and Joey are the most qualified to do the job.

Mike: I told you, I can't work with Joey.

_____

_____

_____

_____

_____

_____

Conversation III:

Ed: Hey, Jill.

Jill: Sir?

Ed: Have you received the latest DOL, EEO requirement to be sure the OFCP administrator finds our records up to date when he reviews them in August?

Jill: DOL, EEO, and OFCP?

Ed: Oh, and don't forget the MSHA, OSHA, and EPA reports are due this afternoon.

Jill: MSHA, OSHA, and EPA?

_____

_____

_____

_____

_____

_____

_____

Conversation IV:

Susan: Hey, Bob, would you do me a favor?

Bob: Okay, Sue. What is it?

Susan: I was reading the concrete inspection report and found the concrete in Bays 4A, 3B, 6C, and 5D didn't meet the 3,000 psi strength requirements. Also, the concrete inspector on the job told me the two batches that came in today had to be refused because they didn't meet the slump requirements as noted on page 16 of the spec. I need to know if any placement problems happened on those bays, how long the ready mix trucks were waiting today, and what we plan to do to stop these problems in the future.

_____

_____

_____

_____

_____

_____

Read the following written memos, and identify any problems:

Memo I:

Let's start with the transformer vault $285.00 due. For what you ask? Answer: practically nothing I admit, but here is the story. Paul the superintendent decided it was not the way good ole Comm Ed wanted it, we took out the ladder and part of the grading (as Paul instructed us to do) we brought it back here to change it. When Comm Ed the architect or Doe found out that everything would still work the way it was, Paul instructed us to reinstall the work. That is the whole story there is please add the $285.00 to my next payout.

_____

_____

_____

_____

_____

_____

_____

Memo II:

Let's take rooms C 307-C-312 and C-313 we made the light track supports and took them to the job to erect them when we tried to put them in we found direct work in the way, my men spent all day trying to find out what to do so ask your Superintendent (Frank) he will verify seven hours pay for these men as he went back and forth while my men waited. Now the Architect has changed the system of hanging and has the gall to say that he has made my work easier, I can't see how. Anyway, we want an extra two (2) men for seven (7) hours for April 21 at $55.00 per hour or $385.00 on April 28th Doe Reference 197 finally resolved this problem. We will have no additional charges on Doe Reference 197, please note.

_____

_____

_____

_____

_____

_____

5.0.0 MOTIVATION

The ability to motivate others is a key skill that leaders must develop. Motivation is the ability to influence. It also describes the amount of effort that a person is willing to put forth to accomplish something. For example, a crew member who skips breaks and lunch in an effort to complete a job on time is thought to be highly motivated, but a crew member who does the bare minimum or just enough to keep his or her job is considered unmotivated.

Employee motivation has dimension because it can be measured. Examples of how motivation can be measured include determining the level of absenteeism, the percentage of employee turnover, and the number of complaints, as well as the quality and quantity of work produced.

5.1.0 Employee Motivators

Different things motivate different people in different ways. Consequently, there is no one-size-fits-all approach to motivating crew members. It is important to recognize that what motivates one crew member may not motivate another. In addition, what works to motivate a crew member once may not motivate that same person again in the future.

Frequently, the needs that motivate individuals are the same as those that create job satisfaction. They include the following:

- Recognition and praise
- Accomplishment
- Opportunity for advancement
- Job importance
- Change
- Personal growth
- Rewards

A crew leader's ability to satisfy these needs increases the likelihood of high morale within a crew. Morale refers to an individual's attitude toward the tasks he or she is expected to perform. High morale, in turn, means that employees will be motivated to work hard, and they will have a positive attitude about coming to work and doing their jobs.

5.1.1 Recognition and Praise

Recognition and praise refer to the need to have good work appreciated, applauded, and acknowledged by others. This can be accomplished by simply thanking employees for helping out on a project, or it can entail more formal praise, such as an award for Employee of the Month.

Some tips for giving recognition and praise include the following:

- Be available on the job site so that you have the opportunity to witness good work.
- Know good work and praise it when you see it.
- Look for good work and look for ways to praise it.
- Give recognition and praise only when truly deserved; otherwise, it will lose its meaning.
- Acknowledge satisfactory performance, and encourage improvement by showing confidence in the ability of the crew members to do above-average work.

5.1.2 Accomplishment

Accomplishment refers to a worker's need to set challenging goals and achieve them. There is nothing quite like the feeling of achieving a goal, particularly a goal one never expected to accomplish in the first place.

Crew leaders can help their crew members attain a sense of accomplishment by encouraging them to develop performance plans, such as goals for the year that will be used in performance evaluations. In addition, crew leaders can provide the support and tools (such as training and coaching) necessary to help their crew members achieve these goals.

5.1.3 Opportunity for Advancement

Opportunity for advancement refers to an employee's need to gain additional responsibility and develop new skills and abilities. It is important that employees know that they are not limited to their current jobs. Let them know that they have a chance to grow with the company and to be promoted as recognition for excelling in their work.

Effective leaders encourage their crew members to work to their full potentials. In addition, they share information and skills with their employees in an effort to help them to advance within the organization.

5.1.4 Job Importance

Job importance refers to an employee's need to feel that his or her skills and abilities are valued and make a difference. Employees who do not feel valued tend to have performance and attendance issues. Crew leaders should attempt to make every crew member feel like an important part of the team, as if the job wouldn't be possible without their help.

5.1.5 Change

Change refers to an employee's need to have variety in work assignments. Change is what keeps things interesting or challenging. It prevents the boredom that results from doing the same task day after day with no variety.

5.1.6 Personal Growth

Personal growth refers to an employee's need to learn new skills, enhance abilities, and grow as a person. It can be very rewarding to master a new competency on the job. Similar to change, personal growth prevents the boredom associated with doing the same thing day after day without developing any new skills.

Crew leaders should encourage the personal growth of their employees as well as themselves. Learning should be a two-way street on the job site; crew leaders should teach their crew members and learn from them as well. In addition, crew members should be encouraged to learn from each other.

5.1.7 Rewards

Rewards are compensation for hard work. Rewards can include a crew member's base salary or go beyond that to include bonuses or other incentives. They can be monetary in nature (salary raises, holiday bonuses, etc.), or they can be non-monetary, such as free merchandise (shirts, coffee mugs, jackets, etc.) or other prizes. Attendance at training courses can be another form of reward.

5.2.0 Motivating Employees

To increase motivation in the workplace, crew leaders must individualize how they motivate different crew members. It is important that crew leaders get to know their crew members and determine what motivates them as individuals. Once again, as diversity increases in the workforce, this becomes even more challenging; therefore, effective communication skills are essential.

Here is a list of some tips for motivating employees:

- Keep jobs challenging and interesting. Boredom is a guaranteed de-motivator.
- Communicate your expectations. People need clear goals in order to feel a sense of accomplishment when the goals are achieved.
- Involve the employees. Feeling that their opinions are valued leads to pride in ownership and active participation.
- Provide sufficient training. Give employees the skills and abilities they need to be motivated to perform.
- Mentor the employees. Coaching and supporting employees boosts their self-esteem, their self-confidence, and ultimately their motivation.
- Lead by example. Become the kind of leader employees admire and respect, and they will be motivated to work for you.
- Treat employees well. Be considerate, kind, caring, and respectful; treat employees the way that you want to be treated.
- Avoid using scare tactics. Threatening employees with negative consequences can backfire, resulting in employee turnover instead of motivation.
- Reward your crew for doing their best by giving them easier tasks from time to time. It is tempting to give your best employees the hardest or dirtiest jobs because you know they will do the jobs correctly.
- Reward employees for a job well done.

Participant Exercise B

You are the crew leader of a masonry crew. Sam Williams is the person whom the company holds responsible for ensuring that equipment is operable and distributed to the jobs in a timely manner.

Occasionally, disagreements with Sam have resulted in tools and equipment arriving late. Sam, who has been with the company 15 years, resents having been placed in the job and feels that he outranks all the crew leaders.

Sam figured it was about time he talked with someone about the abuse certain tools and other items of equipment were receiving on some of the jobs. Saws were coming back with guards broken and blades chewed up, bits were being sheared in half, motor housings were bent or cracked, and a large number of tools were being returned covered with mud. Sam was out on your job when he observed a mason carrying a portable saw by the cord. As he watched, he saw the mason bump the swinging saw into a steel column. When the man arrived at his workstation, he dropped the saw into the mud.

You are the worker's crew leader. Sam approached as you were coming out of the work trailer. He described the incident. He insisted, as crew leader, you are responsible for both the work of the crew and how its members use company property. Sam concluded, "You'd better take care of this issue as soon as possible! The company is sick and tired of having your people mess up all the tools!"

You are aware that some members of your crew have been mistreating the company equipment.

1. How would you respond to Sam's accusations?

_____

_____

_____

_____

2. What action would you take regarding the misuse of the tools?

_____

_____

_____

_____

3. How can you motivate the crew to take better care of their tools? Explain.

_____

_____

_____

_____

 Fundamentals of Crew Leadership

6.0.0 TEAM BUILDING

Organizations are making the shift from the traditional boss-worker mentality to one that promotes teamwork. The manager becomes the team leader, and the workers become team members. They all work together to achieve the common goals of the team.

There are a number of benefits associated with teamwork. They include the ability to complete complex projects more quickly and effectively, higher employee satisfaction, and a reduction in turnover.

6.1.0 Successful Teams

Successful teams are made up of individuals who are willing to share their time and talents in an effort to reach a common goal—the goal of the team. Members of successful teams possess an *Us* or *We* attitude rather than an *I* or *You* attitude; they consider what's best for the team and put their egos aside.

Some characteristics of successful teams include the following:

- Everyone participates and every team member counts.
- There is a sense of mutual trust and interdependence.
- Team members are empowered.
- They communicate.
- They are creative and willing to take risks.
- The team leader develops strong people skills and is committed to the team.

6.2.0 Building Successful Teams

To be successful in the team leadership role, the crew leader should contribute to a positive attitude within the team.

There are several ways in which the team leader can accomplish this. First, he or she can work with the team members to create a vision or purpose of what the team is to achieve. It is important that every team member is committed to the purpose of the team, and the team leader is instrumental in making this happen.

Team leaders within the construction industry are typically assigned a crew. However, it can be beneficial for the team leader to be involved in selecting the team members. Selection should be based on a willingness of people to work on the team and the resources that they are able to bring to the team.

When forming a new team, team leaders should do the following:

- Explain the purpose of the team. Team members need to know what they will be doing, how long they will be doing it (if they are temporary or permanent), and why they are needed.
- Help the team establish goals or targets. Teams need a purpose, and they need to know what it is they are responsible for accomplishing.
- Define team member roles and expectations. Team members need to know how they fit into the team and what is expected of them as members of the team.
- Plan to transfer responsibility to the team as appropriate. Teams should be responsible for the tasks to be accomplished.

7.0.0 GETTING THE JOB DONE

Crew leaders must implement policies and procedures to make sure that the work is done correctly. Construction jobs have crews of people with various experiences and skill levels available to perform the work. The crew leader's job is to draw from this expertise to get the job done well and in a timely manner.

7.1.0 Delegating

Once the various activities that make up the job have been determined, the crew leader must identify the person or persons who will be responsible for completing each activity. This requires that the crew leader be aware of the skills and abilities of the people on the crew. Then, the crew leader must put this knowledge to work in matching the crew's skills and abilities to specific tasks that must be accomplished to complete the job.

After matching crew members to specific activities, the crew leader must then delegate the assignments to the responsible person(s). Delegation is generally communicated verbally by the crew leader talking directly to the person who has been assigned the activity. However, there may be times when work is assigned indirectly through written instructions or verbally through someone other than the crew leader.

When delegating work, remember to:

- Delegate work to a crew member who can do the job properly. If it becomes evident that he or she does not perform to the standard desired, either teach the crew member to do the work correctly or turn it over to someone else who can.

NCCER – *Contren® Learning Series*   46101-11

- Make sure the crew member understands what to do and the level of responsibility. Make sure desired results are clear, specify the boundaries and deadlines for accomplishing the results, and note the available resources.
- Identify the standards and methods of measurement for progress and accomplishment, along with the consequences of not achieving the desired results. Discuss the task with the crew member and check for understanding by asking questions. Allow the crew member to contribute feedback or make suggestions about how the task should be performed in a safe and quality manner.
- Give the crew member the time and freedom to get started without feeling the pressure of too much supervision. When making the work assignment, be sure to tell the crew member how much time there is to complete it, and confirm that this time is consistent with the job schedule.
- Examine and evaluate the result once a task is complete. Then, give the crew member some feedback as to how well it has been done. Get the crew member's comments. The information obtained from this is valuable and will enable the crew leader to know what kind of work to assign that crew member in the future. It will also give the crew leader a means of measuring his or her own effectiveness in delegating work.

7.2.0 Implementing Policies and Procedures

Every company establishes policies and procedures that employees are expected to follow and the crew leaders are expected to implement. Company policies and procedures are essentially guides for how the organization does business. They can also reflect organizational philosophies such as putting safety first or making the customer the top priority. Examples of policies and procedures include safety guidelines, credit standards, and billing processes.

Here are some tips for implementing policies and procedures:

- Learn the purpose of each policy. That way, you can follow it and apply it properly and fairly.
- If you're not sure how to apply a company policy or procedure, check the company manual or ask your supervisor.
- Apply company policies and procedures. Remember that they combine what's best for the customer and the company. In addition, they provide direction on how to handle specific situations and answer questions.

- If you are uncertain how to apply a policy, check with your supervisor.

Crew leaders may need to issue orders to their crew members. Basically, an order initiates, changes, or stops an activity. Orders may be general or specific, written or oral, and formal or informal. The decision of how an order will be issued is up to the crew leader, but it is governed by the policies and procedures established by the company.

When issuing orders:

- Make them as specific as possible.
- Avoid being general or vague unless it is impossible to foresee all of the circumstances that could occur in carrying out the order.
- Recognize that it is not necessary to write orders for simple tasks unless the company requires that all orders be written.
- Write orders for more complex tasks that will take considerable time to complete or orders that are permanent.
- Consider what is being said, the audience to whom it applies, and the situation under which it will be implemented to determine the appropriate level of formality for the order.

8.0.0 PROBLEM SOLVING AND DECISION MAKING

Problem solving and decision making are a large part of every crew leader's daily work. There will always be problems to be resolved and decisions to be made, especially in fast-paced, deadline-oriented industries.

8.1.0 Decision Making vs. Problem Solving

Sometimes, the difference between decision making and problem solving is not clear. Decision making refers to the process of choosing an alternative course of action in a manner appropriate for the situation. Problem solving involves determining the difference between the way things are and the way things should be, and finding out how to bring the two together. The two activities are interrelated because in order to make a decision, you may also have to use problem-solving techniques.

8.2.0 Types of Decisions

Some decisions are routine or simple. Such decisions can be made based on past experiences. An example would be deciding how to get to and from work. If you've worked at the same place for a long time, you are already aware of the options for traveling to and from work (take the bus, drive a car, carpool with a co-worker, take a taxi, etc.). Based on past experiences with the options identified, you can make a decision about how best to get to and from work.

On the other hand, some decisions are more difficult. These decisions require more careful thought about how to carry out an activity by using a formal problem-solving technique. An example is planning a trip to a new vacation spot. If you are not sure how to get there, where to stay, what to see, etc., one option is to research the area to determine the possible routes, hotel accommodations, and attractions. Then, you will have to make a decision about which route to take, what hotel to choose, and what sites to visit, without the benefit of direct past experience.

8.3.0 Problem Solving

The ability to solve problems is an important skill in any workplace. It's especially important for craftworkers, whose workday is often not predictable or routine. In this section, you will learn a five-step process for solving problems, which you can apply to both workplace and personal issues. Review the following steps and then see how they can be applied to a job-related problem. Keep in mind that a problem will not be solved until everyone involved admits that there is a problem.

Step 1 **Define the problem.** This isn't as easy as it sounds. Thinking through the problem often uncovers additional problems.

Step 2 **Think about different ways to solve the problem.** There is often more than one solution to a problem, so you must think through each possible solution and pick the best one. The best solution might be taking parts of two different solutions and combining them to create a new solution.

Step 3 **Pick the solution that seems best and figure out an action plan.** It is best to receive input both from those most affected by the problem and from those who will be most affected by any potential solution.

Step 4 **Test the solution to determine whether it actually works.** Many solutions sound great in theory but in practice don't turn out to be effective. On the other hand, you might discover from trying to apply

a solution that it is acceptable with a little modification. If a solution does not work, think about how you could improve it, and then implement your new plan.

Step 5 **Evaluate the process.** Review the steps you took to discover and implement the solution. Could you have done anything better? If the solution turns out to be satisfactory, you can add the solution to your knowledge base.

Next, you will see how to apply the problem-solving process to a workplace problem. Read the following situation and apply the five-step problem-solving process to come up with a solution to the issues posed by the situation.

Situation:

You are part of a team of workers assigned to a new shopping mall project. The project will take about 18 months to complete. The only available parking is half a mile from the job site. The crew has to carry heavy toolboxes and safety equipment from their cars and trucks to the work area at the start of the day, and then carry them back at the end of their shifts.

Step 1 **Define the problem.** Workers are wasting time and energy hauling all their equipment to and from the work site.

Step 2 **Think about different ways to solve the problem.** Several solutions have been proposed:
- Install lockers for tools and equipment closer to the work site.
- Have workers drive up to the work site to drop off their tools and equipment before parking.
- Bring in another construction trailer where workers can store their tools and equipment for the duration of the project.
- Provide a round-trip shuttle service to ferry workers and their tools.

> **NOTE**
> Each solution will have pros and cons, so it is important to receive input from the workers affected by the problem. For example, workers will probably object to any plan (like the drop-off plan) that leaves their tools vulnerable to theft.

Step 3 Pick the solution that seems best and figure out an action plan. The workers decide that the shuttle service makes the most sense. It should solve the time and energy problem, and workers can keep their tools with them. To put the plan into effect, the project supervisor arranges for a large van and driver to provide the shuttle service.

Step 4 Test the solution to determine whether it actually works. The solution works, but there is a problem. All the workers are scheduled to start and leave at the same time, so there is not enough room in the van for all the workers and their equipment. To solve this problem, the supervisor schedules trips spaced 15 minutes apart. The supervisor also adjusts worker schedules to correspond with the trips. That way, all the workers will not try to get on the shuttle at the same time.

Step 6 Evaluate the process. This process gave both management and workers a chance to express an opinion and discuss the various solutions. Everyone feels pleased with the process and the solution.

8.4.0 Special Leadership Problems

Because they are responsible for leading others, it is inevitable that crew leaders will encounter problems and be forced to make decisions about how to respond to the problem. Some problems will be relatively simple to resolve, like covering for a sick crew member who has taken a day off from work. Other problems will be complex and much more difficult to handle.

Some complex problems are relatively common. A few of the major employee problems include:

- Inability to work with others
- Absenteeism and turnover
- Failure to comply with company policies and procedures

8.4.1 Inability to Work with Others

Crew leaders will sometimes encounter situations where an employee has a difficult time working with others on the crew. This could be a result of personality differences, an inability to communicate, or some other cause. Whatever the reason, the crew leader must address the issue and get the crew working as a team.

The best way to determine the reason for why individuals don't get along or work well together is to talk to the parties involved. The crew leader should speak openly with the employee, as well as the other individual(s) to uncover the source of the problem and discuss its resolution.

Once the reason for the conflict is found, the crew leader can determine how to respond. There may be a way to resolve the problem and get the workers communicating and working as a team again. On the other hand, there may be nothing that can be done that will lead to a harmonious solution. In this case, the crew leader would either have to transfer the employee to another crew or have the problem crew member terminated. This latter option should be used as a last measure and should be discussed with one's superiors or Human Resources Department.

8.4.2 Absenteeism and Turnover

Absenteeism and turnover are big problems. Without workers available to do the work, jobs are delayed, and money is lost.

Absenteeism refers to workers missing their scheduled work time on a job. Absenteeism has many causes, some of which are inevitable. For instance, people get sick, they have to take time off for family emergencies, and they have to attend family events such as funerals. However, there are some causes of absenteeism that can be prevented by the crew leader.

The most effective way to control absenteeism is to make the company's policy clear to all employees. Companies that do this find that chronic absenteeism is reduced. New employees should have the policy explained to them. This explanation should include the number of absences allowed and the reasons for which sick or personal days can be taken. In addition, all workers should know how to inform their crew leaders when they miss work and understand the consequences of exceeding the number of sick or personal days allowed.

Once the policy on absenteeism is explained to employees, crew leaders must be sure to implement it consistently and fairly. If the policy is administered equally, employees will likely follow it. However, if the policy is not administered equally and some employees are given exceptions, then it will not be effective. Consequently, the rate of absenteeism is likely to increase.

Despite having a policy on absenteeism, there will always be employees who are chronically late or miss work. In cases where an employee abuses the absenteeism policy, the crew leader should discuss the situation directly with the employee. The crew leader should confirm that the employee understands the company's policy and insist that the employee comply with it. If the employee's behavior continues, disciplinary action may be in order.

Turnover refers to the loss of an employee that is initiated by that employee. In other words, the employee quits and leaves the company to work elsewhere or is fired for cause.

Like absenteeism, there are some causes of turnover that cannot be prevented and others that can. For instance, it is unlikely that a crew leader could keep an employee who finds a job elsewhere earning twice as much money. However, crew leaders can prevent some employee turnover situations. They can work to ensure safe working conditions for their crew, treat their workers fairly and consistently, and help promote good working conditions. The key is communication. Crew leaders need to know the problems if they are going to be able to successfully resolve them.

Some of the major causes of turnover include the following:

- Unfair/inconsistent treatment by the immediate supervisor
- Unsafe project sites
- Lack of job security

For the most part, the actions described for absenteeism are also effective for reducing turnover. Past studies have shown that maintaining harmonious relationships on the job site goes a long way in reducing both turnover and absenteeism. This requires effective leadership on the part of the crew leader.

8.4.3 Failure to Comply With Company Policies and Procedures

Policies are the rules that define the relationship between the company, its employees, its clients, and its subcontractors. Procedures are the instructions for carrying out the policies. Some companies have dress codes that are reflected in their policies. The dress code may be partly to ensure safety, and partly to define the image a company wants to project to the outside world.

Companies develop procedures to ensure that everyone who performs a task does it safely and efficiently. Many procedures deal with safety. A lockout/tagout procedure is an example. In this procedure, the company defines who may perform a lockout, how it is done, and who has the authority to remove or override it. Workers who fail to follow the procedure endanger themselves, as well as their co-workers.

Among a typical company's policies is the policy on disciplinary action. This policy defines steps to be taken in the event that an employee violates the company's policies or procedures. The steps range from counseling by a supervisor for the first offense, to a written warning, to dismissal for repeat offenses. This will vary from one company to another. For example, some companies will fire an employee for any violation of safety procedures.

The crew leader has the first-line responsibility for enforcing company policies and procedures. The crew leader should take the time with a new crew member to discuss the policies and procedures and show the crew member how to access them. If a crew member shows a tendency to neglect a policy or procedure, it is up to the crew leader to counsel that individual. If the crew member continues to violate a policy or procedure, the crew leader has no choice but to refer that individual to the appropriate authority within the company for disciplinary action.

Case I:

On the way over to the job trailer, you look up and see a piece of falling scrap heading for one of the laborers. Before you can say anything, the scrap material hits the ground about five feet in front of the worker. You notice the scrap is a piece of conduit. You quickly pick it up, assuring the worker you will take care of this matter.

Looking up, you see your crew on the third floor in the area from which the material fell. You decide to have a talk with them. Once on the deck, you ask the crew if any of them dropped the scrap. The men look over at Bob, one of the electricians in your crew. Bob replies, "I guess it was mine. It slipped out of my hand."

It is a known fact that the Occupational Safety and Health Administration (OSHA) regulations state that an enclosed chute of wood shall be used for material waste transportation from heights of 20 feet or more. It is also known that Bob and the laborer who was almost hit have been seen arguing lately.

1. Assuming Bob's action was deliberate, what action would you take?

_____

_____

_____

2. Assuming the conduit accidentally slipped from Bob's hand, how can you motivate him to be more careful?

_____

_____

_____

3. What follow-up actions, if any, should be taken relative to the laborer who was almost hit?

_____

_____

_____

4. Should you discuss the apparent OSHA violation with the crew? Why or why not?

_____

_____

_____

5. What acts of leadership would be effective in this case? To what leadership traits are they related?

_____

_____

_____

Case II:

Mike has just been appointed crew leader of a tile-setting crew. Before his promotion into management, he had been a tile setter for five years. His work had been consistently of superior quality.

Except for a little good-natured kidding, Mike's co-workers had wished him well in his new job. During the first two weeks, most of them had been cooperative while Mike was adjusting to his supervisory role.

At the end of the second week, a disturbing incident took place. Having just completed some of his duties, Mike stopped by the job-site wash station. There he saw Steve and Ron, two of his old friends who were also in his crew, washing.

"Hey, Ron, Steve, you should not be cleaning up this soon. It's at least another thirty minutes until quitting time," said Mike. "Get back to your work station, and I'll forget I saw you here."

"Come off it, Mike," said Steve. "You used to slip up here early on Fridays. Just because you have a little rank now, don't think you can get tough with us." To this Mike replied, "Things are different now. Both of you get back to work, or I'll make trouble." Steve and Ron said nothing more, and they both returned to their work stations.

From that time on, Mike began to have trouble as a crew leader. Steve and Ron gave him the silent treatment. Mike's crew seemed to forget how to do the most basic activities. The amount of rework for the crew seemed to be increasing. By the end of the month, Mike's crew was behind schedule.

1. How do you think Mike should have handled the confrontation with Ron and Steve?

_____

_____

2. What do you suggest Mike could do about the silent treatment he got from Steve and Ron?

_____

_____

3. If you were Mike, what would you do to get your crew back on schedule?

_____

_____

4. What acts of leadership could be used to get the crew's willing cooperation?

_____

_____

5. To which leadership traits do they correspond?

_____

_____

1. A crew leader differs from a craftworker in that a _____.

   a. crew leader need not have direct experience in those job duties that a craftworker typically performs
   b. crew leader can expect to oversee one or more craftworkers in addition to performing some of the typical duties of the craftworker
   c. crew leader is exclusively in charge of overseeing, since performing technical work is not part of this role
   d. crew leader's responsibilities do not include being present on the job site

2. Among the many traits effective leaders should have is _____.

   a. the ability to communicate the goals of a project
   b. the drive necessary to carry the workload by themselves in order to achieve a goal
   c. a perfectionist nature, which ensures that they will not make useless mistakes
   d. the ability to make decisions without needing to listen to the opinions of others

3. Of the three styles of leadership, the _____ style would be effective in dealing with a craftworker's negative attitude.

   a. facilitator
   b. commander
   c. partner
   d. dictator

4. One way to overcome barriers to effective communication is to _____.

   a. avoid taking notes on the content of the message, since this can be distracting
   b. avoid reacting emotionally to the message
   c. anticipate the content of the message and interrupt if necessary in order to show interest
   d. think about how to respond to the message while listening

5. Feedback is important in verbal communication because it requires the _____.

   a. sender to repeat the message
   b. receiver to restate the message
   c. sender to avoid technical jargon
   d. sender to concentrate on the message

6. A good way to motivate employees is to use a one-size-fits-all approach, since employees are members of a team with a common goal.

   a. True
   b. False

7. A crew leader can effectively delegate responsibilities by _____.

   a. refraining from evaluating the employee's performance once the task is completed, since it is a new task for the employee
   b. doing the job for the employee to make sure the task is done correctly
   c. allowing the employee to give feedback and suggestions about the task
   d. communicating information to the employee, generally in written form

8. Problem solving differs from decision making in that _____.

   a. problem solving involves identifying discrepancies between the way a situation is and the way it should be
   b. decision making involves separating facts from non-facts
   c. decision making involves eliminating differences
   d. problem solving involves determining an alternative course of action for a given situation

Objectives

Upon completion of this section, you will be able to:

1. Explain the importance of safety.
2. Give examples of direct and indirect costs of workplace accidents.
3. Identify safety hazards of the construction industry.
4. Explain the purpose of OSHA.
5. Discuss OSHA inspection procedures.
6. Identify the key points of a safety program.
7. List steps to train employees on how to perform new tasks safely.
8. Identify a crew leader's safety responsibilities.
9. Explain the importance of having employees trained in first aid and cardiopulmonary resuscitation (CPR).
10. Describe the indications of substance abuse.
11. List the essential parts of an accident investigation.
12. Describe ways to maintain employee interest in safety. Distinguish between company policies and procedures.

1.0.0 SAFETY OVERVIEW

Businesses lose millions of dollars every year because of on-the-job accidents. Work-related injuries, sickness, and deaths have caused untold suffering for workers and their families. Project delays and budget overruns from injuries and fatalities result in huge losses for employers, and work-site accidents erode the overall morale of the crew.

Craftworkers are exposed to hazards as part of the job. Examples of these hazards include falls from heights, working on scaffolds, using cranes in the presence of power lines, operating heavy machinery, and working on electrically-charged or pressurized equipment. Despite these hazards, experts believe that applying preventive safety measures could drastically reduce the number of accidents.

As a crew leader, one of your most important tasks is to enforce the company's safety program and make sure that all workers are performing their tasks safely. To be successful, the crew leader should:

- Be aware of the costs of accidents.
- Understand all federal, state, and local governmental safety regulations.
- Be involved in training workers in safe work methods.
- Conduct training sessions.
- Get involved in safety inspections, accident investigations, and fire protection and prevention.

Crew leaders are in the best position to ensure that all jobs are performed safely by their crew members. Providing employees with a safe working environment by preventing accidents and enforcing safety standards will go a long way towards maintaining the job schedule and enabling a job's completion on time and within budget.

1.1.0 Accident Statistics

Each day, workers in construction and industrial occupations face the risk of falls, machinery accidents, electrocutions, and other potentially fatal occupational hazards.

The National Institute of Occupational Safety and Health (NIOSH) statistics show that about 1,000 construction workers are killed on the job each year, more fatalities than in any other industry. Falls are the leading cause of deaths in the construction industry, accounting for over 40 percent of the fatalities. Nearly half of the fatal falls occurred from roofs, scaffolds, or ladders. Roofers, structural metal workers, and painters experienced the greatest number of fall fatalities.

In addition to the number of fatalities that occur each year, there are a staggering number of work-related injuries. In 2007, for example, more than 135,000 job-related injuries occurred in the construction industry. NIOSH reports that approximately 15 percent of all worker's compensation costs are spent on injured construction workers. The causes of injuries on construction sites include falls, coming into contact with electric current, fires, and mishandling of machinery or equipment. According to NIOSH, back injuries are the leading safety problem in workplaces.

Did you know?

When OSHA inspects a job site, they focus on the types of safety hazards that are most likely to cause fatal injuries. These hazards fall into the following classifications:

- Falls from elevations
- Struck-by hazards
- Caught in/between hazards
- Electrical shock hazards

 Fundamentals of Crew Leadership

2.0.0 COSTS OF ACCIDENTS

Occupational accidents are estimated to cost more than $100 billion every year. These costs affect the employee, the company, and the construction industry as a whole.

Organizations encounter both direct and indirect costs associated with workplace accidents. Examples of direct costs include workers' compensation claims and sick pay; indirect costs include increased absenteeism, loss of productivity, loss of job opportunities due to poor safety records, and negative employee morale attributed to workplace injuries. There are many other related costs involved with workplace accidents. A company can be insured against some of them, but not others. To compete and survive, companies must control these as well as all other employment-related costs.

2.1.0 Insured Costs

Insured costs are those costs either paid directly or reimbursed by insurance carriers. Insured costs related to injuries or deaths include the following:

- Compensation for lost earnings (known as worker's comp)
- Medical and hospital costs
- Monetary awards for permanent disabilities
- Rehabilitation costs
- Funeral charges
- Pensions for dependents

Insurance premiums or charges related to property damages include:

- Fire
- Loss and damage
- Use and occupancy
- Public liability
- Replacement cost of equipment, material, and structures

2.2.0 Uninsured Costs

The costs related to accidents can be compared to an iceberg, as shown in *Figure 7*. The tip of the iceberg represents direct costs, which are the visible costs. The more numerous indirect costs are not readily measureable, but they can represent a greater burden than the direct costs.

Uninsured costs related to injuries or deaths include the following:

- First aid expenses
- Transportation costs
- Costs of investigations
- Costs of processing reports
- Down time on the job site
- Costs to train replacement workers

Uninsured costs related to wage losses include:

- Idle time of workers whose work is interrupted
- Time spent cleaning the accident area
- Time spent repairing damaged equipment
- Time lost by workers receiving first aid
- Costs of training injured workers in a new career

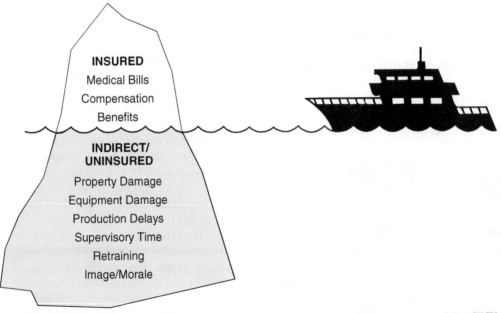

46101-11_F07.EPS

Figure 7 Costs associated with accidents.

NCCER – *Contren® Learning Series* 46101-11

Uninsured costs related to production losses include:

- Product spoiled by accident
- Loss of skill and experience
- Lowered production or worker replacement
- Idle machine time

Associated costs may include the following:

- Difference between actual losses and amount recovered
- Costs of rental equipment used to replace damaged equipment
- Costs of new workers used to replace injured workers
- Wages or other benefits paid to disabled workers
- Overhead costs while production is stopped
- Impact on schedule
- Loss of bonus or payment of forfeiture for delays

Uninsured costs related to off-the-job activities include:

- Time spent on injured workers' welfare
- Loss of skill and experience of injured workers
- Costs of training replacement workers

Uninsured costs related to intangibles include:

- Lowered employee morale
- Increased labor conflict
- Unfavorable public relations
- Loss of bid opportunities because of poor safety records
- Loss of client goodwill

3.0.0 SAFETY REGULATIONS

To reduce safety and health risks and the number of injuries and fatalities on the job, the federal government has enacted laws and regulations, including the *Occupational Safety and Health Act of 1970*. The purpose of OSHA is "to assure so far as possible every working man and woman in the Nation safe and healthful working conditions and to preserve our human resources."

To promote a safe and healthy work environment, OSHA issues standards and rules for working conditions, facilities, equipment, tools, and work processes. It does extensive research into occupational accidents, illnesses, injuries, and deaths in an effort to reduce the number of occurrences and adverse effects. In addition, OSHA regulatory agencies conduct workplace inspections to ensure that companies follow the standards and rules.

3.1.0 Workplace Inspections

To enforce OSHA regulations, the government has granted regulatory agencies the right to enter public and private properties to conduct workplace safety investigations. The agencies also have the right to take legal action if companies are not in compliance with the Act. These regulatory agencies employ OSHA Compliance Safety and Health Officers (CSHOs), who are chosen for their knowledge in the occupational safety and health field. The CSHOs are thoroughly trained in OSHA standards and in recognizing safety and health hazards.

States with their own occupational safety and health programs conduct inspections. To do so, they enlist the services of qualified state CSHOs.

Companies are inspected for a multitude of reasons. They may be randomly selected, or they may be chosen because of employee complaints, due to an imminent danger, or as a result of major accidents or fatalities.

OSHA can assess significant financial penalties for safety violations. In some cases, a superintendent or crew leader can be held criminally liable for repeat violations.

3.2.0 Penalties for Violations

OSHA has established monetary fines for the violation of their regulations. The penalties as of 2010 are shown in *Table 1*.

In addition to the fines, there are possible criminal charges for willful violations resulting in death or serious injury. There can also be personal liability for failure to comply with OSHA regulations. The attitude of the employer and their safety history can have a significant effect on the outcome of a case.

Table 1 OSHA Penalties for Violations

Violation	Penalty
Willful Violations	Maximum $70,000
Repeated Violations	Minimum $70,000
Serious, Other-than-Serious, Other Specific Violations	Minimum $7,000
OSHA Notice Violation	$1,000
Failure to Post *OSHA 300A Summary of Work-Related Injuries and Illnesses*	$1,000
Failure to Properly Maintain *OSHA 300 Log of Work-Related Injuries and Illnesses*	$1,000
Failure to Promptly and Properly Report Fatality/Catastrophe	$5,000
Failure to Permit Access to Records Under *OSHA 1904* Regulations	$1,000
Failure to Follow Advance Notification Requirements Under *OSHA 1903.6* Regulations	$2,000
Failure to Abate – for Each Calendar Day Beyond Abatement Date	$7,000
Retaliation Against Individual for Filing OSHA Complaint	$10,000

Did you know?

Nearly half the states in the U.S. have state-run OSHA programs. These programs are set up under federal OSHA guidelines and must establish job health and safety standards that are at least as effective as the federal standards. The states have the option of adopting more stringent standards or setting standards for hazards not addressed in the federal program. Of the 22 states with state-run OSHA programs, eight of them limit their coverage to public employees.

4.0.0 EMPLOYER SAFETY RESPONSIBILITIES

Each employer must set up a safety and health program to manage workplace safety and health and to reduce work-related injuries, illnesses, and fatalities. The program must be appropriate for the conditions of the workplace. It should consider the number of workers employed and the hazards to which they are exposed while at work.

To be successful, the safety and health program must have management leadership and employee participation. In addition, training and informational meetings play an important part in effective programs. Being consistent with safety policies is the key. Regardless of the employer's responsibility, however, the individual worker is ultimately responsible for his or her own safety.

4.1.0 Safety Program

The crew leader plays a key role in the successful implementation of the safety program. The crew leader's attitudes toward the program set the standard for how crew members view safety. Therefore, the crew leader should follow all program guidelines and require crew members to do the same.

Safety programs should consist of the following:

- Safety policies and procedures
- Hazard identification and assessment
- Safety information and training
- Safety record system
- Accident investigation procedures
- Appropriate discipline for not following safety procedures
- Posting of safety notices

4.1.1 Safety Policies and Procedures

Employers are responsible for following OSHA and state safety standards. Usually, they incorporate OSHA and state regulations into a safety policies and procedures manual. Such a manual is presented to employees when they are hired.

Basic safety requirements should be presented to new employees during their orientation to the company. If the company has a safety manual, the new employee should be required to read it and sign a statement indicating that it is understood. If the employee cannot read, the employer should have someone read it to the employee and answer any questions that arise. The employee should then sign a form stating that he or she understands the information.

It is not enough to tell employees about safety policies and procedures on the day they are hired and never mention them again. Rather, crew leaders should constantly emphasize and reinforce the importance of following all safety policies and procedures. In addition, employees should play an active role in determining job safety hazards and find ways that the hazards can be prevented and controlled.

4.1.2 Hazard Identification and Assessment

Safety policies and procedures should be specific to the company. They should clearly present the hazards of the job. Therefore, crew leaders should also identify and assess hazards to which employees are exposed. They must also assess compliance with OSHA and state standards.

To identify and assess hazards, OSHA recommends that employers conduct inspections of the workplace, monitor safety and health information logs, and evaluate new equipment, materials, and processes for potential hazards before they are used.

Crew leaders and employees play important roles in identifying hazards. It is the crew leader's responsibility to determine what working conditions are unsafe and to inform employees of hazards and their locations. In addition, they should encourage their crew members to tell them about hazardous conditions. To accomplish this, crew leaders must be present and available on the job site.

The crew leader also needs to help the employee be aware of and avoid the built-in hazards to which craftworkers are exposed. Examples include working at elevations, working in confined spaces such as tunnels and underground vaults, on caissons, in excavations with earthen walls, and other naturally dangerous projects.

In addition, the crew leader can take safety measures, such as installing protective railings to prevent workers from falling from buildings, as well as scaffolds, platforms, and shoring.

4.1.3 Safety Information and Training

The employer must provide periodic information and training to new and long-term employees. This must be done as often as necessary so that all employees are adequately trained. Special training and informational sessions must be provided when safety and health information changes or workplace conditions create new hazards. It is important to note that safety-related information must be presented in a manner that the employee will understand.

Whenever a crew leader assigns an experienced employee a new task, the crew leader must ensure that the employee is capable of doing the work in a safe manner. The crew leader can accomplish this by providing safety information or training for groups or individuals.

The crew leader should do the following:

- Define the task.
- Explain how to do the task safely.
- Explain what tools and equipment to use and how to use them safely.
- Identify the necessary personal protective equipment.
- Explain the nature of the hazards in the work and how to recognize them.
- Stress the importance of personal safety and the safety of others.
- Hold regular safety training sessions with the crew's input.
- Review material safety data sheets (MSDSs) that may be applicable.

4.1.4 Safety Record Systems

OSHA regulations (*CFR 29, Part 1904*) require that employers keep records of hazards identified and document the severity of the hazard. The information should include the likelihood of employee exposure to the hazard, the seriousness of the harm associated with the hazard, and the number of exposed employees.

In addition, the employer must document the actions taken or plans for action to control the hazards. While it is best to take corrective action immediately, it is sometimes necessary to develop a plan for the purpose of setting priorities and deadlines and tracking progress in controlling hazards.

Employers who are subject to the recordkeeping requirements of the *Occupational Safety and Health Act of 1970* must maintain a log of all recordable occupational injuries and illnesses. This is known as the *OSHA 300/300A* form.

An MSDS is designed to provide both workers and emergency personnel with the proper procedures for handling or working with a substance that may be dangerous. An MSDS will include information such as physical data (melting point, boiling point, flash point, etc.), toxicity, health effects, first aid, reactivity, storage, disposal, protective equipment, and spill/leak procedures. These sheets are of particular use if a spill or other accident occurs.

Any company with 11 or more employees must post an *OSHA 300A* form, *Log of Work-Related Injuries and Illnesses,* between February 1 and April 30 of each year. Employees have the right to review this form. Check your company's policies with regard to these reports.

OSHA's Form 300A (Rev. 01/2004)

Summary of Work-Related Injuries and Illnesses

Year 20___

U.S. Department of Labor
Occupational Safety and Health Administration

Form approved OMB no. 1218-0176

All establishments covered by Part 1904 must complete this Summary page, even if no work-related injuries or illnesses occurred during the year. Remember to review the Log to verify that the entries are complete and accurate before completing this summary.

Using the Log, count the individual entries you made for each category. Then write the totals below, making sure you've added the entries from every page of the Log. If you had no cases, write "0."

Employees, former employees, and their representatives have the right to review the OSHA Form 300 in its entirety. They also have limited access to the OSHA Form 301 or its equivalent. See 29 CFR Part 1904.35, in OSHA's recordkeeping rule, for further details on the access provisions for these forms.

Number of Cases

Total number of deaths

(G)

Total number of cases with days away from work

(H)

Total number of cases with job transfer or restriction

(I)

Total number of other recordable cases

(J)

Number of Days

Total number of days away from work

(K)

Total number of days of job transfer or restriction

(L)

Injury and Illness Types

Total number of . . .
(M)

(1) Injuries

(2) Skin disorders

(3) Respiratory conditions

(4) Poisonings

(5) Hearing loss

(6) All other illnesses

Post this Summary page from February 1 to April 30 of the year following the year covered by the form.

Public reporting burden for this collection of information is estimated to average 58 minutes per response, including time to review the instructions, search and gather the data needed, and complete and review the collection of information. Persons are not required to respond to the collection of information unless it displays a currently valid OMB control number. If you have any comments about these estimates or any other aspects of this data collection, contact: US Department of Labor, OSHA Office of Statistical Analysis, Room N-3644, 200 Constitution Avenue, NW, Washington, DC 20210. Do not send the completed forms to this office.

Establishment Information

Your establishment name _____

Street _____

City _____ State _____ ZIP _____

Industry description (e.g., Manufacture of motor truck trailers) _____

Standard Industrial Classification (SIC), if known (e.g., 3715) ___ ___ ___ ___

OR

North American Industrial Classification (NAICS), if known (e.g., 336212) ___ ___ ___ ___ ___ ___

Employment Information (If you don't have these figures, see the Worksheet on the back of this page to estimate.)

Annual average number of employees _____

Total hours worked by all employees last year _____

Sign here

Knowingly falsifying this document may result in a fine.

I certify that I have examined this document and that to the best of my knowledge the entries are true, accurate, and complete.

_____ _____
Company executive   Title

_____ _____
Phone   Date

46101-11_SA01.EPS

Logs must be maintained and retained for five years following the end of the calendar year to which they relate. Logs must be available (normally at the establishment) for inspection and copying by representatives of the Department of Labor, the Department of Health and Human Services, or states accorded jurisdiction under the Act. Employees, former employees, and their representatives may also have access to these logs.

4.1.5 Accident Investigation

In the event of an accident, the employer is required to investigate the cause of the accident and determine how to avoid it in the future. According to OSHA, the employer must investigate each work-related death, serious injury or illness, or incident having the potential to cause death or serious physical harm. The employer should document any findings from the investigation, as well as the action plan to prevent future occurrences. This should be done immediately, with photos or video if possible. It is important that the investigation uncover the root cause of the accident so that it can be avoided in the future. In many cases, the root cause can be traced to a flaw in the system that failed to recognize the unsafe condition or the potential for an unsafe act (*Figure 8*).

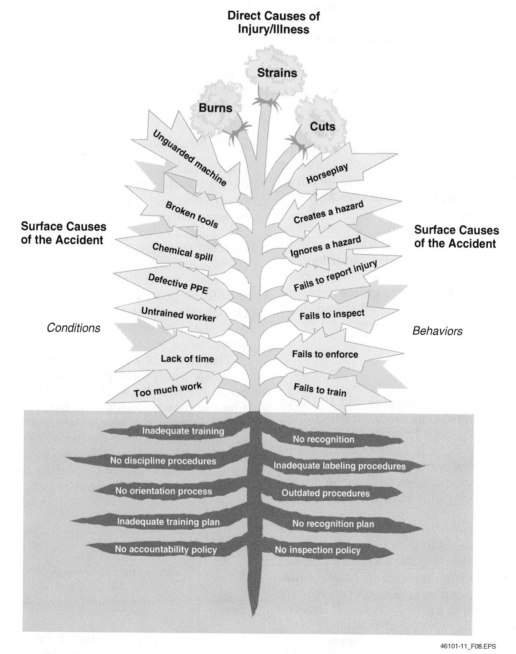

46101-11_F08.EPS

Figure 8 Root causes of accidents.

5.0.0 CREW LEADER INVOLVEMENT IN SAFETY

To be effective leaders, crew leaders must be actively involved in the safety program. Crew leader involvement includes conducting frequent safety training sessions and inspections; promoting first aid and fire protection and prevention; preventing substance abuse on the job; and investigating accidents.

5.1.0 Safety Training Sessions

A safety training session may be a brief, informal gathering of a few employees or a formal meeting with instructional films and talks by guest speakers. The size of the audience and the topics to be addressed determine the format of the meeting. Small, informal safety sessions are typically conducted weekly.

Safety training sessions should be planned in advance, and the information should be communicated to all employees affected. In addition, the topics covered in these training sessions should be timely and practical. A log of each safety session must be kept and signed by all attendees. It must then be maintained as a record and available for inspection.

5.2.0 Inspections

Crew leaders must make regular and frequent inspections to prevent accidents from happening. They must also take steps to avoid accidents. For that purpose, they need to inspect the job sites where their workers perform tasks. It is recommended that these inspections be done before the start of work each day and during the day at different times.

Crew leaders must protect workers from existing or potential hazards in their work areas. Crew leaders are sometimes required to work in areas controlled by other contractors. In these situations, the crew leader must maintain control over the safety exposure of his or her crew. If hazards exist, the crew leader should immediately bring the hazards to the attention of the contractor at fault, their superior, and the person responsible for the job site.

Crew leader inspections are only valuable if action is taken to correct potential hazards. Therefore, crew leaders must be alert for unsafe acts on their work sites. When an employee performs an unsafe action, the crew leader must explain to the employee why the act was unsafe, ask that the employee not do it again, and request cooperation in promoting a safe working environment. The crew leader must document what happened and what the employee was asked to do to correct the situation. It is then very important that crew leaders follow up to make certain the employee is complying with the safety procedures. Never allow a safety violation to go uncorrected. There are three courses of action that you, as a crew leader, can take in an unsafe situation:

- Get the appropriate party to correct the problem.
- Fix the problem yourself.
- Refuse to have the crew work in the area until the problem is corrected.

5.3.0 First Aid

The primary purpose of first aid is to provide immediate and temporary medical care to employees involved in accidents, as well as employees experiencing non-work-related health emergencies, such as chest pains or breathing difficulty. To meet this objective, every crew leader should be aware of the location and contents of first aid kits available on the job site. Emergency numbers should be posted in the job trailer. In addition, OSHA requires that at least one person trained in first aid be present at the job site at all times. Someone on site should also be trained in CPR.

The victim of an accident or sudden illness at a job site poses more problems than normal since he or she may be working in a remote location. The site may be located far from a rescue squad, fire department, or hospital, presenting a problem in the rescue and transportation of the victim to a hospital. The worker may also have been injured by falling rock or other materials, so special rescue equipment or first-aid techniques are often needed.

NOTE

CPR training must be renewed every two years.

The employer benefits by having personnel trained in first aid at each job site in the following ways:

- The immediate and proper treatment of minor injuries may prevent them from developing into more serious conditions. As a result, medical expenses, lost work time, and sick pay may be eliminated or reduced.
- It may be possible to determine if professional medical attention is needed.
- Valuable time can be saved when a trained individual prepares the patient for treatment when professional medical care arrives. As a result, lives can be saved.

The American Red Cross, Medic First Aid, and the United States Bureau of Mines provide basic and advanced first aid courses at nominal costs. These courses include both first aid and CPR. The local area offices of these organizations can provide further details regarding the training available.

5.4.0 Fire Protection and Prevention

Fires and explosions kill and injure many workers each year, so it is important that crew leaders understand and practice fire-prevention techniques as required by company policy.

The need for protection and prevention is increasing as new building materials are introduced. Some building materials are highly flammable. They produce great amounts of smoke and gases, which cause difficulties for fire fighters, and can quickly overcome anyone present. Other materials melt when ignited and may spread over floors, preventing fire-fighting personnel from entering areas where this occurs.

OSHA has specific standards for fire safety. They require that employers provide proper exits, fire-fighting equipment, and employee training on fire prevention and safety. For more information, consult OSHA guidelines.

5.5.0 Substance Abuse

Unfortunately, drug and alcohol abuse is a continuing problem in the workplace. Drug abuse means inappropriately using drugs, whether they are legal or illegal. Some people use illegal street drugs, such as cocaine or marijuana. Others use legal prescription drugs incorrectly by taking too many pills, using other people's medications, or self-medicating. Others consume alcohol to the point of intoxication.

It is essential that crew leaders enforce company policies and procedures regarding substance abuse. Crew leaders must work with management to deal with suspected drug and alcohol abuse and should not handle these situations themselves. These cases are normally handled by the Human Resources Department or designated manager. There are legal consequences of drug and alcohol abuse and the associated safety implications. If you suspect that an employee is suffering from drug or alcohol abuse, immediately contact your supervisor and/or Human Resources Department for assistance. That way, the business and the employee's safety are protected.

It is the crew leader's responsibility to make sure that safety is maintained at all times. This may include removing workers from a work site where they may be endangering themselves or others.

For example, suppose several crew members go out and smoke marijuana or have a few beers during lunch. Then, they return to work to erect scaffolding for a concrete pour in the afternoon. If you can smell marijuana on the crew member's clothing or alcohol on their breath, you must step in and take action. Otherwise, they might cause an accident that could delay the project or cause serious injury or death to themselves or others.

It is often difficult to detect drug and alcohol abuse because the effects can be subtle. The best way is to look for identifiable effects, such as those mentioned above or sudden changes in behavior that are not typical of the employee. Some examples of such behaviors include the following:

- Unscheduled absences; failure to report to work on time
- Significant changes in the quality of work
- Unusual activity or lethargy
- Sudden and irrational temper flare-ups
- Significant changes in personal appearance, cleanliness, or health

There are other more specific signs that should arouse suspicion, especially if more than one is exhibited. Among them are:

- Slurring of speech or an inability to communicate effectively
- Shiftiness or sneaky behavior, such as an employee disappearing to wooded areas, storage areas, or other private locations
- Wearing sunglasses indoors or on overcast days to hide dilated or constricted pupils
- Wearing long-sleeved garments, particularly on hot days, to cover marks from needles used to inject drugs
- Attempting to borrow money from co-workers
- The loss of an employee's tools and company equipment

5.6.0 Job-Related Accident Investigations

There are two times when a crew leader may be involved with an accident investigation. The first time is when an accident, injury, or report of work-connected illness takes place. If present on site, the crew leader should proceed immediately to the accident location to ensure that proper first aid is being provided. He or she will also want to make sure that other safety and operational measures are taken to prevent another incident.

If mandated by company policy, the crew leader will need to make a formal investigation and submit a report after an incident. An investigation looks for the causes of the accident by examining the situation under which it occurred and talking to the people involved. Investigations are perhaps the most useful tool in the prevention of future accidents.

There are four major parts to an accident investigation. The crew leader will be concerned with each one. They are:

- Describing the accident
- Determining the cause of the accident
- Determining the parties involved and the part played by each
- Determining how to prevent re-occurrences

Case Study

For years, a prominent safety engineer was confused as to why sheet metal workers fractured their toes frequently. The crew leader had not performed thorough accident investigations, and the injured workers were embarrassed to admit how the accidents really occurred. It was later discovered they used the metal reinforced cap on their safety shoes as a "third hand" to hold the sheet metal vertically in place when they fastened it. The sheet metal was inclined to slip and fall behind the safety cap onto the toes, causing fractures. Several injuries could have been prevented by performing a proper investigation after the first accident.

6.0.0 PROMOTING SAFETY

The best way for crew leaders to encourage safety is through example. Crew leaders should be aware that their behavior sets standards for their crew members. If a crew leader cuts corners on safety, then the crew members may think that it is okay to do so as well.

The key to effectively promote safety is good communication. It is important to plan and coordinate activities and to follow through with safety programs. The most successful safety promotions occur when employees actively participate in planning and carrying out activities.

Some activities used by organizations to help motivate employees on safety and help promote safety awareness include:

- Safety training sessions
- Contests
- Recognition and awards
- Publicity

6.1.0 Safety Training Sessions

Safety training sessions can help keep workers focused on safety and give them the opportunity to discuss safety concerns with the crew. This topic was addressed in a previous section.

6.2.0 Safety Contests

Contests are a great way to promote safety in the workplace. Examples of safety-related contests include the following:

- Sponsoring housekeeping contests for the cleanest job site or work area
- Challenging employees to come up with a safety slogan for the company or department
- Having a poster contest that involves employees or their children creating safety-related posters
- Recording the number of accident-free workdays or person-hours
- Giving safety awards (hats, T-shirts, other promotional items or prizes)

One of the positive aspects of contests is their ability to encourage employee participation. It is important, however, to ensure that the contest has a valid purpose. For example, the posters or slogans created in a poster contest can be displayed throughout the organization as safety reminders.

6.3.0 Incentives and Awards

Incentives and awards serve several purposes. Among them are acknowledging and encouraging good performance, building goodwill, reminding employees of safety issues, and publicizing the importance of practicing safety standards. There are countless ways to recognize and award safety. Examples include the following:

- Supplying food at the job site when a certain goal is achieved
- Providing a reserved parking space to acknowledge someone for a special achievement
- Giving gift items such as T-shirts or gift certificates to reward employees
- Giving plaques to a department or an individual (*Figure 9*)
- Sending a letter of appreciation
- Publicly honoring a department or an individual for a job well done

Creativity can be used to determine how to recognize and award good safety on the work site. The only precautionary measure is that the award should be meaningful and not perceived as a bribe. It should be representative of the accomplishment.

6.4.0 Publicity

Publicizing safety is the best way to get the message out to employees. An important aspect of publicity is to keep the message accurate and current. Safety posters that are hung for years on end tend to lose effectiveness. It is important to keep ideas fresh.

Examples of promotional activities include posters or banners, advertisements or information on bulletin boards, payroll mailing stuffers, and employee newsletters. In addition, merchandise can be purchased that promotes safety, including buttons, hats, T-shirts, and mugs.

46101-11_F09.EPS

Figure 9 Examples of safety plaques.

Described here are three scenarios that reflect unsafe practices by craft workers. For each of these scenarios write down how you would deal with the situation, first as the crew leader of the craft worker, and then as the leader of another crew.

1. You observe a worker wearing his hard hat backwards and his safety glasses hanging around his neck. He is using a concrete saw.

_____

_____

_____

_____

_____

_____

2. As you are supervising your crew on the roof deck of a building under construction, you notice that a section of guard rail has been removed. Another contractor was responsible for installing the guard rail.

_____

_____

_____

_____

_____

_____

3. Your crew is part of plant shutdown at a power station. You observe that a worker is welding without a welding screen in an area where there are other workers.

_____

_____

_____

_____

_____

_____

1. One of a crew leader's responsibilities is to enforce company safety policies.

   a. True
   b. False

2. Which of the following is an indirect cost of an accident?

   a. Medical bills
   b. Production delays
   c. Compensation
   d. Employee benefits

3. A crew leader can be held criminally liable for repeat safety violations.

   a. True
   b. False

4. OSHA inspection of a business or job site _____.

   a. can be done only by invitation
   b. is done only after an accident
   c. can be conducted at random
   d. can be conducted only if a safety violation occurs

5. The *OSHA 300* form deals with _____.

   a. penalties for safety violations
   b. workplace illnesses and injuries
   c. hazardous material spills
   d. safety training sessions

6. A crew leader's responsibilities include all of the following, *except* _____.

   a. conducting safety training sessions
   b. developing a company safety program
   c. performing safety inspections
   d. participating in accident investigations

7. In order to ensure workplace safety, the crew leader should _____.

   a. hold formal safety training sessions
   b. have crew members conduct on-site safety inspections
   c. notify contractors and their supervisor of hazards in a situation where a job is being performed in an unsafe area controlled by other contractors
   d. hold crew members responsible for making a formal report and investigation following an accident

8. Prohibitions on the abuse of drugs deals only with illegal drugs such as cocaine and marijuana.

   a. True
   b. False

PROJECT CONTROL

Objectives

Upon completion of this section, you will be able to:

1. Describe the three phases of a construction project.
2. Define the three types of project delivery systems.
3. Define planning and describe what it involves.
4. Explain why it is important to plan.
5. Describe the two major stages of planning.
6. Explain the importance of documenting job site work.
7. Describe the estimating process.
8. Explain how schedules are developed and used.
9. Identify the two most common schedules.
10. Explain how the critical path method (CPM) of scheduling is used.
11. Describe the different costs associated with building a job.
12. Explain the crew leader's role in controlling costs.
13. Illustrate how to control the main resources of a job: materials, tools, equipment, and labor.
14. Explain the differences between production and productivity and the importance of each.

Performance Tasks

1. Develop and present a look-ahead schedule.
2. Develop an estimate for a given work activity

1.0.0 PROJECT CONTROL OVERVIEW

The contractor, project manager, superintendent, and crew leader each have management responsibilities for their assigned jobs. For example, the contractor's responsibility begins with obtaining the contract, and it does not end until the client takes ownership of the project. The project manager is generally the person with overall responsibility for coordinating the project. Finally, the superintendent and crew leader are responsible for coordinating the work of one or more workers, one or more crews of workers within the company and, on occasion, one or more crews of subcontractors. The crew leader directs a crew in the performance of work tasks.

This section describes methods of effective and efficient project control. It examines estimating, planning and scheduling, and resource and cost control. All the workers who participate in the job are responsible at some level for controlling cost and schedule performance and for ensuring that the project is completed according to plans and specifications.

> **NOTE**
>
> The material in this section is based largely on building-construction projects. However, the project control principles described here apply generally to all types of projects.

Construction projects are made up of three phases: the development phase, the planning phase, and the construction phase.

1.1.0 Development Phase

A new building project begins when an owner has decided to build a new facility or add to an existing facility. The development process is the first stage of planning for a new building project. This process involves land research and feasibility studies to ensure that the project has merit. Architects or engineers develop the conceptual drawings that define the project graphically. They then provide the owner with sketches of room layouts and elevations and make suggestions about what construction materials should be used.

During the development phase, an estimate for the proposed project is developed in order to establish a preliminary budget. Once that budget has been established, the financing of the project is discussed with lending institutions. The architects/engineers and/or the owner begins preliminary reviews with government agencies. These reviews include zoning, building restrictions, landscape requirements, and environmental impact studies.

Also during the development phase, the owner must analyze the project's cost and potential return on investment (ROI) to ensure that its costs will not exceed its market value and that the project provides a good return on investment. If the project passes this test, the architect/engineer will proceed to the planning phase.

1.2.0 Planning Phase

When the architect/engineer begins to develop the project drawings and specifications, other design professionals such as structural, mechanical, and electrical engineers are brought in. They perform the calculations, make a detailed technical analysis, and check details of the project for accuracy.

 Fundamentals of Crew Leadership

The design professionals create drawings and specifications. These drawings and specifications are used to communicate the necessary information to the contractors, subcontractors, suppliers, and workers that contribute to a project.

During the planning phase, the owners hold many meetings to refine estimates, adjust plans to conform to regulations, and secure a construction loan. If the project is a condominium, an office building, or a shopping center, then a marketing program must be developed. In such cases, the selling of the project is often started before actual construction begins.

Next, a complete set of drawings, specifications, and bid documents is produced. Then the owner will select the method to obtain contractors. The owner may choose to negotiate with several contractors or select one through competitive bidding. Note that safety must also be considered as part of the planning process. A safety crew leader may walk through the site as part of the pre-bid process.

Contracts can take many forms. The three basic types from which all other types are derived are firm fixed price, cost reimbursable, and guaranteed maximum price.

- *Firm fixed price* – In this type of contract, the buyer generally provides detailed drawings and specifications, which the contractor uses to calculate the cost of materials and labor. To these costs, the contractor adds a percentage representing company overhead expenses such as office rent, insurance, and accounting/payroll costs. On top of all this, the contractor adds a profit factor. When submitting the bid, the contractor will state very specifically the conditions and assumptions on which the bid is based. These conditions and assumptions form the basis from which changes can be priced. Because the price is established in advance, any changes in the job requirements once the job is started will impact the contractor's profit margin. This is where the crew leader can play an important role by identifying problems that increase the amount of labor or material that was planned. By passing this information up the chain of command, the crew leader allows the company to determine if the change is outside the scope of the bid. If so, they can submit a change order request to cover the added cost.

- *Cost reimbursable* – In this type of contract, the buyer reimburses the contractor for labor, materials, and other costs encountered in the performance of the contract. Typically, the contractor and buyer agree in advance on hourly or daily labor rates for different categories of worker.

These rates include an amount representing the contractor's overhead expense. The buyer also reimburses the contractor for the cost of materials and equipment used on the job. The buyer and contractor also negotiate a profit margin. On this type of contract, the profit margin is likely to be lower than that of a fixed-price contract because the contractor's cost risk is significantly reduced. The profit margin is often subject to incentive or penalty clauses that make the amount of profit awarded subject to performance by the contractor. Performance is usually tied to project schedule milestones.

- *Guaranteed maximum price (GMP)* – This form of contract, also called a not-to-exceed contract, is most often used on projects that have been negotiated with the owner. Involvement in the process usually includes preconstruction, and the entire team develops the parameters that define the basis for the work. In some instances, the owner will require a competitively-bid GMP. In such cases, the scope of work has not been fully defined, but bids are taken for general conditions (direct costs) and fee based on an assumed volume of work. The advantages of the GMP contract vehicle are:
  - Reduced design time
  - Allows for phased construction
  - Uses a team approach to a project
  - Reduction in changes related to incomplete drawings

1.3.0 Construction Phase

The designated contractor enlists the help of mechanical, electrical, elevator, and other specialty subcontractors to complete the construction phase. The contractor may perform one or more parts of the construction, and rely on subcontractors for the remainder of the work. However, the general contractor is responsible for managing all the trades necessary to complete the project.

As construction nears completion, the architect/engineer, owner, and government agencies start their final inspections and acceptance of the project. If the project has been managed by the general contractor, the subcontractors have performed their work, and the architect/ engineers have regularly inspected the project to ensure that local codes have been followed, then the inspection procedure can be completed in a timely manner. This results in a satisfied client and a profitable project for all.

On the other hand, if the inspection reveals faulty workmanship, poor design or use of materials, or violation of codes, then the inspection and acceptance will become a lengthy battle and may result in a dissatisfied client and an unprofitable project.

Figure 10 shows the flow of a typical project.

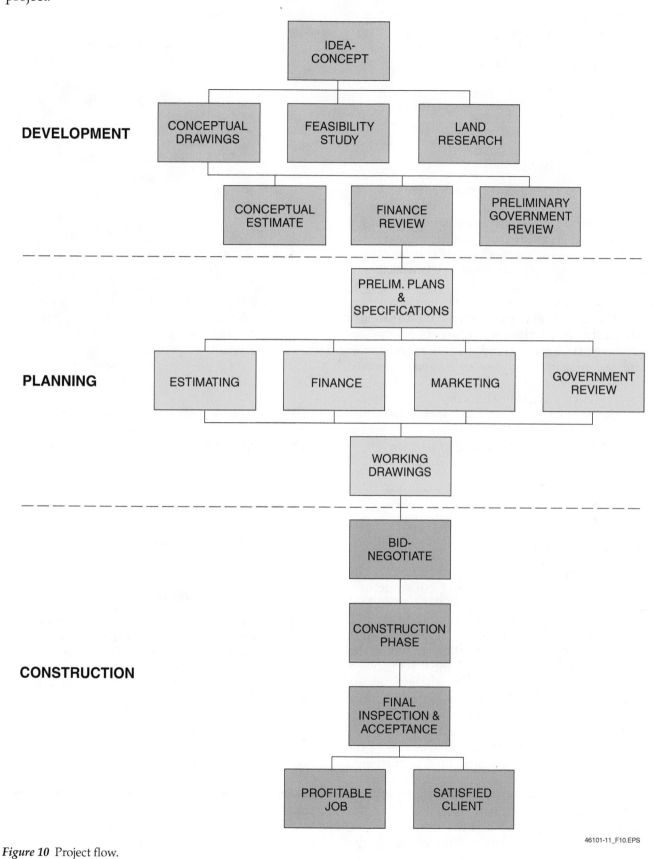

Figure 10 Project flow.

LEED stands for Leadership in Energy and Environmental Design. It is an initiative started by the U.S. Green Building Council (USGBC) to encourage and accelerate the adoption of sustainable construction standards worldwide through a Green Building Rating System™. USGBC is a non-government, not-for-profit group. Their rating system addresses six categories:

1. Sustainable Sites (SS)
2. Water Efficiency (WE)
3. Energy and Atmosphere (EA)
4. Materials and Resources (MR)
5. Indoor Environmental Quality (EQ)
6. Innovation in Design (ID)

46101-11_SA02.EPS

   LEED is a voluntary program that is driven by building owners. Construction crew leaders may not have input into the decision to seek LEED certification for a project, or what materials are used in the project's construction. However, these crew leaders can help to minimize material waste and support recycling efforts, both of which are factors in obtaining LEED certification.
   An important question to ask is whether your project is seeking LEED certification. If the project is seeking certification, the next step is to ask what your role will be in getting the certification. If you are procuring materials, what information is needed and who should receive it? What specifications and requirements do the materials need to meet? If you are working outside the building or inside in a protected area, what do you need to do to protect the work area? How should waste be managed? Are there any other special requirements that will be your responsibility? Do you see any opportunities for improvement? LEED principles are described in more detail in the NCCER publications, *Your Role in the Green Environment* and *Sustainable Construction Supervisor*.

1.3.1 As-Built Drawings

A set of drawings for a construction project reflects the completed project as conceived by the architect and engineers. During construction, changes usually are necessary because of factors unforeseen during the design phase. For example, if cabling or conduit is re-routed, or equipment is installed in a different location than shown on the original drawing, such changes must be marked on the drawings. Without this record, technicians called to perform maintenance or modify the equipment at a later date will have trouble locating all the cabling and equipment.

   Any changes made during construction or installation must be documented on the drawings as the changes are made. Changes are usually made using a colored pen or pencil, so the change can be readily spotted. These drawings are commonly called redlines. When the drawings have been revised to reflect the redline changes, the final drawings are called as-builts, and are so marked. These become the drawings of record for the project.

2.0.0 PROJECT DELIVERY SYSTEMS

Project delivery systems refer to the process by which projects are delivered, from development through construction. Project delivery systems focus on three main systems: general contracting, design-build, and construction management (*Figure 11*).

2.1.0 General Contracting

The traditional project delivery system uses a general contractor. In this type of project, the owner determines the design of the project, and then solicits proposals from general contractors. After selecting a general contractor, the owner contracts directly with that contractor, who builds the project as the prime, or controlling, contractor.

	GENERAL CONTRACTING	DESIGN-BUILD	CONSTRUCTION MANAGEMENT
OWNER	Designs project (or hires architect)	Hires general contractor	Hires construction management company
GENERAL CONTRACTOR	Builds project (with owner's design)	Involved in project design, builds project	Builds, may design (hired by construction management company)
CONSTRUCTION MANAGEMENT COMPANY			Hires and manages general contractor and architect

46101-11_F11.EPS

Figure 11 Project delivery systems.

2.2.0 Design-Build

The design-build project delivery system is different from the general contracting delivery system. In the design-build system, both the design and construction of a project are managed by a single entity. GMP contracts are commonly used in these situations.

2.3.0 Construction Management

The construction management project delivery system uses a construction manager to facilitate the design and construction of a project. Construction managers are very involved in project control; their main concerns are controlling time, cost, and the quality of the project.

3.0.0 COST ESTIMATING AND BUDGETING

Before a project is built, an estimate must be prepared. An estimate is the process of calculating the cost of a project. There are two types of costs to consider, including direct and indirect costs. Direct costs, also known as general conditions, are those that can clearly be assigned to a budget. Indirect costs are overhead costs that are shared by all projects. These costs are generally applied as an overhead percentage to labor and material costs.

Direct costs include the following:

• Materials
• Labor
• Tools
• Equipment

Indirect costs refer to overhead items such as:

• Office rent
• Utilities
• Telecommunications
• Accounting
• Office supplies, signs

The bid price includes the estimated cost of the project as well as the profit. Profit refers to the amount of money that the contractor will make after all of the direct and indirect costs have been paid. If the direct and indirect costs exceed those estimated to perform the job, the difference between the actual and estimated costs must come out of the company's profit. This reduces what the contractor makes on the job.

Profit is the fuel that powers a business. It allows the business to invest in new equipment and facilities, provide training, and to maintain a reserve fund for times when business is slow. In the case of large companies, profitability attracts investors who provide the capital necessary for the business to grow. For these reasons, contractors cannot afford to lose money on a consistent basis. Those who cannot operate profitably are forced out of business. Crew leaders can help their companies remain profitable by managing budget, schedule, quality, and safety adhering to the drawings, specifications, and project schedule.

3.1.0 The Estimating Process

The cost estimate must consider a number of factors. Many companies employ professional cost estimators to do this work. They also maintain performance data for previous projects. This data can be used as a guide in estimating new projects. A complete estimate is developed as follows:

Step 1 Using the drawings and specifications, an estimator records the quantity of the materials needed to construct the job. This is called a quantity takeoff. The information is placed on a takeoff sheet like the one shown in *Figure 12*.

Step 2 Productivity rates are used to estimate the amount of labor required to complete the project. Most companies keep records of these rates for the type and size of the jobs that they perform. The company's estimating department keeps these records.

Step 3 The amount of work to be done is divided by the productivity rate to determine labor hours. For example, if the productivity rate for concrete finishing is 40 square feet per hour, and there are 10,000 square feet of concrete to be finished, then 250 hours of concrete finishing labor is required. This number would be multiplied by the hourly rate for concrete finishing to determine the cost of that labor category. If this work is subcontracted, then the subcontractor's cost estimate, raised by an overhead factor, would be used in place of direct labor cost.

Step 4 The total material quantities are taken from the quantity takeoff sheet and placed on a summary or pricing sheet, an example of which is shown in *Figure 13*. Material prices are obtained from local suppliers, and the total cost of materials is calculated.

Step 5 The cost of equipment needed for the project is determined. This number could reflect rental cost or a factor used by the company when their own equipment is to be used.

Step 6 The total cost of all resources—materials, equipment, tools, and labor—is then totaled on the summary sheet. The unit cost—the total cost divided by the total number of units of material to be put into place—can also be calculated.

Step 7 The cost of taxes, bonds, insurance, subcontractor work, and other indirect costs are added to the direct costs of the materials, equipment, tools, and labor.

Step 8 Direct and indirect costs are summed to obtain the total project cost. The contractor's expected profit is then added to that total.

> **NOTE**
>
> There are software programs available to simplify the cost estimating process. Many of them are tailored to specific industries such as construction and manufacturing, and to specific trades within the industries. For example, there are programs available for electrical and HVAC contractors. Estimating programs are typically set up to include a takeoff form and a form for estimating labor by category. Most of these programs include a data base that contains current prices for labor and materials, so they automatically price the job and produce a bid. Once the job is awarded, the programs can generate purchase orders for materials.

3.1.1 Estimating Material Quantities

The crew leader may be required to estimate quantities of materials.

A set of construction drawings and specifications is needed in order to estimate the amount of a certain type of material required to perform a job. The appropriate section of the technical specifications and page(s) of drawings should be carefully reviewed to determine the types and quantities of materials required. The quantities are then placed on the worksheet. For example, the specification section on finished carpentry should be reviewed along with the appropriate pages of drawings before taking off the linear feet of door and window trim.

If an estimate is required because not enough materials were ordered to complete the job, the estimator must also determine how much more work is necessary. Once this is known, the crew leader can then determine the materials needed. The construction drawings will also be used in this process.

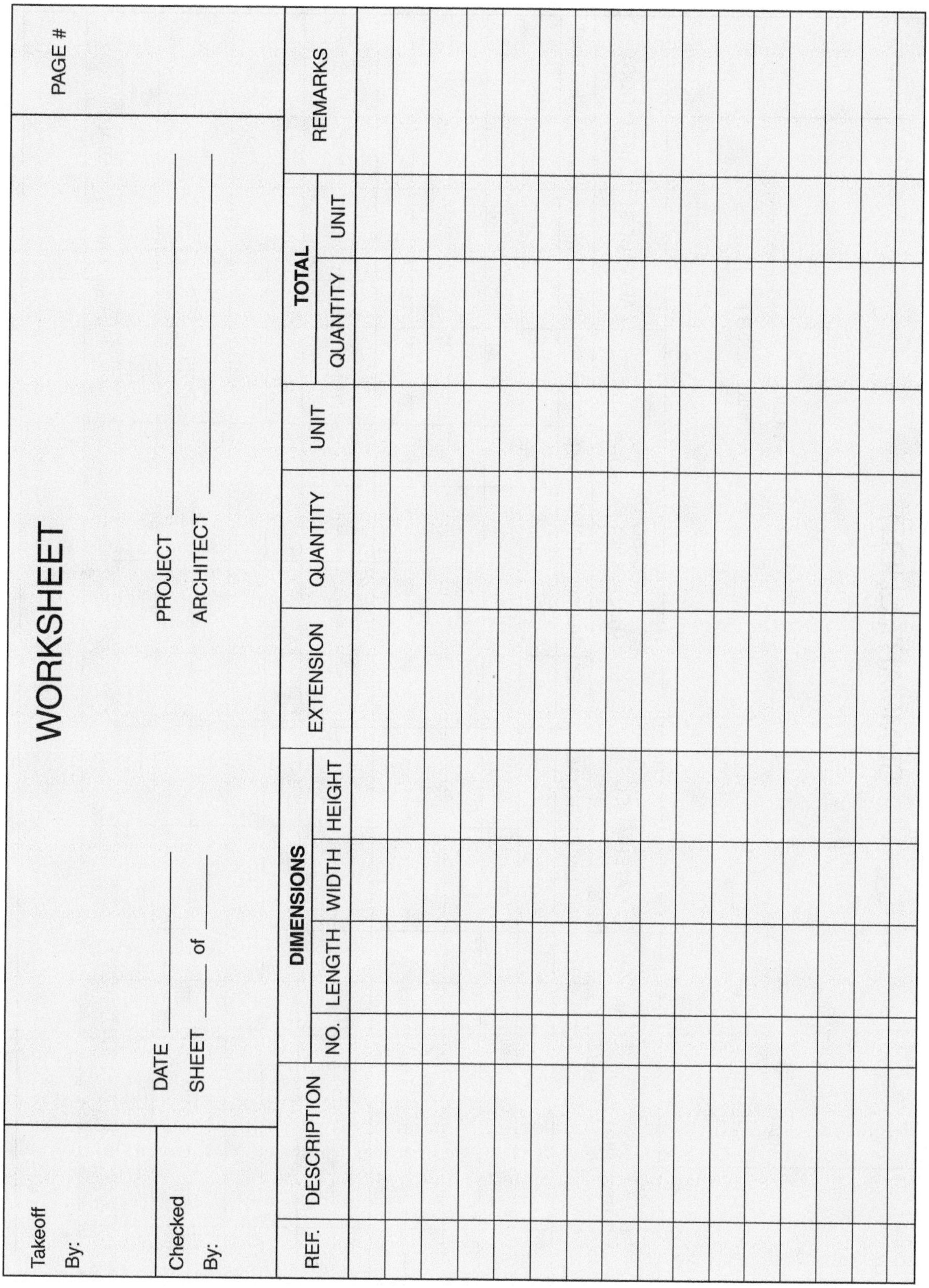

WORKSHEET

Takeoff By: _____

Checked By: _____

PAGE #

PROJECT _____

ARCHITECT _____

DATE _____

SHEET _____ of _____

REF.	DESCRIPTION	DIMENSIONS				EXTENSION				TOTAL		REMARKS
		NO.	LENGTH	WIDTH	HEIGHT		QUANTITY		UNIT	QUANTITY	UNIT	

Figure 12 Quantity takeoff sheet.

46101-11_F12.EPS

SUMMARY SHEET

By:

DATE _____

SHEET _____ of _____

TITLE: _____

PROJECT _____

WORK ORDER # _____

PAGE #

DESCRIPTION	QUANTITY		MATERIAL COST		LABOR MAN HOURS FACTORS					LABOR COST		ITEM COST	
	TOTAL	UT	PER UNIT	TOTAL	CRAFT	PR UNIT	TOTAL	RATE	COST PR	PER	TOTAL	TOTAL	PER UNIT
MATERIAL										LABOR		TOTAL	

46101-11_F13.EPS

Figure 13 Summary sheet.

NCCER – *Contren® Learning Series*   46101-11

Assume you are the leader of a crew building footing formwork for the construction shown in *Figure 14*. You have used all of the materials provided for the job, yet you have not completed it. You study the drawings and see that the formwork consists of two side forms, each 12" high. The total length of footing for the entire project is 115'-0". You have completed 88'-0" to date; therefore, you have 27'-0" remaining (115'-0" – 88'-0" = 27'-0"). Your job is to prepare an estimate of materials that you will need to complete the job. In this case, only the side forms will be estimated (the miscellaneous materials will not be considered here).

- Footing length to complete: 27'-0"
- Footing height: 1'-0"

Refer to the worksheet in *Figure 15* for a final tabulation of the side forms needed to complete the job.

1. Using the same footing as described in the example above, calculate the quantity (square feet) of formwork needed to finish 203 linear feet of the footing. Place this information directly on the worksheet.
2. You are the crew leader of a carpentry crew whose task is to side a warehouse with plywood sheathing. The wall height is 16 feet, and there is a total of 480 linear feet of wall to side. You have done 360 linear feet of wall and have run out of materials. Calculate how many more feet of plywood you will need to complete the job. If you are using 4' × 8' plywood panels, how many will you need to order, assuming no waste? Write your estimate on the worksheet.

Show your calculations to the instructor.

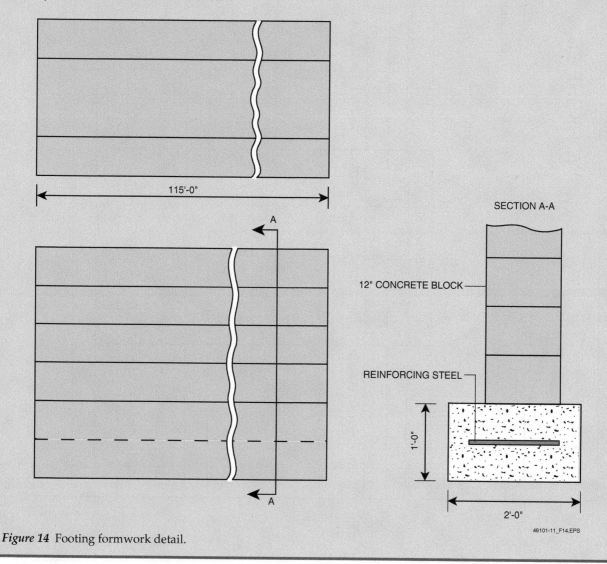

Figure 14 Footing formwork detail.

46101-11_F14.EPS

WORKSHEET

PAGE #1

Takeoff
By: RWH

DATE ___2/1/15___

Checked
By:

SHEET _01_ of _01_

PROJECT ___Sam's Diner___

ARCHITECT ___654b___

| REF. | DESCRIPTION | DIMENSIONS | | | | EXTENSION | QUANTITY | UNIT | TOTAL | | REMARKS |
		NO	LENGTH	WIDTH	HEIGHT				QUANTITY	UNIT	
	Footing Side Forms	2	27'0"		1'0"	2x27x1	54	SF	54	SF	

46101-11_F15.EPS

Figure 15 Worksheet with entries.

4.0.0 PLANNING

Planning can be defined as determining the method used to carry out the different tasks to complete a project. It involves deciding what needs to be done and coming up with an organized sequence of events or plan for doing the work.

Planning involves the following:

- Determining the best method for performing the job
- Identifying the responsibilities of each person on the work crew
- Determining the duration and sequence of each activity
- Identifying what tools and equipment are needed to complete a job
- Ensuring that the required materials are at the work site when needed
- Making sure that heavy construction equipment is available when required

- Working with other contractors in such a way as to avoid interruptions and delays

4.1.0 Why Plan?

With a plan, a crew leader can direct work efforts efficiently and can use resources such as personnel, materials, tools, equipment, work area, and work methods to their full potential.

Some reasons for planning include the following:

- Controlling the job in a safe manner so that it is built on time and within cost
- Lowering job costs through improved productivity
- Preparing for bad weather or unexpected occurrences
- Promoting and maintaining favorable employee morale
- Determining the best and safest methods for performing the job

---

Participant Exercise F

1. In your own words, define planning, and describe how a job can be done better if it is planned. Give an example.

   _____

   _____

   _____

   _____

2. Consider a job that you recently worked on to answer the following:
   a. List the material(s) used.
   b. List each member of the crew with whom you worked and what each person did.
   c. List the kinds of equipment used.

   _____

   _____

   _____

   _____

   _____

3. List some suggestions for how the job could have been done better, and describe how you would plan for each of the suggestions.

   _____

   _____

   _____

   _____

 Fundamentals of Crew Leadership

4.2.0 Stages of Planning

There are various times when planning is done for a construction job. The two most important occur in the pre-construction phase and during the construction work.

4.2.1 Pre-Construction Planning

The pre-construction stage of planning occurs before the start of construction. Except in a fairly small company or for a relatively small job, the crew leader usually does not get directly involved in the pre-construction planning process, but it is important to understand what it involves.

There are two phases of pre-construction planning. The first is when the proposal, bid, or negotiated price for the job is being developed. This is when the estimator, the project manager, and the field superintendent develop a preliminary plan for how the work will be done. This is accomplished by applying experience and knowledge from previous projects. It involves determining what methods, personnel, tools, and equipment will be used and what level of productivity they can achieve.

The second phase occurs after the contract is awarded. This phase requires a thorough knowledge of all project drawings and specifications. During this stage, the actual work methods and resources needed to perform the work are selected. Here, crew leaders might get involved, but their planning must adhere to work methods, production rates, and resources that fit within the estimate prepared before the contract was awarded. If the project requires a method of construction different from what is normal, the crew leader will usually be informed of what method to use.

4.2.2 Construction Planning

During construction, the crew leader is directly involved in planning on a daily basis. This planning consists of selecting methods for completing tasks before beginning work. Effective planning exposes likely difficulties, and enables the crew leader to minimize the unproductive use of personnel and equipment. Effective planning also provides a gauge to measure job progress. Effective crew leaders develop what is known as look-ahead (short-term) schedules. These schedules consider actual circumstances as well as projections two to three weeks into the future. Developing a look-ahead schedule helps ensure that all resources are available on the project when needed.

One of the characteristics of an effective crew leader is the ability to reduce each job to its simpler parts and organize a plan for handling each task.

Project planners establish time and cost limits for the project; the crew leader's planning must fit within those constraints. Therefore, it is important to consider the following factors that may affect the outcome:

- Site and local conditions, such as soil types, accessibility, or available staging areas
- Climate conditions that should be anticipated during the project
- Timing of all phases of work
- Types of materials to be installed and their availability
- Equipment and tools required and their availability
- Personnel requirements and availability
- Relationships with the other contractors and their representatives on the job

On a simple job, these items can be handled almost automatically. However, larger or more complex jobs require the planner to give these factors more formal consideration and study.

5.0.0 THE PLANNING PROCESS

The planning process consists of the following five steps:

Step 1 Establish a goal.

Step 2 Identify the work activities that must be completed in order to achieve the goal.

Step 3 Identify the tasks that must be done to accomplish those activities.

Step 4 Communicate responsibilities.

Step 5 Follow up to see that the goal is achieved.

NCCER – *Contren® Learning Series* 46101-11

5.1.0 Establish a Goal

The term *goal* has different meanings for different people. In general, a goal is a specific outcome that one works toward. For example, the project superintendent of a home construction project could establish the goal to have a house dried-in by a certain date. (Dried-in means ready for the application of roofing and siding.) In order to meet that goal, the leader of the framing crew and the superintendent would need to agree to a goal to have the framing completed by a given date. The crew leader would then establish sub-goals (objectives) for the crew to complete each element of the framing (floors, walls, roof) by a set time. The superintendent would need to set similar goals with the crews that install sheathing, building wrap, windows, and exterior doors. However, if the framing crew does not meet its goal, the other crews will be delayed.

5.2.0 Identify the Work to be Done

The second step in planning is to identify the work to be done to achieve the goal. In other words, it is a series of activities that must be done in a certain sequence. The topic of breaking down a job into activities is covered later in this section. At this point, the crew leader should know that, for each activity, one or more objectives must be set.

An objective is a statement of what is desired at a specific time. An objective must:

- Mean the same thing to everyone involved
- Be measurable, so that everyone knows when it has been reached
- Be achievable
- Have everyone's full support

Examples of practical objectives include the following:

- By 4:30 P.M. today, the crew will have completed installation of the floor joists.
- By closing time Friday, the roof framing will be complete.

Notice that both examples meet the first three requirements of an objective. In addition, it is assumed that everyone involved in completing the task is committed to achieving the objective. The advantage in developing objectives for each work activity is that it allows the crew leader to evaluate whether or not the plan and schedules are being followed. In addition, objectives serve as sub-goals that are usually under the crew leader's control.

Some construction work activities, such as installing 12"-deep footing forms, are done so often that they require little planning. However, other jobs, such as placing a new type of mechanical equipment, require substantial planning. This type of job requires that the crew leader set specific objectives.

Whenever faced with a new or complex activity, take the time to establish objectives that will serve as guides for accomplishing the job. These guides can be used in the current situation, as well as in similar situations in the future.

5.3.0 Identify Tasks to be Performed

To plan effectively, the crew leader must be able to break a work activity assignment down into smaller tasks. Large jobs include a greater number of tasks than small ones, but all jobs can be broken down into manageable components.

When breaking down an assignment into tasks, make each task identifiable and definable. A task is identifiable when the types and amounts of resources it requires are known. A task is definable if it has a specific duration. For purposes of efficiency, the job breakdown should not be too detailed or complex, unless the job has never been done before or must be performed with strictest efficiency.

For example, a suitable breakdown for the work activity to install 12" × 12" vinyl floor tile in a cafeteria might be the following:

Step 1   Prepare the floor.

Step 2   Lay out the tile.

Step 3   Spread the adhesive.

Step 4   Lay the tile.

Step 5   Clean the tile.

Step 6   Wax the floor.

The crew leader could create even more detail by breaking down any one of the tasks, such as lay the tile, into subtasks. In this case, however, that much detail is unnecessary and wastes the crew leader's time and the project's money. However, breaking tasks down further might be necessary in a case where the job is very complex or the analysis of the job needs to be very detailed.

 Fundamentals of Crew Leadership

Every work activity can be divided into three general parts:

- Preparing
- Performing
- Cleaning up

One of the most frequent mistakes made in the planning process is forgetting to prepare and to clean up. The crew leader must be certain that preparation and cleanup are not overlooked.

After identifying the various tasks that make up the job and developing an objective for each task, the crew leader must determine what resources the job requires. Resources include labor, equipment, materials, and tools. In most jobs, these resources are identified in the job estimate. The crew leader must make sure that these resources are available on the site when needed.

5.4.0 Communicating Responsibilities

A crew leader is unable to complete all of the activities within a job independently. Other people must be relied upon to get everything done. Therefore, most jobs have a crew of people with various experiences and skill levels to assist in the work. The crew leader's job is to draw from this expertise to get the job done well and in a safe and timely manner.

Once the various activities that make up the job have been determined, the crew leader must identify the person or persons responsible for completing each activity. This requires that the crew leader be aware of the skills and abilities of the people on the crew. Then, the crew leader must put this knowledge to work in matching the crew's skills and abilities to specific tasks that must be performed to complete the job.

After matching crew members to specific activities, the crew leader must then communicate the assignments to the crew. Communication of responsibilities is generally handled verbally; the crew leader often talks directly to the person to which the activity has been assigned. There

may be times when work is assigned indirectly through written instructions or verbally through someone other than the crew leader. Either way, the crew members should know what it is they are responsible for accomplishing on the job.

5.5.0 Follow-Up Activities

Once the activities have been delegated to the appropriate crew members, the crew leader must follow up to make sure that they are completed effectively and efficiently. Follow-up work involves being present on the job site to make sure all the resources are available to complete the work; ensuring that the crew members are working on their assigned activities; answering any questions; and helping to resolve any problems that occur while the work is being done. In short, follow-up activity means that the crew leader is aware of what's going on at the job site and is doing whatever is necessary to make sure that the work is completed on schedule.

Figure 16 reviews the planning steps.

The crew leader should carry a small note pad or electronic device to be used for planning and note taking. That way, thoughts about the project can be recorded as they occur, and pertinent details will not be forgotten. The crew leader may also choose to use a planning form such as the one illustrated in *Figure 17*.

As the job is being built, refer back to these plans and notes to see that the tasks are being done in sequence and according to plan. This is referred to as analyzing the job. Experience shows that jobs that are not built according to work plans usually end up costing more and taking more time; therefore, it is important that crew leaders refer back to the plans periodically.

The crew leader is involved with many activities on a day-to-day basis. As a result, it is easy to forget important events if they are not documented. To help keep track of events such as job changes, interruptions, and visits, the crew leader should keep a job diary.

Figure 16 Steps to effective planning.

DAILY WORK PLAN		
"PLAN YOUR WORK AND WORK YOUR PLAN = EFFICIENCY"		
Plan of _____		Date _____
PRIORITY	DESCRIPTION	✓ When Completed ✗ Carried Forward

46101-11_F17.EPS

Figure 17 Planning form.

A job diary is a notebook in which the crew leader records activities or events that take place on the job site that may be important later. When recording in a job diary, make sure that the information is accurate, factual, complete, consistent, organized, and up to date. Follow company policy in determining which events should be recorded. However, if there is a doubt about what to include, it is better to have too much information than too little.

Figure 18 shows a sample page from a job diary.

6.0.0 PLANNING RESOURCES

Once a job has been broken down into its tasks or activities, the next step is to assign the various resources needed to perform them.

6.1.0 Safety Planning

Using the company safety manual as a guide, the crew leader must assess the safety issues associated with the job and take necessary measures to minimize any risk to the crew. This may involve working with the company or site safety officer and may require a formal hazard analysis.

6.2.0 Materials Planning

The materials required for the job are identified during pre-construction planning and are listed on the job estimate. The materials are usually ordered from suppliers who have previously provided quality materials on schedule and within estimated cost.

July 8, 2015

Weather: Hot and Humid

Project: Company XYZ Building

- The paving contractor crew arrived late (10 am).

- The owner representative inspected the footing foundation at approximately 1 pm.

- The concrete slump test did not pass. Two trucks had to be ordered to return to the plant, causing a delay.

- John Smith had an accident on the second floor. I sent him to the doctor for medical treatment. The cause of the accident is being investigated.

46101-11_F18.EPS

Figure 18 Sample page from a job diary.

The crew leader is usually not involved in the planning and selection of materials, since this is done in the pre-construction phase. The crew leader does, however, have a major role to play in the receipt, storage, and control of the materials after they reach the job site.

The crew leader is also involved in planning materials for tasks such as job-built formwork and scaffolding. In addition, the crew leader may run out of a specific material, such as fasteners, and need to order more. In such cases, a higher authority should be consulted, since most companies have specific purchasing policies and procedures.

6.3.0 Site Planning

There are many planning elements involved in site work. The following are some of the key elements:

- Emergency procedures
- Access roads
- Parking
- Stormwater runoff
- Sedimentation control
- Material and equipment storage
- Material staging
- Site security

6.4.0 Equipment Planning

Much of the planning for use of construction equipment is done during the pre-construction phase. This planning includes the types of equipment needed, the use of the equipment, and the length of time it will be on the site. The crew leader must work with the home office to make certain that the equipment reaches the job site on time. The crew leader must also ensure that sure equipment operators are properly trained.

Coordinating the use of the equipment is also very important. Some equipment is used in combination with other equipment. For example, dump trucks are generally required when loaders and excavators are used. The crew leader should also coordinate equipment with other contractors on the job. Sharing equipment can save time and money and avoid duplication of effort.

Habitat for Humanity, a charitable organization that builds homes for disadvantaged families, accepts surplus building materials for use in their projects. In some cities, they have stores called ReStores, which serve as retail outlets for such materials. They may also accept materials salvaged during demolition of a structure. LEED credits can be obtained through practices such as salvaging building materials, segregating scrap materials for recycling, and taking steps to minimize waste.

46101-11_SA03.EPS

The crew leader must designate time for equipment maintenance in order to prevent equipment failure. In the event of an equipment failure, the crew leader must know who to contact to resolve the problem. An alternate plan must be ready in case one piece of equipment breaks down, so that the other equipment does not sit idle. This planning should be done in conjunction with the home office or the crew leader's immediate superior.

6.5.0 Tool Planning

A crew leader is responsible for planning what tools will be used on the job. This task includes:

- Determining the tools required
- Informing the workers who will provide the tools (company or worker)
- Making sure the workers are qualified to use the tools safely and effectively
- Determining what controls to establish for tools

6.6.0 Labor Planning

All jobs require some sort of labor because the crew leader cannot complete all the work alone. When planning for labor, the crew leader must:

- Identify the skills needed to perform the work.
- Determine how many people having those specific skills are needed.
- Decide who will actually be on the crew.

In many companies, the project manager or job superintendent determines the size and makeup of the crew. Then, the crew leader is expected to accomplish the goals and objectives with the crew provided. Even though the crew leader may not be involved in staffing the crew, the crew leader is responsible for training the crew members to ensure that they have the skills needed to do the job.

In addition, the crew leader is responsible for keeping the crew adequately staffed at all times so that jobs are not delayed. This involves dealing with absenteeism and turnover, two common problems that affect industry today.

7.0.0 SCHEDULING

Planning and scheduling are closely related and are both very important to a successful job. Planning involves determining the activities that must be completed and how they should be accomplished. Scheduling involves establishing start and finish times or dates for each activity.

A schedule for a project typically shows:

- Operations listed in sequential order
- Units of construction
- Duration of activities
- Estimated date to start and complete each activity
- Quantity of materials to be installed

There are different types of schedules used today. They include the bar chart; the network schedule, which is sometimes called the critical path method (CPM) or precedence diagram; and the short-term, or look-ahead schedule.

7.1.0 The Scheduling Process

The following is a brief summary of the steps a crewleader must complete to develop a schedule.

Step 1   Make a list of all of the activities that will be performed to build the job, including individual work activities and special tasks, such as inspections or the delivery of materials.

At this point, the crew leader should just be concerned with generating a list, not with determining how the activities will be accomplished, who will perform them, how long they will take, or in what sequence they will be completed.

Step 2   Use the list of activities created in Step 1 to reorganize the work activities into a logical sequence.

When doing this, keep in mind that certain steps cannot happen until others have been completed. For example, footings must be excavated before concrete can be placed.

Step 3   Assign a duration or length of time that it will take to complete each activity and determine the start time for each. Each activity will then be placed into a schedule format. This step is important because it helps the crew leader ensure that the activities are being completed on schedule.

The crew leader must be able to read and interpret the job schedule. On some jobs, the beginning and expected end date for each activity, along with the expected crew or worker's production rate, is provided on the form. The crew leader can use this

information to plan work more effectively, set realistic goals, and measure whether or not they were accomplished within the scheduled time.

Before starting a job, the crew leader must:

- Determine the materials, tools, equipment, and labor needed to complete the job.
- Determine when the various resources are needed.
- Follow up to ensure that the resources are available on the job site when needed.

Availability of needed resources should be verified three to four working days before the start of the job. It should be done even earlier for larger jobs. This advance preparation will help avoid situations that could potentially delay the job or cause it to fall behind schedule.

7.2.0 Bar Chart Schedule

Bar chart schedules, also known as Gantt charts, can be used for both short-term and long-term jobs. However, they are especially helpful for jobs of short duration.

Bar charts provide management with the following:

- A visual concept of the overall time required to complete the job through the use of a logical method rather than a calculated guess
- A means to review each part of the job
- Coordination requirements between crafts
- Alternative methods of performing the work

A bar chart can be used as a control device to see whether the job is on schedule. If the job is not on schedule, immediate action can be taken in the office and the field to correct the problem and ensure that the activity is completed on schedule.

A bar chart is illustrated in *Figure 19*.

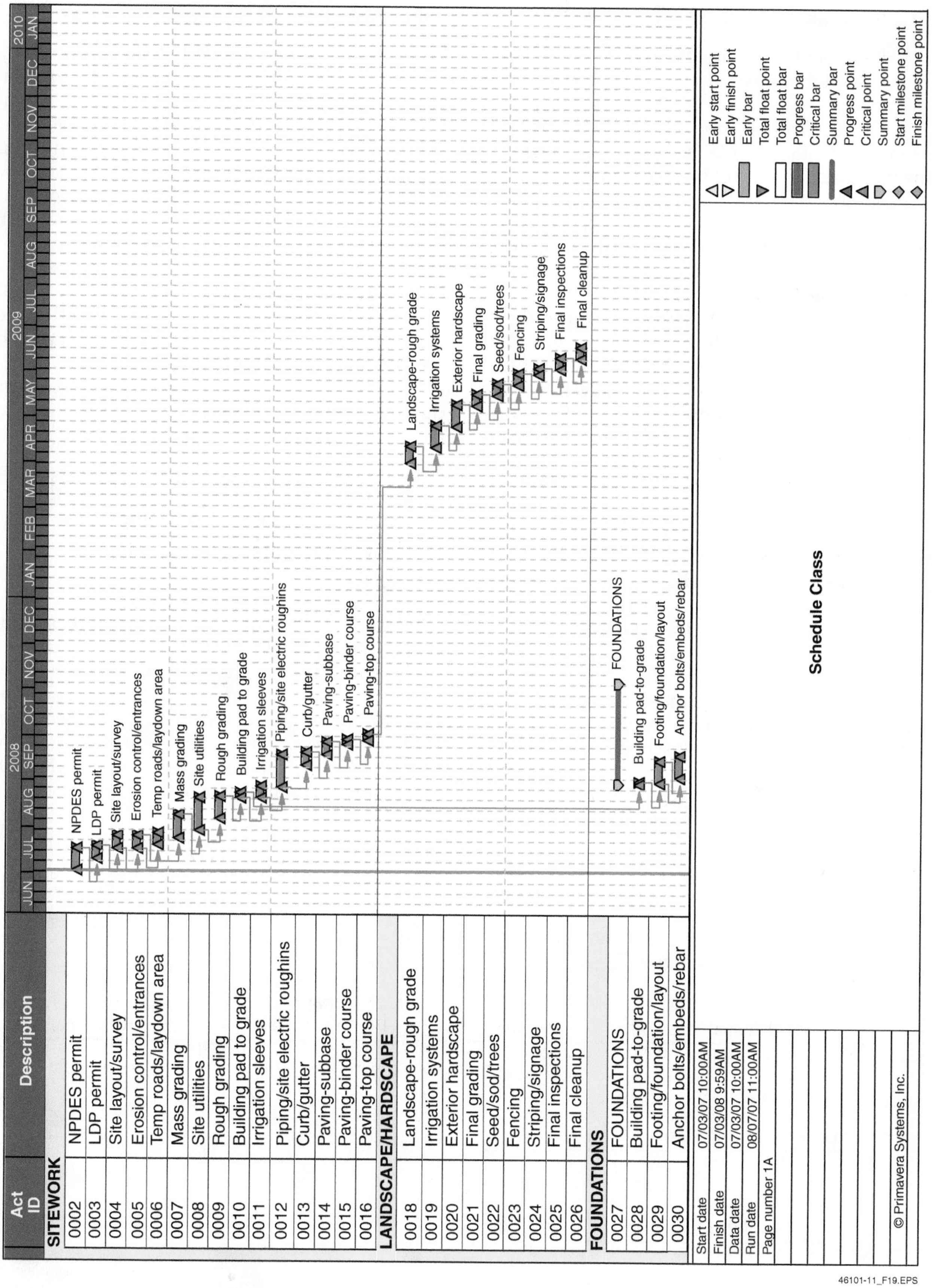

Figure 19 Example of a bar chart schedule.

7.3.0 Network Schedule

Network schedules are an effective project management tool because they show dependent (critical path) activities and activities that can be performed in parallel. In *Figure 20,* for example, reinforcing steel cannot be set until the concrete forms have been built and placed. Other activities are happening in parallel, but the forms are in the critical path. When building a house, drywall cannot be installed and finished until wiring, plumbing, and HVAC ductwork have been roughed-in. Because other activities, such as painting and trim work, depend on drywall completion, the drywall work is a critical-path function. That is, until it is complete, the other tasks cannot be started, and the project itself is likely to be delayed by the amount of delay in any dependent activity. Likewise, drywall work can't even be started until the rough-ins are complete. Therefore, the project superintendent is likely to focus on those activities when evaluating schedule performance.

The advantage of a network schedule is that it allows project leaders to see how a schedule change on one activity is likely to affect other activities and the project in general. A network schedule is laid out on a timeline and usually shows the estimated duration for each activity. Network schedules are generally used for complex jobs that take a long time to complete. The PERT (program evaluation and review technique) schedule is a form of network schedule.

7.4.0 Short-Term Scheduling

Since the crew leader needs to maintain the job schedule, he or she needs to be able to plan daily production. Short-term scheduling is a method used to do this. An example is shown in *Figure 21.*

The information to support short-term scheduling comes from the estimate or cost breakdown. The schedule helps to translate estimate data and the various job plans into a day-to-day schedule of events. The short-term schedule provides the crew leader with visibility over the project. If actual production begins to slip behind estimated production, the schedule will warn the crew leader that a problem lies ahead and that a schedule slippage is developing.

Short-term scheduling can be used to set production goals. It is generally agreed that production can be improved if workers:

- Know the amount of work to be accomplished
- Know the time they have to complete the work
- Can provide input when setting goals

Consider the following example:

Situation:

A carpentry crew on a retaining wall project is about to form and pour catch basins and put up wall forms. The crew has put in a number of catch basins, so the crew leader is sure that they can perform the work within the estimate. However, the crew leader is concerned about their production of the wall forms. The crew will work on both the basins and the wall forms at the same time.

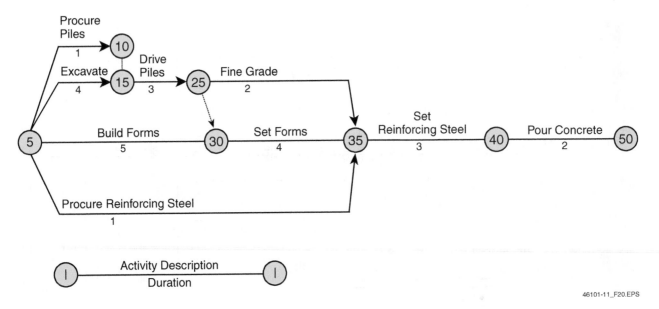

46101-11_F20.EPS

Figure 20 Example of a network schedule.

NCCER – *Contren® Learning Series* 46101-11

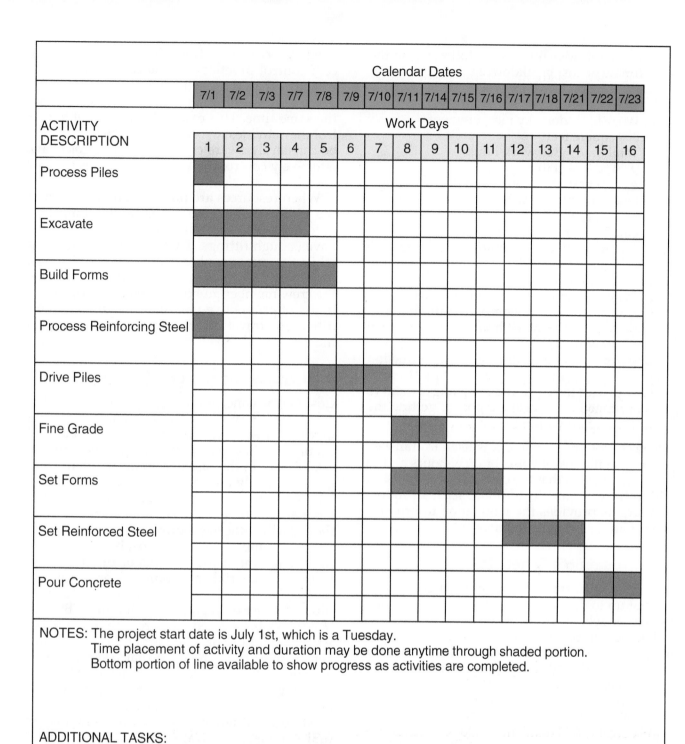

ACTIVITY DESCRIPTION	Calendar Dates															
	7/1	7/2	7/3	7/7	7/8	7/9	7/10	7/11	7/14	7/15	7/16	7/17	7/18	7/21	7/22	7/23
Work Days	1	2	3	4	5	6	7	8	9	10	11	12	13	14	15	16
Process Piles	■															
Excavate	■	■	■	■												
Build Forms	■	■	■	■	■											
Process Reinforcing Steel	■															
Drive Piles					■	■	■									
Fine Grade								■	■							
Set Forms								■	■	■	■					
Set Reinforced Steel												■	■	■		
Pour Concrete															■	■

NOTES: The project start date is July 1st, which is a Tuesday.
Time placement of activity and duration may be done anytime through shaded portion.
Bottom portion of line available to show progress as activities are completed.

ADDITIONAL TASKS:

Figure 21 Short-term schedule.

1. The crew leader notices the following in the estimate or cost breakdown:
   a. Production factor for wall forms = 16 work-hours per 100 square feet
   b. Work to be done by measurement = 800 square feet
   c. Total time = 128 work-hours (800 × 16 ÷ 100)
2. The carpenter crew consists of the following:
   a. One carpenter crew leader
   b. Four carpenters
   c. One laborer
3. The crew leader determines the goal for the job should be set at 128 work-hours (from the cost breakdown).
4. If the crew remains the same (six workers), the work should be completed in about 21 crew-hours (128 work hours ÷ 6 workers = 21.33 crew-hours).
5. The crew leader then discusses the production goal (completing 800 square feet in 21 crew-hours) with the crew and encourages them to work together to meet the goal of getting the forms erected within the estimated time.

The short-term schedule was used to translate production into work-hours or crew-hours and to schedule work so that the crew can accomplish it within the estimate. In addition, setting production targets provides the motivation to produce more than the estimate requires.

7.5.0 Updating a Schedule

No matter what type of schedule is used, it must be kept up to date to be useful to the crew leader. Inaccurate schedules are of no value.

The person responsible for scheduling in the office handles the updates. This person uses information gathered from job field reports to do the updates.

The crew leader is usually not directly involved in updating schedules. However, he or she may be responsible for completing field or progress reports used by the company. It is critical that the crew leader fill out any required forms or reports completely and accurately so that the schedule can be updated with the correct information.

8.0.0 COST CONTROL

Being aware of costs and controlling them is the responsibility of every employee on the job. It is the crew leader's job to ensure that employees uphold this responsibility. Control refers to the comparison of estimated performance against actual performance and following up with any

needed corrective action. Crew leaders who use cost-control practices are more valuable to the company than those who do not.

On a typical job, many activities are going on at the same time. This can make it difficult to control the activities involved. The crew leader must be constantly aware of the costs of a project and effectively control the various resources used on the job.

When resources are not controlled, the cost of the job increases. For example, a plumbing crew of four people is installing soil pipe and does not have enough fittings. Three crew members wait while one crew member goes to the supply house for a part that costs only a few dollars. It takes the crew member an hour to get the part, so four hours of productive work have been lost. In addition, the travel cost for retrieving the supplies must be added.

8.1.0 Assessing Cost Performance

Cost performance on a project is determined by comparing actual costs to estimated costs. Regardless of whether the job is a contract bid project or an in-house project, a budget must first be established. In the case of a contract bid, the budget is generally the cost estimate used to bid the job. For an in-house job, participants will submit labor and material forecasts, and someone in authority will authorize a project budget.

It is common to estimate cost by either breaking the job into funded tasks or by forecasting labor and materials expenditures on a timeline. Many companies create a work breakdown structure (WBS) for each project. Within the WBS, each major task is assigned a discrete charge number. Anyone working on that task charges that number on their time sheet, so that project managers can readily track cost performance. However, knowing how much has been spent does not necessarily determine cost performance.

Although financial reports can show that actual expenses are tracking forecast expenses, they don't show if the work is being done at the required rate. Thus it is possible to have spent half the budget, but have less than half of the work compete. When the project is broken down into funded tasks related to schedule activities and events, there is far greater control over cost performance.

8.2.0 Field Reporting System

The total estimated cost comes from the job estimates, but the actual cost of doing the work is obtained from an effective field reporting system.

A field reporting system is made up of a series of forms, which are completed by the crew leader and others. Each company has its own forms and methods for obtaining information. The general information and the process of how they are used are described here.

First, the number of hours each person worked on each task must be known. This information comes from daily time cards. Once the accounting department knows how many hours each employee worked on an activity, it can calculate the total cost of the labor by multiplying the number of hours worked by the wage rate for each worker. The cost for the labor to do each task can be calculated as the job progresses. This cost will be compared with the estimated cost. This comparison will also be done at the end of the job.

When material is put in place, a designated person will measure the quantities from time to time and send this information to the home office and, possibly, the crew leader. This information, along with the actual cost of the material and the amount of hours it took the workers to install it, is compared to the estimated cost. If the cost is greater than the estimate, management and the crew leader will have to take action to reduce the cost.

A similar process is used to determine if the costs to operate equipment or the production rate are comparable to the estimated cost and production rate.

For this comparison process to be of use, the information obtained from field personnel must be correct. It is important that the crew leader be accurate in reporting. The crew leader is responsible for carrying out his or her role in the field reporting system. One of the best ways to do this is to maintain a daily diary, using a notebook or electronic device. In the event of a legal/contractual conflict with the client, such diaries are considered as evidence in court proceedings, and can be helpful in reaching a settlement.

Here is an example. You are running a crew of five concrete finishers for a subcontractor. When you and your crew show up to finish a slab, the GC says, "We're a day behind on setting the forms, so I need you and your crew to stand down until tomorrow." What do you do?

Of course, you would first call your office to let them know about the delay. Then, you would immediately record it in your job diary. A six-man crew for one day represents 48 labor hours. If your company charges $30 an hour, that's a potential loss of $1,440, which the company would want to recover from the GC. If there is a dispute, your entry in the job diary could result in a favorable decision for your employer.

8.3.0 Crew Leader's Role in Cost Control

The crew leader is often the company representative in the field, where the work takes place. Therefore, the crew leader has a great deal to do with determining job costs. When work is assigned to a crew, the crew leader should be given a budget and schedule for completing the job. It is then up to the crew leader to make sure the job is done on time and stays within budget. This is done by actively managing the use of labor, materials, tools, and equipment.

If the actual costs are at or below the estimated costs, the job is progressing as planned and scheduled, and the company will realize the expected profit. However, if the actual costs exceed the estimated costs, one or more problems may result in the company losing its expected profit, and maybe more. No company can remain in business if it continually loses money. One of the factors that can increase cost is client-related changes. The crew leaders must be able to assess the potential impact of such changes and, if necessary, confer with their employer to determine the course of action. If losses are occurring, the crew leader and superintendent will need to work together to get the costs back in line.

Noted below are some of the reasons why actual costs can exceed estimated costs and suggestions on what the crew leader can do to bring the costs back in line. Before starting any action, however, the crew leader should check with his or her superior to see that the action proposed is acceptable and within the company's policies and procedures.

- *Cause* – Late delivery of materials, tools, and/or equipment
  Corrective Action: Plan ahead to ensure that job resources will be available when needed
- *Cause* – Inclement weather
  Corrective Action: Work with the superintendent and have alternate plans ready
- *Cause* – Unmotivated workers
  Corrective Action: Counsel the workers
- *Cause* – Accidents
  Corrective Action: Enforce the existing safety program

There are many other methods to get the job done on time if it gets off schedule. Examples include working overtime, increasing the size of the crew, pre-fabricating assemblies, or working staggered shifts. However, these examples may increase the cost of the job, so they should not be done without the approval of the project manager.

9.0.0 RESOURCE CONTROL

The crew leader's job is to ensure that assigned tasks are completed safely according to the plans and specifications, on schedule, and within the scope of the estimate. To accomplish this, the crew leader must closely control how resources of materials, equipment, tools, and labor are used. Waste must be minimized whenever possible.

Control involves measuring performance and correcting deviations from plans and specifications to accomplish objectives. Control anticipates deviation from plans and specifications and takes measures to prevent it from occurring.

An effective control process can be broken down into the following steps:

Step 1   Establish standards and divide them into measurable units.

For example, a baseline can be created using experience gained on a typical job, where 2,000 LF of 1¼" copper water tube was installed in five days. Dividing 2,000 by 5 gives 400. Thus, the average installation rate in this case for 1¼" copper water tube was 400 LF/day.

Step 2   Measure performance against a standard.

On another job, 300 square feet of the same tube was placed during an average day. Thus, this average production of 300 LF/day did not meet the average rate of 400 LF/day.

Step 3   Adjust operations to ensure that the standard is met.

In Step 2 above, if the plan called for the job to be completed in five days, the crew leader would have to take action to ensure that this happens. If 300 LF/day is the actual average daily rate, it will have to be increased by 100 LF/day to meet the standard.

9.1.0 Materials Control

The crew leader's responsibility in materials control depends on the policies and procedures of the company. In general, the crew leader is responsible for ensuring on-time delivery, preventing waste, controlling delivery and storage, and preventing theft of materials.

9.1.1 Ensuring On-Time Delivery

It is essential that the materials required for each day's work be on the job site when needed. The crew leader should confirm in advance that all materials have been ordered and will be delivered on schedule. A week or so before the delivery date, follow-up is needed to make sure there will be no delayed deliveries or items on backorder.

To be effective in managing materials, the crew leader must be familiar with the plans and specifications to be used, as well as the activities to be performed. He or she can then determine how many and what types of materials are needed.

If other people are responsible for providing the materials for a job, the crew leader must follow up to make sure that the materials are available when they are needed. Otherwise, delays occur as crew members stand around waiting for the materials to be delivered.

9.1.2 Preventing Waste

Waste in construction can add up to loss of critical and costly materials and may result in job delays. The crew leader needs to ensure that every crew member knows how to use the materials efficiently. The crew should be monitored to make certain that no materials are wasted.

An example of waste is a carpenter who saws off a piece of lumber from a fresh piece, when the size needed could have been found in the scrap pile. Another example of waste involves installing a fixture or copper tube incorrectly. The time spent installing the item incorrectly is wasted because the task will need to be redone. In addition, the materials may need to be replaced if damaged during installation or removal.

Under LEED, waste control is very important. Credits are given for finding ways to reduce waste and for recycling waste products. Waste material should be separated for recycling if feasible (*Figure 22*).

Did you know?

Just-in-time (JIT) delivery is a strategy in which materials are delivered to the job site when they are needed. This means that the materials may be installed right off the truck. This method reduces the need for on-site storage and staging. It also reduces the risk of loss or damage as products are moved about the site. Other modern material management methods include the use of radio frequency identification tags (RFIDs) that make it easy to locate material in crowded staging areas.

9.1.3 Verifying Material Delivery

A crew leader may be responsible for the receipt of materials delivered to the work site. When this happens, the crew leader should require a copy of the shipping ticket and check each item on the shipping ticket against the actual materials to see that the correct amounts were received.

46101-11_F22.EPS

Figure 22 Waste material separated for recycling.

The crew leader should also check the condition of the materials to verify that nothing is defective before signing the shipping ticket. This can be difficult and time consuming because it means that cartons must be opened and their contents examined. However, it is necessary, because a signed shipping ticket indicates that all of the materials were received undamaged. If the crew leader signs for the materials without checking them, and then finds damage, no one will be able to prove that the materials came to the site in that condition.

Once the shipping ticket is checked and signed, the crew leader should give the original or a copy to the superintendent or project manager. The shipping ticket will then be filed for future reference because it serves as the only record the company has to check bills received from the supply house.

9.1.4 Controlling Delivery and Storage

Another very important element of materials control is where the materials will be stored on the job site. There are two factors in determining the appropriate storage location. The first is convenience. If possible, the materials should be stored near where they will be used. The time and effort saved by not having to carry the materials long distances will greatly reduce the installation costs.

Next, the materials must be stored in a secure area where they will not be damaged. It is important that the storage area suit the materials being stored. For instance, materials that are sensitive to temperature, such as chemicals or paints, should be stored in climate-controlled areas. Otherwise, waste may occur.

9.1.5 Preventing Theft and Vandalism

Theft and vandalism of construction materials increase costs because these materials are needed to complete the job. The replacement of materials and the time lost because the needed materials are missing can add significantly to the cost. In addition, the insurance that the contractor purchases will increase in cost as the theft and vandalism rate grows.

The best way to avoid theft and vandalism is a secure job site. At the end of each work day, store unused materials and tools in a secure location, such as a locked construction trailer. If the job site is fenced or the building can be locked, the materials can be stored within. Many sites have security cameras and/or intrusion alarms to help minimize theft and vandalism.

9.2.0 Equipment Control

The crew leader may not be responsible for long-term equipment control. However, the equipment required for a specific job is often the crew leader's responsibility. The first step is to identify when the required equipment must be transported from the shop or rental yard. The crew leader is responsible for informing the shop where it is being used and seeing that it is returned to the shop when the job is done.

It is common for equipment to lay idle at a job site because the job has not been properly planned and the equipment arrived early. For example, if wire-pulling equipment arrives at a job site before the conduit is in place, this equipment will be out of service while awaiting the conduit installation. In addition, it is possible that this unused equipment could be damaged, lost, or stolen.

The crew leader needs to control equipment use, ensure that the equipment is operated in accordance with its design, and that it is being used within time and cost guidelines. The crew leader must also ensure that equipment is maintained and repaired as indicated by the preventive maintenance schedule. Delaying maintenance and repairs can lead to costly equipment failures. The crew leader must also ensure that the equipment operators have the necessary credentials to operate the equipment, including applicable licenses.

The crew leader is responsible for the proper operation of all other equipment resources, including cars and trucks. Reckless or unsafe operation of vehicles will likely result in damaged equipment and a delayed or unproductive job. This, in turn, could affect the crew leader's job security.

The crew leader should also ensure that all equipment is secured at the close of each day's work in an effort to prevent theft. If the equipment is still being used for the job, the crew leader should ensure that it is locked in a safe place; otherwise, it should be returned to the shop.

9.3.0 Tool Control

Among companies, various policies govern who provides hand and power tools to employees. Some companies provide all the tools, while others furnish only the larger power tools. The crew leader should find out about and enforce any company policies related to tools.

Tool control is a twofold process. First, the crew leader must control the issue, use, and maintenance of all tools provided by the company. Next, the crew leader must control how the tools are being used to do the job. This applies to tools that are issued by the company as well as tools that belong to the workers.

Using the proper tools correctly saves time and energy. In addition, proper tool use reduces the chance of damage to the tool being used. Proper use also reduces injury to the worker using the tool, and to nearby workers.

Tools must be adequately maintained and properly stored. Making sure that tools are cleaned, dried, and lubricated prevents rust and ensures that the tools are in the proper working order.

In the event that tools are damaged, it is essential that they be repaired or replaced promptly. Otherwise, an accident or injury could result.

> **NOTE**
>
> Regardless of whether a tool is owned by a worker or the company, OSHA will hold the company responsible for it when it is used on a job site. The company can be held accountable if an employee is injured by a defective tool. Therefore, the crew leader needs to be aware of any defects in the tools the crew members are using.

Company-issued tools should be taken care of as if they are the property of the user. Workers should not abuse tools simply because they are not their own.

One of the major causes of time lost on a job is the time spent searching for a tool. To prevent this from occurring, a storage location for company-issued tools and equipment should be established. The crew leader should make sure that all company-issued tools and equipment are returned to this designated location after use. Similarly, workers should make sure that their personal toolboxes are organized so that they can readily find the appropriate tools and return their tools to their toolboxes when they are finished using them.

Studies have shown that the key to an effective tool control system lies in:

- Limiting the number of people allowed access to stored tools
- Limiting the number of people held responsible for tools
- Controlling the ways in which a tool can be returned to storage
- Making sure tools are available when needed

9.4.0 Labor Control

Labor typically represents more than half the cost of a project, and therefore has an enormous impact on profitability. For that reason, it is essential to manage a crew and their work environment in a way that maximizes their productivity. One of the ways to do that is to minimize the time spent on unproductive activities such as:

- Engaging in bull sessions
- Correcting drawing errors
- Retrieving tools, equipment, and materials
- Waiting for other workers to finish

If crew members are habitually goofing off, it is up to the crew leader to counsel those workers. The counseling should be documented in the crew leader's daily diary. Repeated violations will need to be referred to the attention of higher management as guided by company policy.

Errors will occur and will need to be corrected. Some errors, such as mistakes on drawings, may be outside of the crew leader's control. However, some drawing errors can be detected by carefully examining the drawings before work begins.

If crew members are making mistakes due to inexperience, the crew leader can help avoid these errors by providing on-the-spot training and by checking on inexperienced workers more often.

The availability and location of tools, equipment, and materials can have a profound effect on a crew's productivity. If the crew has to wait for these things, or travel a distance to get them, it reduces their productivity. The key to minimizing such problems is proactive management of these resources. As discussed earlier, practices such as checking in advance to be sure equipment and materials will be available when scheduled and placing materials close to the work site will help minimize unproductive time.

Delays caused by others can be avoided by carefully tracking the project schedule. By doing so, crew leaders can anticipate delays that will affect the work of their crews and either take action to prevent the delay or redirect the crew to another task.

Participant Exercise G

1. List the methods your company uses to minimize waste.

_____

_____

_____

_____

2. List the methods your company uses to control small tools on the job.

_____

_____

_____

_____

3. List five ways that you feel your company could control labor to maximize productivity.

_____

_____

_____

_____

 Fundamentals of Crew Leadership

10.0.0 PRODUCTION AND PRODUCTIVITY

Production is the amount of construction put in place. It is the quantity of materials installed on a job, such as 1,000 linear feet of waste pipe installed in a given day. On the other hand, productivity depends on the level of efficiency of the work. It is the amount of work done per hour or day by one worker or a crew.

Production levels are set during the estimating stage. The estimator determines the total amount of materials to be put in place from the plans and specifications. After the job is complete, the actual amount of materials installed can be assessed, and the actual production can be compared to the estimated production.

Productivity relates to the amount of materials put in place by the crew over a certain time period. The estimator uses company records during the estimating stage to determine how much time and labor it will take to place a certain quantity of materials. From this information, the estimator calculates the productivity necessary to complete the job on time.

For example, it might take a crew of two people ten days to paint 5,000 square feet. The crew's productivity per day is obtained by dividing 5,000 square feet by ten days. The result is 500 square feet per day. The crew leader can compare the daily production of any crew of two painters doing similar work with this average, as discussed previously.

Planning is essential to productivity. The crew must be available to perform the work, and have all of the required materials, tools, and equipment in place when the job begins.

The time on the job should be for business, not for taking care of personal problems. Anything not work-related should be handled after hours, away from the job site. Planning after-work activities, arranging social functions, or running personal errands should be handled after work or during breaks.

Organizing field work can save time. The key to effectively using time is to work smarter, not necessarily harder. For example, most construction projects require that the contractor submit a set of as-built plans at the completion of the work. These plans describe how the materials were actually installed. The best way to prepare these plans is to mark a set of working plans as the work is in progress. That way, pertinent details will not be forgotten and time will not be wasted trying to remember how the work was done.

The amount of material actually used should not exceed the estimated amount. If it does, either the estimator has made a mistake, undocumented changes have occurred, or rework has caused the need for additional materials. Whatever the case, the crew leader should use effective control techniques to ensure the efficient use of materials.

When bidding a job, most companies calculate the cost per labor hour. For example, a ten-day job might convert to 160 labor hours (two painters for ten days at eight hours per day). If the company charges a labor rate of $30/hour, the labor cost would be $4,800. The estimator then adds the cost of materials, equipment, and tools, along with overhead costs and a profit factor, to determine the price of the job.

After a job has been completed, information gathered through field reporting allows the home office to calculate actual productivity and compare it to the estimated figures. This helps to identify productivity issues and improves the accuracy of future estimates.

The following labor-related practices can help to ensure productivity:

- Ensure that all workers have the required resources when needed.
- Ensure that all personnel know where to go and what to do after each task is completed.
- Make reassignments as needed.
- Ensure that all workers have completed their work properly.

1. Which of these activities occurs during the development phase of a project?

    a. Architect/engineer sketches are prepared and a preliminary budget is developed.
    b. Government agencies give a final inspection of the design, adherence to codes, and materials used.
    c. Project drawings and specifications are prepared.
    d. Contracts for the project are awarded.

2. The type of contract in which the client pays the contractor for their actual labor and material expenses they incur is known as a _____.

    a. firm fixed-price contract
    b. time-spent contract
    c. cost-reimbursable contract
    d. performance-based contract

3. On-site changes in the original design that are made during construction are recorded in the _____.

    a. as-built plans
    b. takeoff sheet
    c. project schedule
    d. job specifications

4. On a design-build project, _____.

    a. the owner is responsible for providing the design
    b. the architect does the design and the general contractor builds the project
    c. the same contractor is responsible for both design and construction
    d. a construction manager is hired to oversee the project

5. One example of a direct cost when bidding a job is _____.

    a. office rent
    b. labor
    c. accounting
    d. utilities

6. The control method that a crew leader uses to plan a few weeks in advance is a _____.

    a. network schedule
    b. bar chart schedule
    c. daily diary
    d. look-ahead schedule

7. A job diary should typically indicate _____.

    a. items such as job interruptions and visits
    b. changes needed to project drawings
    c. the estimated time for each job task related to a particular project
    d. the crew leader's ideas for improving employee morale

8. Gantt charts can help crew leaders in the field by _____.

    a. offering a comparison of actual production to estimated production
    b. providing short-term and long-term schedule information
    c. stating the equipment and materials necessary to complete a task
    d. providing the information needed to develop an estimate or an estimate breakdown

9. What is the crew leader's responsibility with regard to cost control?

    a. Cost control is outside the scope of a crew leader's responsibility.
    b. The crew leader is responsible only for minimizing material waste.
    c. The crew leader must ensure that all team members are aware of project costs and how to control them.
    d. The crew leader typically prepares the company's cost estimate and is therefore responsible for cost performance.

10. Which of the following is a correct statement regarding project cost?

    a. Cost is handled by the accounting department and is not a concern of the crew leader.
    b. A company's profit on a project is affected by the difference between the estimated cost and the actual cost.
    c. Wasted material is factored into the estimate, and is therefore not a concern.
    d. The contractor's overhead costs are not included in the cost estimate.

11. The crew leader is responsible for ensuring that equipment used by his or her crew is properly maintained.

    a. True
    b. False

12. Who is responsible if a defect in an employee's tool results in an accident?

    a. The employee
    b. The company
    c. The crew leader
    d. The tool manufacturer

13. Productivity is defined as the amount of work accomplished.

    a. True
    b. False

14. If a crew of masons is needed to lay 1,000 concrete blocks, and the estimator determined that two masons could complete the job in one eight-hour day, what is the estimated productivity rate?

    a. 125 blocks per hour
    b. 62.5 blocks per hour
    c. 31.25 blocks per hour
    d. 16 blocks per hour

Additional Resources

This module presents thorough resources for task training. The following resource material is suggested for further study.

Aging Workforce News, www.agingworkforce-news.com.

American Society for Training and Development (ASTD), www.astd.org.

Architecture, Engineering, and Construction Industry (AEC), www.aecinfo.com.

CIT Group, www.citgroup.com.

Equal Employment Opportunity Commission (EEOC), www.eeoc.gov.

National Association of Women in Construction (NAWIC), www.nawic.org.

National Census of Fatal Occupational Injuries (NCFOI), www.bls.gov.

National Center for Construction Education and Research, www.nccer.org.

National Institute of Occupational Safety and Health (NIOSH), www.cdc.gov/niosh.

National Safety Council, www.nsc.org.

NCCER Publications:

Your Role in the Green Environment

Sustainable Construction Supervisor

Occupational Safety and Health Administration (OSHA), www.osha.gov.

Society for Human Resources Management (SHRM), www.shrm.org.

United States Census Bureau, www.census.gov.

United States Department of Labor, www.dol.gov.

USA Today, www.usatoday.com.

Figure Credits

© 2010 Photos.com, a division of Getty Images. All rights reserved., Module opener

United States Department of Labor, Occupational Safety and Health Administration, SA01

plaquemaster.com, Figure 9

© U.S. Green Building Council, SA02

Sushil Shenoy, SA03 and Figure 22

John Ambrosia, Figure 19

CONTREN® LEARNING SERIES — USER UPDATE

NCCER makes every effort to keep its textbooks up-to-date and free of technical errors. We appreciate your help in this process. If you find an error, a typographical mistake, or an inaccuracy in NCCER's Contren® materials, please fill out this form (or a photocopy), or complete the online form at www.nccer.org/olf. Be sure to include the exact module number, page number, a detailed description, and your recommended correction. Your input will be brought to the attention of the Authoring Team. Thank you for your assistance.

Instructors – If you have an idea for improving this textbook, or have found that additional materials were necessary to teach this module effectively, please let us know so that we may present your suggestions to the Authoring Team.

NCCER Product Development and Revision

3600 NW 43rd Street, Building G, Gainesville, FL 32606

Fax: 352-334-0932
Email: curriculum@nccer.org
Online: www.nccer.org/olf

❏ Trainee Guide   ❏ AIG   ❏ Exam   ❏ PowerPoints   Other _____

Craft / Level: _____ Copyright Date: _____

Module Number / Title: _____

Section Number(s): _____

Description: _____

_____

_____

_____

_____

Recommended Correction: _____

_____

_____

_____

_____

Your Name: _____

Address: _____

_____

Email: _____ Phone: _____

Rack Assembly

33305-11

Trainees with successful module completions may be eligible for credentialing through NCCER's National Registry. To learn more, go to **www.nccer.org** or contact us at **1.888.622.3720**. Our website has information on the latest product releases and training, as well as online versions of our *Cornerstone* newsletter and Pearson's Contren® product catalog.

Your feedback is welcome. You may email your comments to **curriculum@nccer.org**, send general comments and inquiries to **info@nccer.org**, or use the User Update form at the back of this module.

 V.1 4/11

RACK ASSEMBLY

Objectives

When you have completed this module, you will be able to do the following:

1. Identify various types of electronic equipment racks.
2. Select the appropriate rack for a given application.
3. Establish proper grounding of racks to ensure personnel safety and minimize signal interference.
4. Explain proper ventilation techniques to avoid overheating of rack-mounted electronic equipment.
5. Describe the installation practices for rack-mounted equipment.
6. Prepare a rack layout drawing.
7. Calculate power requirements and heat dissipation requirements for a rack installation.
8. Assemble a rack, including lacing rails.
9. Populate a rack with equipment.

Performance Tasks

Under the supervision of the instructor, you should be able to do the following:

1. Select a rack unit for a given application.
2. Prepare a rack layout drawing.
3. Calculate power requirements and Btu dissipation for a rack installation.
4. Properly install electronic equipment in a rack.
5. Assemble a rack, including lacing rails.

Trade Terms

Backplane
British thermal units per
  hour (Btu/h)

Convection
Ground loops

Headroom
Rack unit (RU)

Separately derived system
Static pressure

Contents

Topics to be presented in this module include:

Figures and Tables

Figures and Tables (continued)

1.0.0 INTRODUCTION

Complex electronic systems require many equipment components. In order to organize and protect the equipment, manage cabling effectively, and use space efficiently, this equipment is usually housed in equipment racks designed for this purpose (*Figure 1*).

One common use for equipment racks is to house equipment in computer data centers, which are also known as server farms. Server farms typically contain a large number of PCs. In comparison to common office PCs, the PCs used as servers are configured with high-reliability components to reduce the risk of failure. In addition, they usually have larger, much faster hard drives. The equipment in these centers is commonly used for high-volume data processing, data storage, and backup. Data centers also may contain network management equipment such as routers and switches that provide access to users outside the center. Some data centers have hundreds of servers housed in air-conditioned equipment racks. These centers are used by institutions, banks and financial services, government agencies, and large businesses. They may contain classified information, personnel records, or individual financial and personal data. Because of that, security is very important. Racks used to house such servers are designed with special security features, including heavy-gauge frames and lockable doors (*Figure 2*).

Racks used to house telecommunications equipment (*Figure 3*) are of a different design. They are typically not as deep as server racks and do not generally have the security features needed for servers. Telecommunications equipment rooms (*Figure 4*) contain the routers and switching devices needed to manage phone and network equipment.

Many years ago, the electronics industry decided to standardize the size of commercial equipment so that it could be mounted in standard equipment racks. Standard widths of 19", 24", and 30" were established for equipment panels and equipment racks. The 19" rack is the universally accepted size for audio equipment. Racks are equipped with mounting rails to which the equipment components or panels are attached.

The overwhelming majority of the industry uses the standard 19" racks, which are available from many manufacturers. Occasionally, specialized equipment cannot be manufactured to the standard. In those cases, custom racks are created, or a standard rack is adapted using accessories made for that purpose. As an example, *Figure 5* shows a VCR mounted in a special carrier.

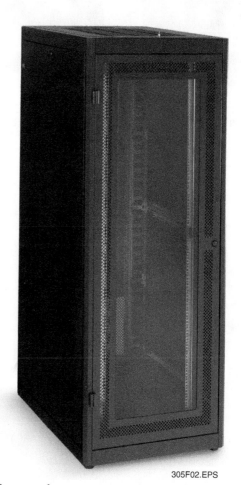

305F02.EPS

Figure 2 Server rack.

305F01.EPS

Figure 1 Equipment racks.

Although the 19" rack is the most common, there is also a standard 23" rack used primarily for telecommunications and data network equipment. Racks are available in a variety of depths to accommodate the wide variety of electronic devices on the market.

In this module, we will focus on racks used in commercial applications. However, there are also racks designed for home use, such as the one shown in *Figure 6*.

2.0.0 TYPES OF RACKS

Many kinds of racks have been designed to meet a variety of needs. The types covered here include the following:

- Free-standing vertical racks
- Consoles
- Wall-mount racks
- Mobile racks
- Sliding racks
- NEMA enclosures

305F05.EPS

Figure 5 VCR mounted in rack adapter.

FRONT VIEW
PLEXI GLASS DOOR
SOLID SIDE PANEL

REAR VIEW
SPLIT MESH DOOR
VENTED SIDE PANEL

305F03.EPS

Figure 3 Telecommunications equipment rack.

305F04.EPS

Figure 4 Telecommunications equipment room.

305F06.EPS

Figure 6 Rack system for home use.

Free-standing racks may be either open or closed (*Figure 7*). Open racks can be used in rooms where the equipment is not exposed to heavy foot traffic or airborne dirt. An example would be an equipment room accessed primarily by equipment technicians and operators who are likely to respect the equipment. Some open racks are not self-supporting and may need to be bolted to the floor to ensure equipment and personnel safety.

Closed free-standing racks have closed sides. They can also be equipped with optional doors for protection of equipment and people. Free-standing racks are used in applications where there is room to get behind the unit for maintenance and troubleshooting. Free-standing racks can be either the welded type or the knockdown type. The latter is shipped in four sections (top, bottom, and two sides), along with equipment mounting hardware. Knockdown racks are less expensive and easier to ship and handle, but require more labor at the work site because they must be assembled.

Console racks are designed primarily for display and control equipment, although many consoles accommodate other electronic equipment as well. They are available in a variety of forms. Equipment mounting rails (*Figure 8*) are incorporated into the console for convenient mounting of equipment.

Operator consoles like the one shown in *Figure 9* are designed for ease of viewing. Upper sections are tilted downward, while lower sections are tilted slightly upward.

Wall-mount racks (*Figure 10*), as the name implies, are designed to be attached to a wall. There is no direct access to the rear of the rack as there is with free-standing racks. For that reason, wall-mount racks consist of two sections: a backplane, also called a backbox, that is attached to the wall; and a mounting section that is hinged so it can be swung away from the backplane. The mounting section can be hinged to open on either side. Front doors are optional, as they are with free-standing racks. The backplane has knockouts for cable and conduit access at top and bottom. The mounting section may have knockouts for ventilating fans and/or antennas. When planning an installation, it is important to leave enough clearance to allow the rack to be fully opened.

(A) ENCLOSED RACK   (B) OPEN RACK   (C) OPEN FRAME

305F07.EPS

Figure 7 Free-standing racks.

Mobile or portable racks are designed specifically to be moved from place to place. A mobile rack can be as basic as a frame on wheels or as complex as a ruggedized container designed for shipping electronic equipment from city to city by air, truck, or rail. The interior of this shipping container is padded with foam to provide protection against shock. Additional foam inserts are used to protect equipment components during shipping.

The rolling cabinet (*Figure 11*) is used to house equipment that may have to be moved from one location to another within a building or complex. It does not have foam padding.

Like other racks that have been covered, all these containers have equipment mounting rails for mounting standard electronic components.

Sliding or slideout racks (*Figure 12*) contain drawer slides that allow electronic components to be pulled out from the housing. In some cases, the entire rack can be pulled out. In others, a selected segment of the rack can be pulled out. The slideout rack is an alternative to the wall-mount rack.

The National Electrical Manufacturers Association (NEMA) publishes manufacturing standards for electrical equipment. The expressed purpose of NEMA with regard to standards is to provide "a forum for the standardization of electrical equipment, enabling customers to select from a range of safe, effective, and compatible electrical products." *NEMA Publication 250-2008 Enclosures for Electrical Equipment*, identifies 13 types of enclosures designed to meet various requirements in non-hazardous indoor or outdoor environments. This publication also identifies four enclosure types for hazardous locations.

305F09.EPS

Figure 9 Console rack application.

305F08.EPS

Figure 8 Console rack with equipment mounting rails.

(A)

(B)

305F10.EPS

Figure 10 Wall-mount racks.

3.0.0 RACK DIMENSIONS

The height of an equipment panel is measured in rack units (RU). A rack unit is 1.75". Therefore, the smallest panel is 1.75" high. The largest panel allowed is 5 RU, or 8.75". Standard racks are designed to accept a 19"- or 23"-wide equipment panel. The panels are attached to predrilled mounting rails. *Figure 13* shows the typical hole spacing for standard mounting rails.

A distinction needs to be made between the terms *equipment* and *panel*. Equipment means an equipment package, sometimes referred to as a chassis, with an extended front panel that allows it to be attached to standard mounting rails in equipment racks. The term *panel* can apply to blank panels, baffle panels, and vent panels.

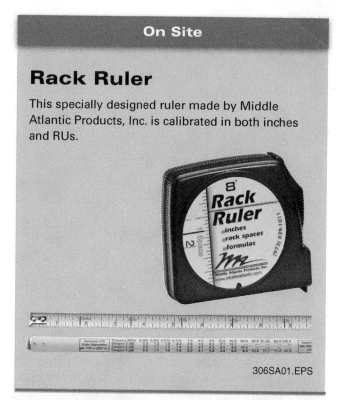

305F11.EPS

Figure 11 Examples of mobile racks.

Blank panels (*Figure 14*) are simply inserts installed to cover unused spaces. Filling the opening with blank panels gives the installation a more professional appearance and avoids problems with ventilation airflow. If future expansion is likely, small blank panels are better than large ones. Large panels may have to be discarded when equipment is added. In any case, the height of a blank panel should not exceed 5 RU.

305F12.EPS

Figure 12 Slideout rack.

Vent panels have louvers cut into them to allow ventilating air to be pulled into the unit. A baffle panel is like a vent panel, but it has a rear baffle that traps heat and expels it through the louvers on the front of the panel.

4.0.0 POWER AND GROUNDING

Equipment racks receive AC power from the building distribution system. Because line voltage is involved, safety is a primary concern. Proper grounding of the rack and its components is essential to avoid electrocution hazards and to protect equipment from stray voltage and electrostatic discharge (ESD). AC voltage is likely to be anywhere in the rack. In some cases, rack equipment is plugged directly into AC outlets mounted on power strips in the rack. In others, the AC power is hard-wired directly to the rack equipment, which contains its own power supplies, to produce the DC voltages needed to operate the unit. In still other situations, the AC line voltage is supplied to a power supply mounted in the rack. The power supply provides DC voltages to other equipment in the rack.

In audio and video equipment, static, buzzing, hums, and signal distortion can be caused by improper grounding and by ground loops. For that reason, special precautions must be taken to ensure that the equipment is properly grounded and that there is a single, effective ground path.

On Site

Invention of the Battery

A disagreement with another scientist led Alessandro Volta to the discovery of the battery, which originally was known as the voltaic pile. Volta believed that when metal objects came together, they would produce an electric current. To demonstrate his theory, Volta placed two discs, one of zinc and the other of silver, into a weak acidic solution. When he connected the discs with wire, a current flowed through the wire, and the battery was born.

NCCER – *Contren® Learning Series*   33305-11

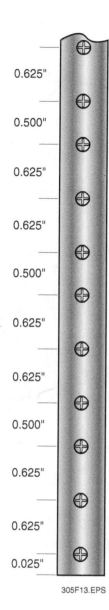

0.625"
0.500"
0.625"
0.625"
0.625"
0.500"
0.625"
0.625"
0.625"
0.500"
0.625"
0.625"
0.625"
0.025"

305F13.EPS

Figure 13 Hole spacing for standard equipment mounting rails.

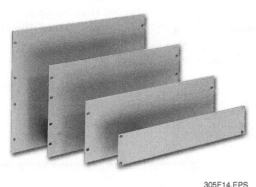

305F14.EPS

Figure 14 Examples of blank panels.

4.1.0 Power

As *Figure 15* shows, the AC voltage used to power electronic equipment originates as a very high voltage at the power generating plant. The voltage is stepped down by a series of transformers to obtain the levels needed for operation.

Residential and light commercial applications use single-phase AC voltage. An Edison hookup (*Figure 16*) is a common wiring arrangement for the power transformer that delivers the single-phase voltage. The neutral line is connected to the building ground, which may be a copper water pipe that runs underground, and/or a copper-clad steel rod driven into the ground. The two hot legs are commonly designated L1 and L2. The transformer secondary is center-tapped, and this leg is called the neutral. The neutral is electrically grounded at the power transformer. Other grounds exist at the electric meter or the main service panel, depending on local codes. The grounding conductor (ground) and the grounded conductor (neutral) are electrically common, but are bonded together only at the main service panel, or at the source of a separately derived system. They are not to be connected on the load side of the main panel.

Three-phase power is common in commercial and industrial applications. The 120-volt power needed to operate the electronic equipment is tapped from the three-phase source. Three-phase power is connected in several ways. The type used in a given situation depends on the amount of voltage and current needed. The common three-phase distribution methods are: the four-wire closed delta, the four-wire open delta, and the four-wire wye.

> **NOTE**
>
> The majority of the material in the remainder of this section was adapted, with permission, from a technical paper provided by Middle Atlantic Products, Inc. See the references page at the end of this module for complete information on this source.

4.1.1 *Load Estimating*

If an electrical load is normally operated for three hours or more, it is considered continuous operation. In that case, the wiring and over-current protection must be sized for 125 percent of the load. The *NEC®* permits 100 percent sizing for non-continuous operation, but that is not recommended because of the headroom needed to faithfully reproduce peaks in amplifiers. *Figure 17* shows an example of wire and circuit breaker sizing for a given load.

Rack Assembly

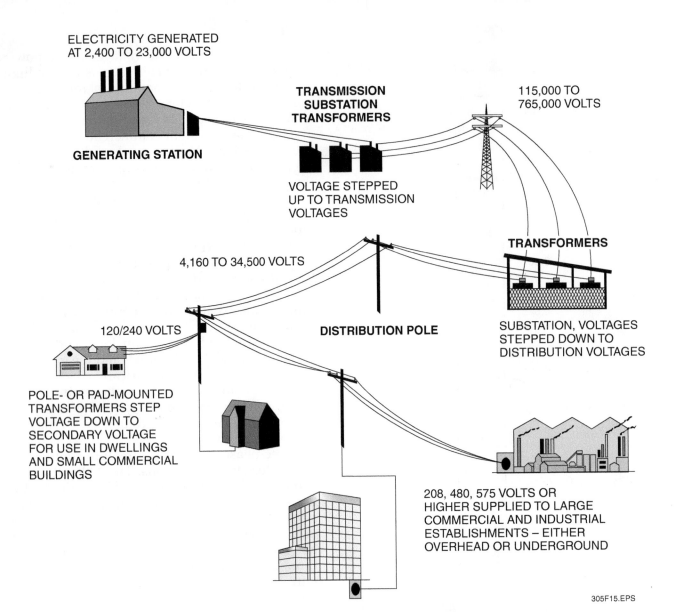

ELECTRICITY GENERATED
AT 2,400 TO 23,000 VOLTS

GENERATING STATION

TRANSMISSION
SUBSTATION
TRANSFORMERS

VOLTAGE STEPPED
UP TO TRANSMISSION
VOLTAGES

115,000 TO
765,000 VOLTS

TRANSFORMERS

4,160 TO 34,500 VOLTS

DISTRIBUTION POLE

SUBSTATION, VOLTAGES
STEPPED DOWN TO
DISTRIBUTION VOLTAGES

120/240 VOLTS

POLE- OR PAD-MOUNTED
TRANSFORMERS STEP
VOLTAGE DOWN TO
SECONDARY VOLTAGE
FOR USE IN DWELLINGS
AND SMALL COMMERCIAL
BUILDINGS

208, 480, 575 VOLTS OR
HIGHER SUPPLIED TO LARGE
COMMERCIAL AND INDUSTRIAL
ESTABLISHMENTS – EITHER
OVERHEAD OR UNDERGROUND

305F15.EPS

Figure 15 AC power distribution.

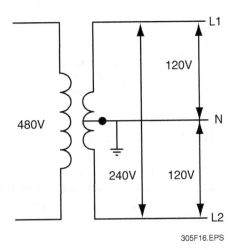

305F16.EPS

Figure 16 Edison hookup.

Other factors to be considered in sizing wire include length of run, ambient temperature, and the number of conductors in the conduit.

Amplifier loads are especially tricky because they depend on the connection method and the application. *Table 1* shows the variations in current draw for the same amplifier, depending on use and output.

> **NOTE**
>
> Gross oversizing of branch circuits may be restricted by the *NEC®*. Consult the manufacturer's literature, and be aware that modifying or changing the line cord connector in a manner that conflicts with the installation instructions may affect the warranty and UL listing.

NCCER – *Contren® Learning Series* 33305-11

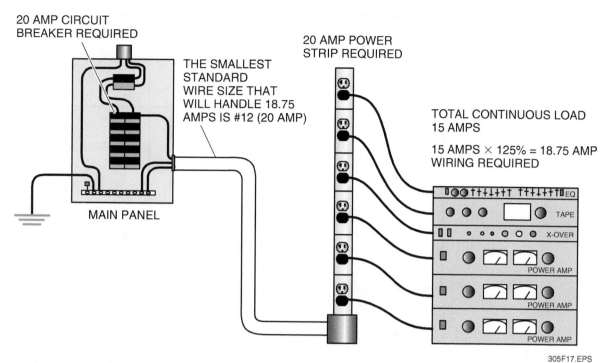

20 AMP CIRCUIT BREAKER REQUIRED

THE SMALLEST STANDARD WIRE SIZE THAT WILL HANDLE 18.75 AMPS IS #12 (20 AMP)

MAIN PANEL

20 AMP POWER STRIP REQUIRED

TOTAL CONTINUOUS LOAD 15 AMPS

15 AMPS × 125% = 18.75 AMP WIRING REQUIRED

305F17.EPS

Figure 17 Circuit sizing example.

Table 1 Example of Amplifier Load Variations

Program Type	Speaker Ohms (Stereo)	Current Draw (Amps)
Individual speech	8	4.1
Individual speech	2	5.8
Compressed rock music	8	13.4
Compressed rock music	2	20.1

4.1.2 Common Wiring Errors

The wiring errors shown in *Figure 18* may not easily be detected, so you should always check the outlets with a meter. The equipment may seem to work okay, but there may be electrical noise. In addition, these errors can create electrical hazards.

If neutral and ground are reversed, it can be detected with a clamp-on ammeter installed on the ground conductor with a load on the circuit. No current should flow on the grounding conductor if the wiring is correct. Reversal of the neutral and hot conductors can be detected using a voltmeter. The reading between neutral and ground should be zero. Between the hot lead and ground, the meter should read 120 volts.

4.1.3 Surge Suppression and EMI Filter

Metal oxide varistors (MOVs) conduct electricity when a predetermined voltage, usually around 300 volts, is reached. Surges and spikes are normally well over this avalanche voltage. When a surge exceeds the avalanche voltage, the MOV

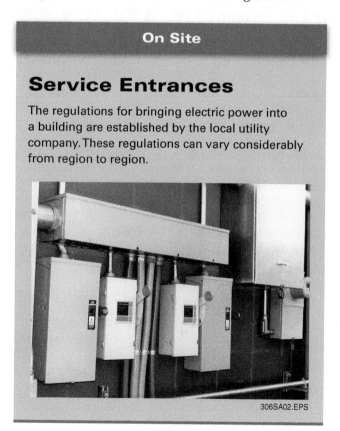

On Site

Service Entrances

The regulations for bringing electric power into a building are established by the local utility company. These regulations can vary considerably from region to region.

306SA02.EPS

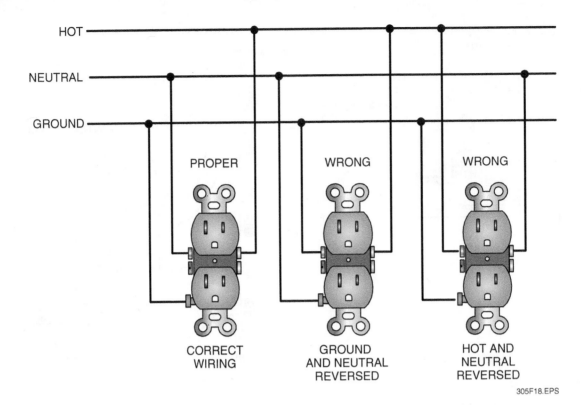

HOT

NEUTRAL

GROUND

PROPER

WRONG

WRONG

CORRECT
WIRING

GROUND
AND NEUTRAL
REVERSED

HOT AND
NEUTRAL
REVERSED

305F18.EPS

Figure 18 Common wiring errors.

conducts, shorting the surge to ground. An individual MOV has limited capacity. Therefore, several MOVs may be wired in parallel (*Figure 19*) to increase capacity.

A disadvantage of the MOV is that it provides a path for high-frequency noise (EMI) during normal operation. For that reason, a capacitor or filter network can be added to help filter out the EMI. However, some installations with high-gain structures may still be affected by this noise path.

4.1.4 Balanced Power Transformer

One way to reduce noise is to use a balanced power transformer (*Figure 20*). In a 60/120 power system, there is no neutral, so the return path is not a grounded conductor. Most noise is induced on the return side and hot side at the same time, but because they are out of phase, the noise is cancelled. The 60/120 system is an effective but expensive way of reducing noise.

4.1.5 Rack Power

Power for equipment mounted in racks typically comes from power strips mounted in the rack. Listed power strips must be used; there are different types of power strips. The power panel (*Figure 21*) is attached to the rack rails just like an

equipment chassis. It provides front panel control of power and is used for situations where power control is needed at the rack. Equipment components are plugged in at the back of the panel. The power strip (*Figure 22*) is designed to be mounted inside the rack. Supply wiring to the power strip or panel can be done with a molded plug, junction box, or flexible metal conduit whip.

4.2.0 Grounding

Most medium to large video, audio, and data system installations will encounter low-voltage signal interference caused by grounding problems. Equipment connected by coaxial cable in rack-mounted video head-ends and rack-mounted distribution amplifiers are susceptible to interference caused by ground loops. In video, this is usually a 60-cycle hum that shows up as slowly rolling, wide horizontal bands (hum bars). In audio systems, interference may show up as a hum or buzz, white noise, or popping sounds. In data systems, it causes bit errors that render the system less effective.

Ground loops can be caused when equipment and enclosures are grounded at two or more points that are at different electrical potentials and are not bonded together. This section covers grounding and bonding methods that will help prevent ground loops.

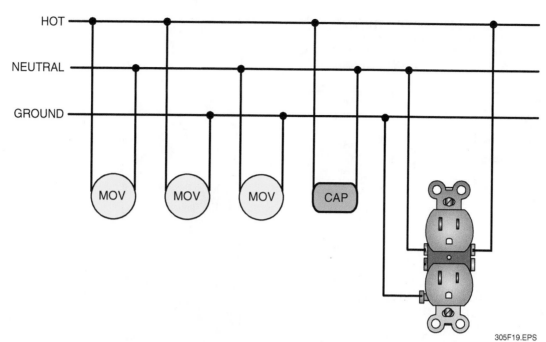

Figure 19 Surge suppression circuit with filter capacitor.

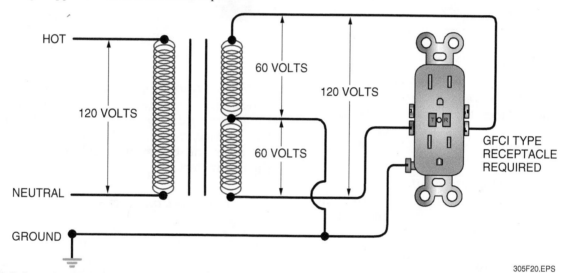

Figure 20 Balanced power transformer.

305F21.EPS

Figure 21 Power panel.

4.2.1 Separately Derived System

A separately derived system (*Figure 23*) is one in which a transformer is used to isolate or step down the voltage from the main power source. In such cases, the secondary service panel is considered the main panel when terminating isolated grounds. The system ground is derived from the transformer secondary, which must be grounded locally.

305F22.EPS

Figure 22 Power strips.

 Rack Assembly

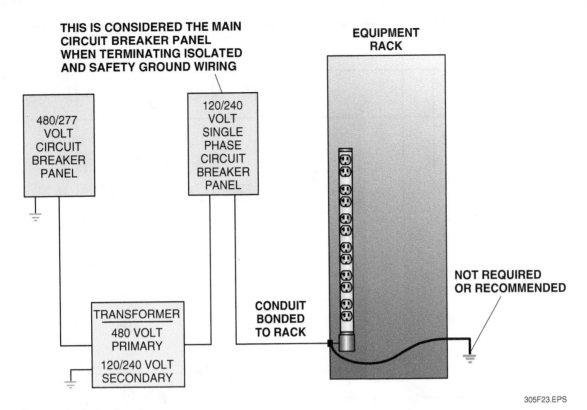

THIS IS CONSIDERED THE MAIN
CIRCUIT BREAKER PANEL
WHEN TERMINATING ISOLATED
AND SAFETY GROUND WIRING

EQUIPMENT
RACK

480/277
VOLT
CIRCUIT
BREAKER
PANEL

120/240
VOLT
SINGLE
PHASE
CIRCUIT
BREAKER
PANEL

TRANSFORMER
480 VOLT
PRIMARY
120/240 VOLT
SECONDARY

CONDUIT
BONDED
TO RACK

NOT REQUIRED
OR RECOMMENDED

305F23.EPS

Figure 23 Separately derived system.

NOTE

Metallic conduit is considered to be a grounding conductor; however, a supplemental ground wire in the conduit is strongly recommended.

A supplemental safety ground via a ground rod or building steel is permitted by the *NEC®*, but it must be bonded to the equipment ground coming from the service panel. Such a grounding system is not recommended because it may cause ground loops if the main grounding system is not properly terminated.

WARNING

Never bond an isolated ground wire to a ground rod or building steel.

4.2.2 Isolated Racks

The *NEC®* requires that all enclosures and raceways be effectively grounded (*Figure 24*). Although isolated grounds are permitted, this does not relieve the requirement to ground equipment and raceways. The isolated ground may be used in place of the normal safety ground, but may not be bonded to the safety ground, except at the service panel. The rack is isolated by using a non-conducting electrical fitting to connect the conduit to the rack.

When terminating a rack ground, remove nonconductive coatings or use a bonding jumper. If the rack is sitting on concrete or another conductive surface, make sure the rack is isolated from the floor and the mounting bolts with non-conductive material.

4.2.3 Flexible Connections

In the examples in *Figure 25*, both racks are isolated from the building safety ground. In example #2, however, the rack is not properly grounded because there is no definite bonding connection from the rack to the power strip. The power strip chassis mounting clips cannot be relied on to conduct fault current.

4.2.4 Isolated-Ground Power Strip

The conduit may contain both the supplemental equipment ground and the isolated outlet ground. Bond the rack and all the equipment to the normal equipment ground and use an isolated-ground power strip (*Figure 26*). The isolated-ground wire is connected to only the outlet ground screw. The power strip chassis and rack are to be bonded to the equipment ground.

A problem can occur when the power cord ground conductor is bonded to the chassis in most of the utilization equipment. In such cases, a ground loop can occur between the isolated

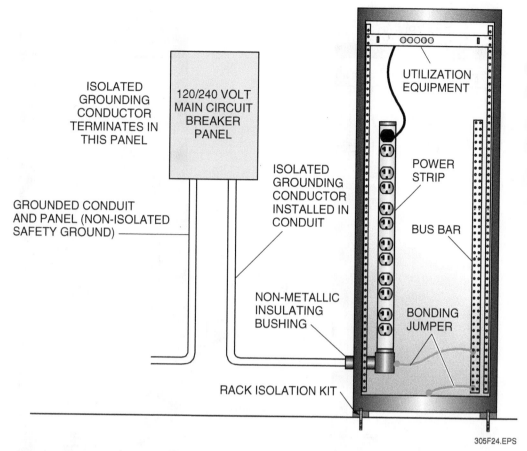

ISOLATED GROUNDING CONDUCTOR TERMINATES IN THIS PANEL

120/240 VOLT MAIN CIRCUIT BREAKER PANEL

UTILIZATION EQUIPMENT

POWER STRIP

ISOLATED GROUNDING CONDUCTOR INSTALLED IN CONDUIT

BUS BAR

GROUNDED CONDUIT AND PANEL (NON-ISOLATED SAFETY GROUND)

NON-METALLIC INSULATING BUSHING

BONDING JUMPER

RACK ISOLATION KIT

305F24.EPS

Figure 24 Completely isolated equipment rack.

ground and the safety ground. Insulating rack ears (*Figure 27*) and screws may be needed. Non-conductive shoulder washers are used to ensure that the isolated technical ground is segregated from the safety ground.

> **WARNING**
>
> Bypassing or lifting an equipment safety ground may reduce noise problems, but it is a dangerous practice and violates the *NEC®*.

4.2.5 Star Grounding Method

When racks are bonded together in a daisy chain fashion, impedances can build up at each bond connection. By the time the daisy chain reaches its end, a voltage drop may exist, causing a ground loop for induced noise currents. While this may not be an unsafe condition, it has an effect on the signal quality. Star grounding, in which each rack is connected to a common ground point (*Figure 28*), reduces the voltage drop and the additive bonding impedances.

4.2.6 Signal Reference Grounding Grid

Star grounding is effective for standard audio equipment, but a different approach may be needed for highly sensitive digital, broadband, and RF transmitting equipment. This equipment is sensitive to ground wire length as well as to high-impedance ground connections.

The signal reference grounding grid (*Figure 29*) is used in such situations. This method can be compared with an antenna system. In an antenna system, frequencies are received best by the antenna that most closely matches their wavelength.

> **On Site**
>
> ## Isolating Grounding Problems
>
> A process of elimination is a good way to determine which piece of equipment is causing noise problems. Disconnect each piece of equipment, one at a time, until the problem is located. This approach will not correct the problem, but will reduce the time it takes to isolate the problem.

Rack Assembly

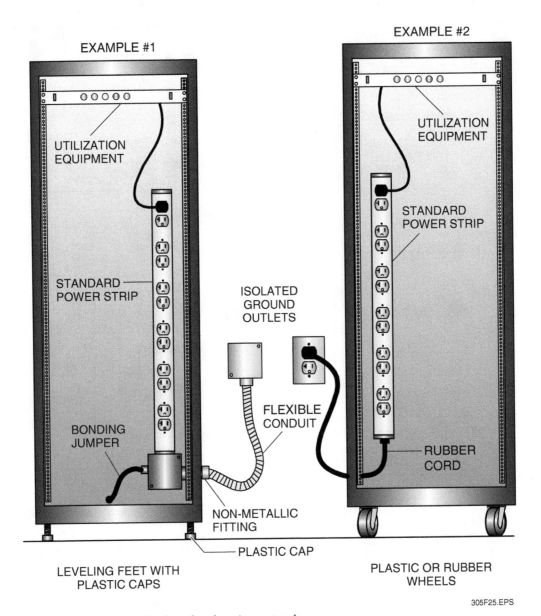

EXAMPLE #1

UTILIZATION EQUIPMENT

STANDARD POWER STRIP

BONDING JUMPER

ISOLATED GROUND OUTLETS

FLEXIBLE CONDUIT

NON-METALLIC FITTING

PLASTIC CAP

LEVELING FEET WITH PLASTIC CAPS

EXAMPLE #2

UTILIZATION EQUIPMENT

STANDARD POWER STRIP

RUBBER CORD

PLASTIC OR RUBBER WHEELS

305F25.EPS

Figure 25 Flexible connections to completely isolated equipment rack.

The same principle applies to high-frequency ground noise. That is, in a grounding grid, the noise signal will seek the path of least impedance for each of the frequencies it contains.

The grid is made of intersecting runs of #6 bare copper wire or $\frac{1}{8}$" × 2" flat copper. The pattern may be of any convenient size. Generally, squares measuring 1' × 1' or 2' × 2' are used. The ideal location for a grounding grid is under a raised floor, in which case, a 2' × 2' pattern would be the most convenient.

4.2.7 Signal Wiring

In *Figure 30*, the signal cable must be shielded at both ends. A balanced-type cable is recommended. If the outlets that feed the preamp and power supply receive their power from two dif-

ferent service panels, it may be necessary to lift the ground from one end of the signal cable in order to eliminate a ground loop. This only applies to a balanced cable. Neither end of the shield of an unbalanced cable should be disconnected.

The shield is not intended to act as a safety ground. The safety ground comes from the grounding conductor in the power cord. When grounding jumpers are used on several pieces of equipment in the same location, all grounds must terminate at the same point.

WARNING

Never lift or otherwise bypass the power cord ground. It could result in a fatal shock.

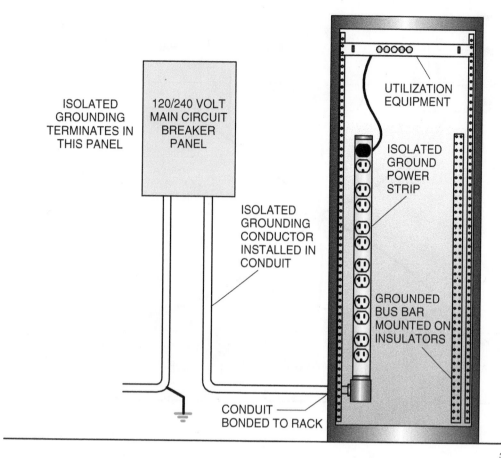

Figure 26 Isolated-ground power strip.

305F26.EPS

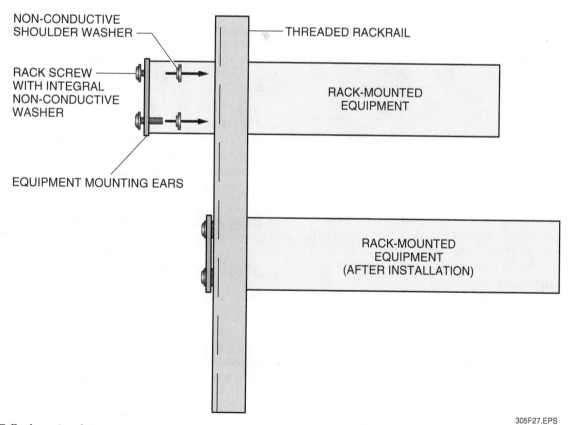

Figure 27 Rack ear insulators.

305F27.EPS

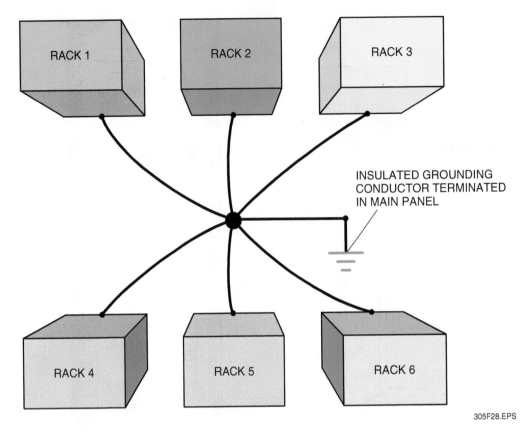

RACK 1

RACK 2

RACK 3

INSULATED GROUNDING
CONDUCTOR TERMINATED
IN MAIN PANEL

RACK 4

RACK 5

RACK 6

305F28.EPS

Figure 28 Star grounding method.

5.0.0 RACK VENTILATION AND COOLING

NOTE

The majority of the material in this section was adapted, with permission, from a technical paper provided by Middle Atlantic Products, Inc. See the references page at the end of this module for complete information on this source.

Electronic equipment generates heat as it works. This is especially true of power amplifiers when they are being driven. In a closed rack, this heat must be removed because excessive heat can damage sensitive electronic components or cause them to fail prematurely. As a rule of thumb, the heat in a rack containing computer and A/V equipment should be no higher than 85°F. Power amplifiers are less likely to be affected by high temperatures. They may operate well at 110°F.

Heat can be removed either by natural (passive) or forced (active) convection. What is convection? Everyone knows that warm air rises. As it does so, it gives up heat to the cooler air around it and becomes cooler. The cooled air falls toward the floor, and the process is repeated. Natural convection may not be enough to remove all the heat generated by the electronics in a rack. For that reason, ventilating fans are sometimes used to

help the process along. These fans draw cool air into the rack. As the cool air flows through the rack, it picks up heat generated by the rack equipment, then exhausts the heated air. In certain circumstances, even forced ventilating air may not be enough. In those cases, it may be necessary to use a water-cooled heat exchanger or an air conditioner to provide the necessary cooling. Heat exchangers and air conditioners are covered later in this section.

The room in which the racks are located needs to have adequate ventilation to exhaust the heat and may need to be cooled with air conditioning in order for heat to be efficiently removed. This

On Site

EMI Traps

Some equipment inherently produces hum. There is not much that can be done to prevent it, but experimental grounding of the metal chassis may help. One area of concern is equipment that contains EMI traps, which are small modules containing capacitors and inductors to filter EMI. These modules may pass noise onto the ground circuit.

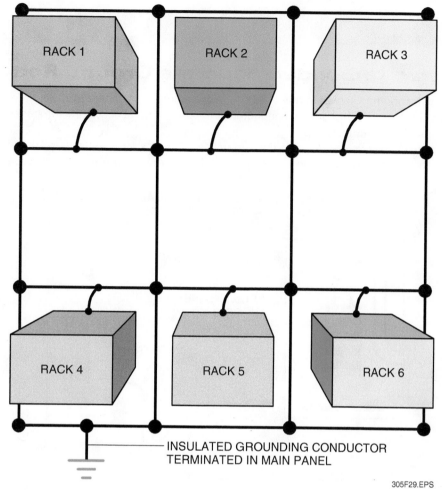

Figure 29 Signal reference grounding grid.

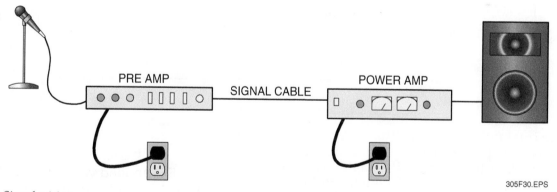

Figure 30 Signal wiring.

will depend on the heat load. The heat load is the amount of heat that must be removed in order to maintain the desired temperature. A room temperature of 75°F is typically needed to maintain the air inside the racks at 85°F. When the outdoor air temperature is more than 75°F, the indoor temperature will likewise be higher than 75°F, so some artificial cooling of the equipment spaces is going to be needed.

A rack can be configured for ventilation in a number of ways. First it is necessary to know how much heat needs to be removed.

5.1.0 Heat Load Calculation

The electrical power consumed by electronic equipment is converted to waste heat. The amount of heat that needs to be removed is measured in

Never Use an Unbonded Separate Ground Rod

During the ground fault condition shown in this diagram, the high-impedance path through the earth will prevent the fault current from being safely and quickly conducted to ground. In most cases, the impedance will be high enough to keep the circuit breaker from tripping.

The soil resistance between two ground rods that are only a few feet apart can be 10 to 20 ohms. This resistance limits the fault current to less than 12 amps, which is not enough to trip a 15-amp breaker.

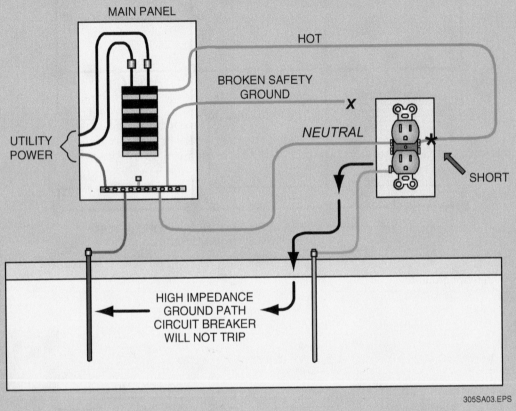

DO NOT SUBSTITUTE THE NORMAL SAFETY GROUND WITH A SEPARATE GROUND ROD!

305SA03.EPS

British thermal units per hour (Btu/h). A Btu is directly related to power consumption, which, as you know from earlier lessons, is related to voltage and current draw. To calculate Btu/h, multiply watts by 3.41.

In a rack containing only passive equipment, the amount of waste heat produced by that equipment is a function of the current drawn by the equipment. At 117 volts, each ampere of current produces 400 Btu/h of heat output. The rack heat is therefore the sum of all the power dissipated by the rack equipment. Sometimes the wattage is provided on the equipment nameplate. In other cases, it will be necessary to calculate the wattage using the power formula:

$$P = EI$$

Where:

$$P = \text{watts}$$
$$E = \text{applied voltage}$$
$$I = \text{rated current}$$

In all cases, it is best to check the current draw of the actual equipment. Don't rely on the fuse rating, as the fuse is normally oversized.

If the rack contains amplifiers, the calculation is a little more complex. First of all, the amount of waste heat is a function of the type of power supply, the type of program material to be played, the speaker load, and other variables. The power consumption of amplifiers also varies with the type of amplifier. Class A amplifiers, the least efficient, are only about 20 percent efficient, which means

---

80 percent of their current draw is converted into waste heat. At the other end of the spectrum, Class D amplifiers are 90 percent efficient. Neither of these two types of amplifiers is in common use.

The Btu/h factor is useful in determining the size of a heat exchanger or air conditioner. Fans, on the other hand, are rated in terms of the cubic feet per minute (cfm) of air they will move and the static pressure they are designed to overcome. For that reason, a fan that has failed must be replaced with a fan having the same characteristics.

If the characteristics of the existing fan are not known, you can determine them using the temperature rise (ΔT) in the rack. This is the difference between the temperature of the inlet air and outlet air. A typical (ΔT) for equipment at room temperature is 20°F. In a warmer operating environment, a lower temperature rise factor should be used. To calculate cfm, use the formula:

$$cfm = (3,160 \times kW) \div (\Delta T)$$

Where:

3,160 represents air volume at sea level

Static pressure is the amount of resistance the fan must overcome to provide the required airflow. This resistance is created by the rack vents and any obstruction in the rack. The original fan selection was based on a particular static pressure range, so the replacement fan should be in the same range. If the static pressure range is unknown, one common practice is to include an allowance of 25 percent over the calculated cfm to compensate for airflow resistance and hot spots.

To see the effect of static pressure, compare the two product data sheets in *Figure 31*. You will see that the cfm is significantly increased by a reduction in static pressure.

5.2.0 Rack Ventilation

Rack ventilation is classified as active or passive. Active ventilation requires ventilating fans to be mounted on the rack to draw in and exhaust ventilating air. Passive ventilation relies on natural convection, but can include fans built into the electronic equipment.

In an environment at normal room temperature, a rack is able to dissipate 300 to 500 watts of waste heat through natural convection. This requires vent openings at the bottom and top of the unit, as well as space around the equipment and unimpeded airflow.

There are some basic best practices for ventilating racks. Knowing them will help you recognize ventilation problems that could be causing equipment failures. Sometimes the initial ventilation design is incorrect. More often, equipment has been added to the rack, making the original design inadequate. Your ability to recognize and correct such problems can make you a hero to a frantic customer.

The following are some important guidelines:

- Rack placement is important. If the rack is placed under an air conditioning supply duct, the cold air can impede the flow of exhaust air from the rack. *Figure 32* shows proper placement.
- On a new installation, it is a good idea to place a thermometer near the warmest expected spot in the rack. Check the reading at 12 and 24 hours with the equipment operating to make sure the temperature is within the required range. If hot spots occur, fan panels can be placed near the spot to provide additional cooling.
- The most common airflow arrangement found in equipment with a high current draw is one in which air is drawn into the front of the cabinet and heated air is exhausted through the rear or sides. You may also encounter equipment in which the air is drawn from the rear and exhausted from the front.
- Some equipment components that use passive ventilation have intake vents on the bottom and outlet vents on the top. Make sure intake and outlet vents are not blocked by stacked equipment or other obstructions. If the components have front intakes, it is okay to stack the components (*Figure 33*).
- The components that generate the most heat should be placed near the top of the cabinet, especially where passive ventilation is being used. The exceptions to this rule occur when a loaded rack is being transported to the job site, or in a seismic installation.

> **WARNING**
> Be careful to avoid making the rack top-heavy when placing components. Also be alert to the fact that power amplifiers are heavy. Placing them at the top of the rack may create a tipping hazard.

- In a passive-ventilation application, don't install vented rack panels between components with front-intake fans. The vents will short-circuit the airflow as shown in *Figure 34*.

Fan and Blower Sizing

To cool this application you will need 356 CFM
at a static pressure of 0.75" I.W.G.

```
Internal Heat Load: 900 Watts
Max. Cabinet Temp.: 85°F
Max. Ambient Temp.: 75°F
Static Pressure:    0.75" I.W.G.
Volts/Hz:           115/60
CFM required:       356 CFM
```

Recommended Models
(CFMs shown below are at the required static pressure of 0.75" I.W.G.)

Single Centrifugal Blowers

```
KBB49          8.19"H x  10.5"W x  7.31"D  115/60    380 CFM    Stocked
```

High Pressure Centrifugal Blowers

```
KBB65          11.31"H x    9"W x  9.63"D  115/60    441 CFM    Stocked
```

Double Centrifugal Blowers

```
KBB37-37       7.19"H x  13.5"W x  6.44"D  115/60    375 CFM
KBB60-60       11.06"H x   12"W x  9.25"D  115/60    490 CFM    Stocked
KBB64-64       11.38"H x   14"W x 10.63"D  115/60    710 CFM    Stocked
KBB80-80       13.75"H x   14"W x 12.13"D  115/60   1020 CFM    Stocked
```

Quadruplex Centrifugal Blowers

```
KBB435         7.44"H x   16"W x   6.5"D   115/60    415 CFM    Stocked
KBB443         7.94"H x   17"W x  7.38"D   115/60    775 CFM    Stocked
KBB451         9.31"H x 14.88"W x 8.31"D   115/60    670 CFM    Stocked
```

2-Pole Backward-Curved Impeller Blowers

```
KBB2E225/63    8.88"H x  8.88"W x  3.88"D  115/60    400 CFM - NEW!
KBB2E250/56    9.88"H x  9.88"W x    4"D   115/60    660 CFM - NEW!
KBB2E280/40   11.06"H x 11.06"W x  4.44"D  115/60    500 CFM - NEW!
```

4-Pole Backward-Curved Impeller Blowers

```
KBB4E315/101 12.44"H x 12.44"W x    6"D    115/60    370 CFM    Stocked
```

(Stocked = Normally in stock at the factory)

305F31A.EPS

Figure 31 Examples of fan product data sheets (1 of 2).

Fan and Blower Sizing

To cool this application you will need 356 CFM at a static pressure of 0.25" I.W.G.

```
Internal Heat Load: 900 Watts
Max. Cabinet Temp.:  85°F
Max. Ambient Temp.:  75°F
Static Pressure:     0.25" I.W.G.
Volts/Hz:            115/60
CFM required:        356 CFM
```

Recommended Models
(CFMs shown below are at the required static pressure of 0.25" I.W.G.)

Single Centrifugal Blowers

KBB49	8.19"H x 10.5"W x 7.31"D	115/60	415 CFM	Stocked

High Pressure Centrifugal Blowers

KBB63	11.19"H x 8.75"W x 9.69"D	115/60	377 CFM	Stocked
KBB65	11.31"H x 9"W x 9.63"D	115/60	463 CFM	Stocked

Double Centrifugal Blowers

KBB36-36	7.13"H x 12.06"W x 6.44"D	115/60	360 CFM	Stocked
KBB37-37	7.19"H x 13.5"W x 6.44"D	115/60	432 CFM	
KBB43-43	8.25"H x 11.63"W x 7.31"D	115/60	412 CFM	
KBB50-50	9.25"H x 13.63"W x 8.25"D	115/60	430 CFM	Stocked
KBB57-57	9.75"H x 13.88"W x 8.44"D	115/60	495 CFM	
KBB60-60	11.06"H x 12"W x 9.25"D	115/60	600 CFM	Stocked
KBB64-64	11.38"H x 14"W x 10.63"D	115/60	820 CFM	Stocked
KBB80-80	13.75"H x 14"W x 12.13"D	115/60	1140 CFM	Stocked

Quadruplex Centrifugal Blowers

KBB435	7.44"H x 16"W x 6.5"D	115/60	510 CFM	Stocked
KBB443	7.94"H x 17"W x 7.38"D	115/60	870 CFM	Stocked
KBB451	9.31"H x 14.88"W x 8.31"D	115/60	760 CFM	Stocked

2-Pole Backward-Curved Impeller Blowers

KBB2E220/45	11.06"H x 11.06"W x 4.44"D	115/60	490 CFM - NEW!
KBB2E225/40	8.88"H x 8.88"W x 3.13"D	115/60	520 CFM Stocked
KBB2E225/63	8.88"H x 8.88"W x 3.88"D	115/60	630 CFM - NEW!
KBB2E250/56	9.88"H x 9.88"W x 4"D	115/60	885 CFM - NEW!
KBB2E280/40	11.06"H x 11.06"W x 4.44"D	115/60	850 CFM - NEW!

4-Pole Backward-Curved Impeller Blowers

KBB4E250/56	9.94"H x 9.94"W x 4.13"D	115/60	400 CFM - NEW!
KBB4E280/80	11.06"H x 11.06"W x 4.94"D	115/60	705 CFM Stocked

305F31B.EPS

Figure 31 Examples of fan product data sheets (2 of 2).

 Rack Assembly

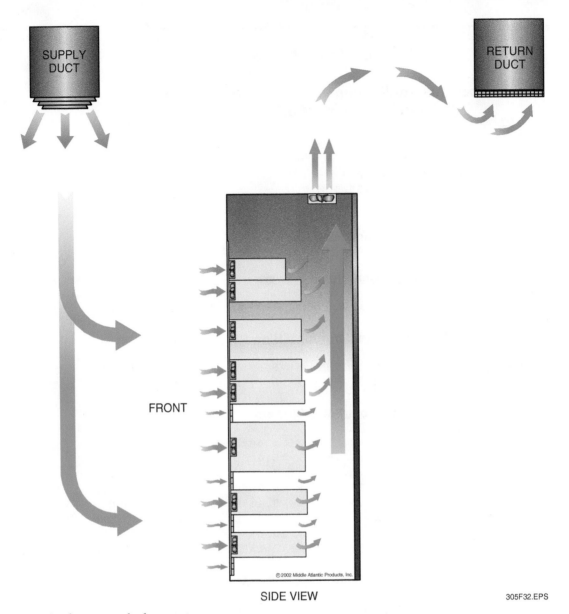

SUPPLY DUCT

RETURN DUCT

FRONT

SIDE VIEW

© 2002 Middle Atlantic Products, Inc.

305F32.EPS

Figure 32 Example of correct rack placement.

- When forced-air ventilation is used, vent panels can be placed between equipment components with front intakes, but it is not necessary to do so. Most front-intake fans produce 25 to 50 cfm, so the rack blower cfm must equal at least the sum of all the front-intake fans. Vented rack panels should not be placed in the upper six spaces of the rack (*Figure 35*).
- Filters must be placed over the intakes if the operating environment contains dust or airborne particles (*Figure 36*).
- Many designers prefer to use a pressurized rack in a dirty environment (*Figure 37*). In this approach, there are no front vents to allow dirty air to enter. Filtered air escapes through cracks around equipment and doors.

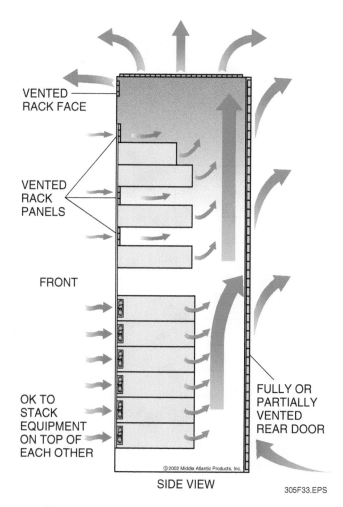

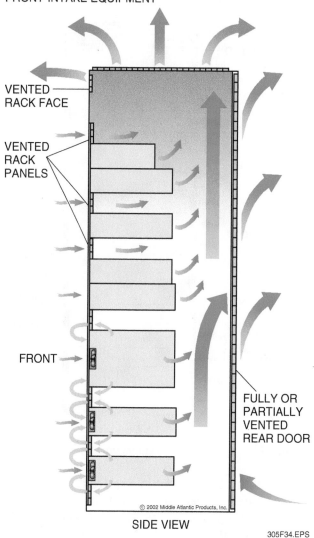

Figure 33 Proper passive convection.

5.3.0 Heat Exchangers and Air Conditioners

In dirty environments, such as factories, rack filters quickly become clogged, and heat rapidly builds inside the rack. In these situations, it is often necessary to use sealed and gasketed NEMA-rated racks. When this occurs, either air conditioning or heat exchangers must be used to provide internal rack cooling. These solutions prevent the dirty ambient air from mixing with the clean air inside the racks.

5.3.1 Heat Exchangers

Heat exchangers can be used in conditions where the internal rack temperature can be higher than that of the ambient air. The heat exchanger (*Figure 38*) is mounted on the rack, and openings in the rack are used for airflow. *Figure 39* shows basic setups for air-cooled and water-cooled heat exchangers. The heat from the rack is drawn out of the rack by a heat pipe in the heat exchanger assembly. This air flows over the heat exchanger coil, which absorbs the heat. The heat is exhausted

Figure 34 Vent panels short-circuiting airflow.

from the heat exchanger by the external blower, and the cooled air is recirculated in the rack.

Depending on the amount of heat, the coil may simply be made of aluminum fins that dissipate the heat as the heated air flows over them. When more heat removal is needed, a refrigerant-filled coil may be used. Refrigerant under pressure boils at a very low temperature because of the relationship between pressure and temperature. For example, water boils at 212°F at atmospheric pressure. Lowering the pressure reduces the boiling point. As you approach a vacuum, the boiling point approaches 0°.

Refrigerants are chemicals that have low boiling points. When heated air flows over the refrigerant-filled heat exchanger coil, the refrigerant absorbs heat from the air and begins to boil, turning into a vapor. This vapor rises to the top of the coil, where the heat is transferred to the cooler

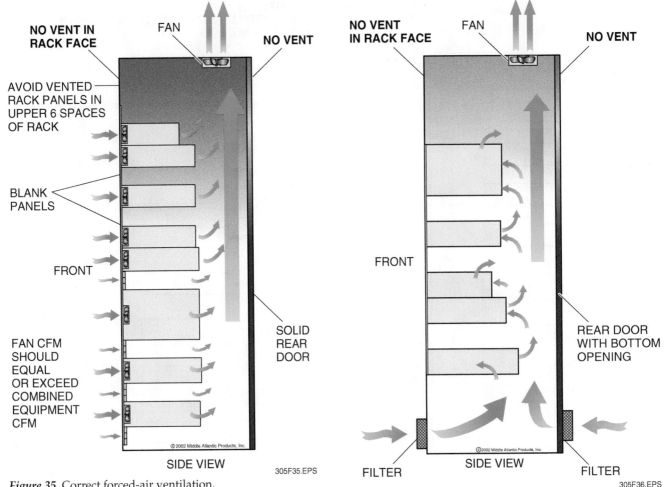

NO VENT IN RACK FACE

FAN

NO VENT

AVOID VENTED RACK PANELS IN UPPER 6 SPACES OF RACK

BLANK PANELS

FRONT

FAN CFM SHOULD EQUAL OR EXCEED COMBINED EQUIPMENT CFM

© 2002 Middle Atlantic Products, Inc.

SIDE VIEW

SOLID REAR DOOR

305F35.EPS

Figure 35 Correct forced-air ventilation.

NO VENT IN RACK FACE

FAN

NO VENT

FRONT

REAR DOOR WITH BOTTOM OPENING

© 2002 Middle Atlantic Products, Inc.

SIDE VIEW

FILTER

FILTER

305F36.EPS

Figure 36 Use of air filters.

ambient air. As the heat is removed, the refrigerant is converted back to a liquid.

If ambient air is too hot for cooling, a water-to-air system can be chosen. In such a system, cold water flows through a cooling coil, absorbs the heat from the rack air, and then is carried away. If the water is cold enough, below-ambient cooling can be achieved.

5.3.2 Air Conditioners

When it is necessary to keep the equipment operating temperature below the ambient room temperature in a sealed rack, an air conditioning system (*Figure 40*) is used. Air conditioners for rack systems are available in air-cooled and water-cooled configurations.

An air conditioner can provide cooling air at much lower temperatures than a heat exchanger. In addition to the evaporator and condenser coils, a refrigerant compressor is used to raise the temperature of the heated refrigerant so that it is much higher than the temperature of the ambient air. In that way, large amounts of heat can be transferred (heat flows from warmer to cooler).

Figure 41 shows a simplified schematic of a room air conditioner to illustrate how an air conditioning system works. Liquid refrigerant at a very low temperature flows through the evaporator coil, picking up heat from the warm air in the room. The refrigerant boils into a vapor, absorbing more heat as it does so. The hot, vaporized refrigerant is drawn into the refrigerant compressor, which raises the refrigerant pressure, and therefore its temperature. The refrigerant temperature may be two to three times higher than that of the ambient air, so the heat from the refrigerant readily transfers to the relatively cooler outdoor air being blown across the condenser. As the heat is removed, the refrigerant reverts back to a liquid. The expansion device is a tiny orifice that has the effect of lowering the pressure, and therefore the temperature, of the liquid refrigerant. When it reaches the evaporator, the refrigerant is once again a cold, low-pressure refrigerant, and the cycle repeats. *Figure 42* shows a simplified schematic of a rack air conditioner.

ADVANTAGE:
FOR DIRTY ENVIRONMENTS, CRACKS AROUND
EQUIPMENT AND DOORS LET FILTERED AIR
ESCAPE, RATHER THAN SUCKING DIRTY AIR IN.

VENTED
TOP REAR
DOOR

NO VENTS
ANYWHERE

FRONT

FAN

FILTER

©2002 Middle Atlantic Products, Inc.

SIDE VIEW

305F37.EPS

Figure 37 Pressurized rack.

305F38.EPS

Figure 38 A rack heat exchanger.

6.0.0 INSTALLATION PRACTICES

Equipment should be installed in racks in a logical pattern. This means planning ahead and preparing rack layout sketches that show where the equipment goes. These sketches usually become part of the as-built drawings that are turned over to the client at the end of the project. *Figure 43* is an example of a rack layout drawing.

> **NOTE**
>
> Racks must be at least 36" from electrical distribution equipment.

CAD and drawing programs on computers greatly simplify the rack layout process because they allow you make changes without having to redraw your layout every time. The more complex the layout, the more important it is to use a computer. The following are some rules of thumb for installation planning:

- Place heavy equipment at the bottom of the rack and on the rear rails. It avoids making the rack top-heavy and makes the equipment easier to install or remove. However, avoid overpopulating the rear rails, as it can make equipment access difficult.
- Place items containing controls, indicator lights, gauges, and front panel jacks at or near eye level.
- Place monitors at eye level.
- Install DVRs and DVD players within easy reach of the user.
- Leave 20 percent space in each rack for future growth.
- Leave extra space around equipment like power amplifiers that tends to run hot. A baffled vent panel may be needed above such units, as was covered earlier in this module.
- Because of the magnetic fields they generate, some amplifiers can induce a hum in sensitive equipment. To avoid this problem, place sensitive equipment a couple of RUs away from amplifiers.

 Rack Assembly

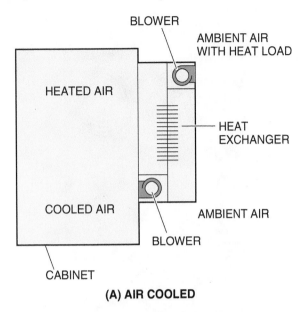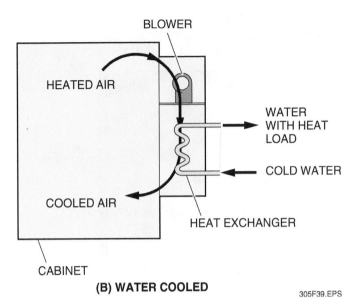

Figure 39 Heat exchanger airflow.

305F40.EPS

Figure 40 Server rack with air conditioner.

Part of the planning process involves determining how to organize the equipment. For example, should you put all of one type of equipment together, or should you put clusters of related equipment together? These decisions are often driven by the amount of rack-to-rack cabling that each method will create. However, the arrangement needs to be logical so that users can easily find equipment. Once you decide on the layout concept, you should stick with it.

If you use a computer, you accumulate files in a logical way. Certain types of files go into certain folders. If you maintain that system, you will always know where to look for those types of files, and in a pinch, someone else would be able to find files on your computer by reviewing the directories. The same principle applies when populating equipment racks: use a logical system and document that system.

6.1.0 Installation Hardware and Accessories

The manufacturers of electronic equipment racks offer a variety of accessories, as well as hardware to support installation of equipment in the racks.

6.1.1 Casters

Casters (*Figure 44*) are required on mobile racks. Caster selection depends on the weight of the unit, as well as the intended use. Material such as hard rubber works on carpet, but may damage wood or vinyl floors. Softer materials such as urethane and polyurethane are used for portable sound systems. Such systems should also have at least two locking casters. If a rack in a fixed installation is located in a tight spot, it may need to be moved to get access to the rear of the rack. Keep in mind, however, that cabling limits the amount of movement.

The quality of the caster depends on the use. If a rack is moved often over considerable distances, a ball-bearing caster should be selected over a sleeve-bearing caster.

NCCER – *Contren® Learning Series* 33305-11

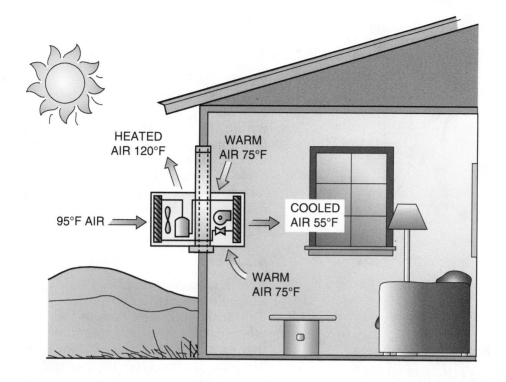

HEATED AIR 120°F

WARM AIR 75°F

95°F AIR

COOLED AIR 55°F

WARM AIR 75°F

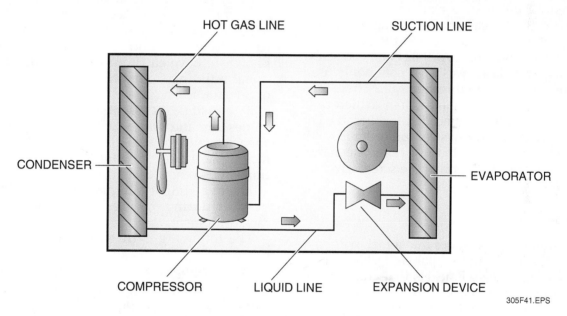

HOT GAS LINE

SUCTION LINE

CONDENSER

EVAPORATOR

COMPRESSOR

LIQUID LINE

EXPANSION DEVICE

305F41.EPS

Figure 41 The air conditioning process.

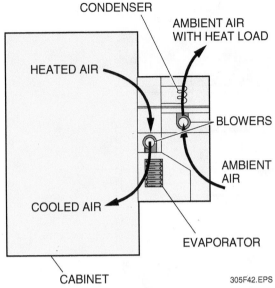

Figure 42 Simplified schematic of a rack air conditioner.

6.1.2 Equipment Mounting Hardware

Equipment mounting rails will be either drilled or tapped for a threaded machine screw. When the rail is drilled, a cage nut or spring nut (*Figure 45*) is attached to the rail behind the hole and is used to secure the screw. A 10-32 pan head screw with a nylon or plastic washer is commonly used to secure equipment to threaded rails. These rails are necessarily heavier gauge than drilled rails because of the thickness needed for threads. For security purposes, screws can be ordered with different types of heads. In those cases, the manufacturer will offer the appropriate screwdriver bit (*Figure 46*).

6.1.3 Rack Ears

Not all equipment that goes into a rack will have the standard 19" mounting panel. Rack ears (*Figure 47*) are right-angle brackets that are attached to the equipment chassis to extend it to the 19" width needed for mounting to the rails.

6.1.4 Rack Support Bases (Plinths)

Some rack manufacturers have optional plinths that are used to insulate the rack from the floor and to prevent the rack from being inadvertently grounded to the building structure or other grounds. Accessory plinths with cable entrance panels are available from rack manufacturers. A plinth can also be made with lumber that is anchored to the floor. The rack would then be anchored to the lumber.

6.2.0 Cable Management

Cabling is routed to and from racks using various devices and methods, including ladder racks (*Figure 48*). The type of device used depends on the type and amount of cable that must be supported.

A variety of accessories for managing cabling within the racks are available from rack manufacturers. These accessories include cable lacing rods, D-rings, cable organizers with clips, and cable channels that attach to the inside of the rack. Cable ties are also an important part of cable management. Examples of cable management accessories for racks are shown in *Figure 49*.

6.3.0 Earthquake Protection

Any rack installation plan should include the possibility of seismic activity. Most people think of California and the infamous San Andreas Fault when earthquakes are mentioned. The fact is, many areas of the U.S. are prone to earthquakes. The New Madrid Fault, for example, runs through parts of Missouri, Tennessee, and Illinois. There are records that a nineteenth-century earthquake in this area changed the course of the Mississippi River. New York and Alabama are other areas known to have earthquakes. The 2000 International Building Code provides a seismic rating, or S-factor, for every area of the country. Any facility classified as essential, such as medical facilities, fire and police facilities, and government agencies must be seismically rated. These same facilities are likely to house critical communications and computer equipment, which makes it all the more important that special consideration be given during the selection and installation of rack equipment.

The following extra precautions need to be taken during the installation of racks and rack-mounted equipment in seismically rated facilities:

- The rack must be securely anchored to the floor using fasteners with a high shear rating and tensile strength.
- The top of the rack should also be anchored. One method of anchoring the top is through the use of a cable ladder.
- Equipment may need to be supported from the rear. This would apply if all the following conditions exist: the unit exceeds 10 pounds per rack space, the center of gravity is toward the rear, and the overall depth of the unit is more than 2.5 times the rack height.

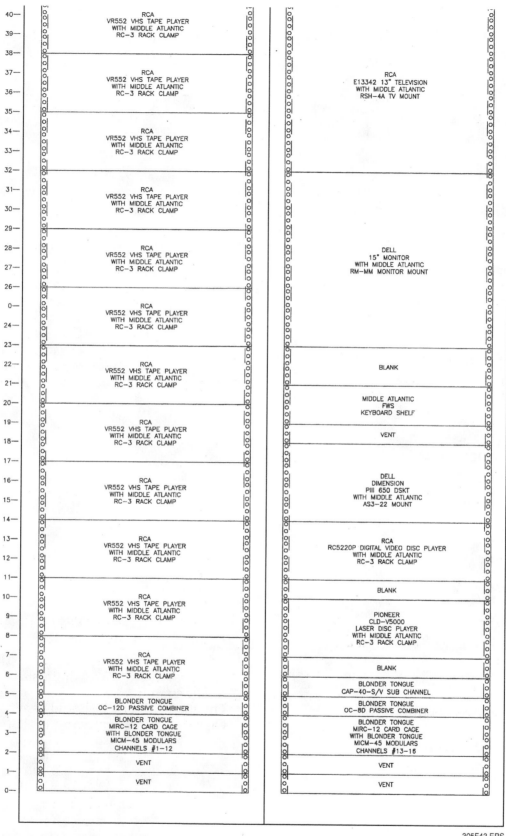

Figure 43 Example of a rack layout drawing.

305F43.EPS

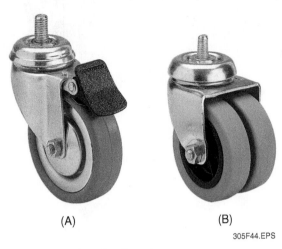

(A)   (B)

305F44.EPS

Figure 44 Examples of rack casters.

305F46.EPS

Figure 46 Square-head screw and driver bit.

SPRING NUT

CAGE NUTS

305F45.EPS

Figure 45 Nuts used in rack systems.

305F47.EPS

Figure 47 Rack ears.

305F48.EPS

Figure 48 Ladder rack.

NCCER – *Contren® Learning Series*   33305-11

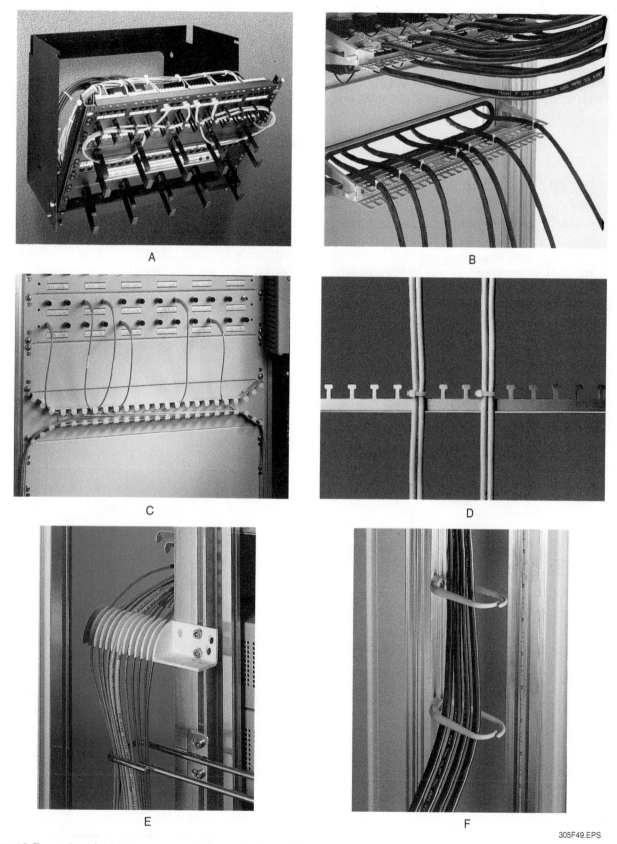

Figure 49 Examples of cable-management accessories for racks.

305F49.EPS

- During installation, keep the center of gravity as low as possible. This means placing heavy units near the bottom, which is a good practice under any circumstances.
- Fill the rack from bottom to top, making sure that every space is filled with a unit or a panel.
- Make sure all screws are tight. This includes screws used to assemble the rack, as well as those used to install equipment and accessories.

SUMMARY

Racks provide a means of organizing and protecting sensitive audio, video, and digital electronic equipment. The electronic equipment is screwed to rack mounting rails and is made specifically to fit the rack. Proper grounding is an important consideration in rack-based electronic systems, both for the protection of personnel and the elimination of noise and other signal degradation that can be caused by ground loops. Electronic equipment can be damaged by excessive heat. For that reason, it is very important to provide proper ventilation for racks, and to be able to adjust ventilation as the rack population changes. Large installations use many racks of equipment. It is therefore important to use a logical approach in laying out the racks, and to document the layout for future use.

1. The width of the rack typically used for tele-communications and data networks is _____.

   a. 23 inches
   b. 25 inches
   c. 30 inches
   d. 36 inches

2. The type of rack that has a backplane is the _____.

   a. console rack
   b. knockdown rack
   c. wall-mount rack
   d. mobile rack

3. A five-RU panel is _____.

   a. 5" high
   b. 7½" high
   c. 8¾" high
   d. 10½" high

4. It is standard practice to bond the grounding conductor to the neutral conductor at each rack.

   a. True
   b. False

NEUTRAL

HOT

GROUND

305RQ01.EPS

Figure 1

5. The wiring arrangement shown in *Figure 1* is _____.

   a. correct
   b. incorrect because the hot and neutral are reversed
   c. incorrect because the ground and neutral are reversed
   d. incorrect because the wrong type of outlet is used

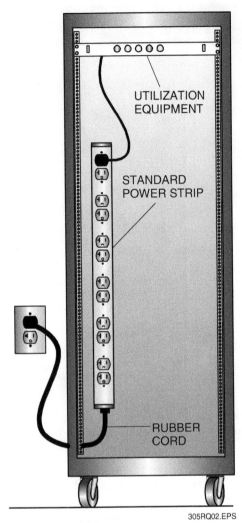

UTILIZATION EQUIPMENT

STANDARD POWER STRIP

RUBBER CORD

305RQ02.EPS

Figure 2

6. The rack shown in *Figure 2* is _____.

   a. correctly grounded
   b. incorrectly grounded because there is no bonding strip between the power strip and the rack
   c. incorrectly grounded because the rack is not isolated from the floor
   d. incorrectly grounded because there is no bonding connection between the power strip and the utilization equipment

7. Star grounding is the preferred method for digital and broadband equipment.

   a. True
   b. False

8. What is the grounding requirement for the shield of audio system cables?

   a. Both ends of the cable shield should be grounded regardless of the type of cable.
   b. The shield of a balanced cable should always be grounded at both ends.
   c. The shield of an unbalanced cable should always be grounded at both ends.
   d. The shields are never grounded.

9. A rack contains five passive equipment chassis that draw a total of 15 amps of current. Assuming 115 volts applied, the heat load in Btu/h for this rack is _____.

   a. 1,725
   b. 2,500
   c. 5,416
   d. 5,882

10. The major factors that control the selection of a fan for rack ventilation are _____.

   a. cfm and static pressure
   b. cfm and current draw
   c. static pressure and current draw
   d. cfm and rack height

11. A rack using passive-convection ventilation is able to handle how many watts of heat at normal room temperature?

   a. 100 to 300
   b. 300 to 500
   c. 500 to 750
   d. 1,000 to 1,500

12. Which type of rack ventilation system relies on natural convection?

   a. Active
   b. Passive
   c. Pressurized
   d. Heat pump

13. An air conditioner works on the principle that _____.
    a. rack heat is transferred from the warm rack air to the refrigerant
    b. the air conditioning creates cold air that is blown across the rack equipment
    c. the racks are cooled by contact with refrigerant piping
    d. cold air is transferred from the condenser to the rack

14. When populating a rack, it is a good idea to leave about _____ of the space for future growth.
    a. 5 percent
    b. 10 percent
    c. 20 percent
    d. 30 percent

15. A sleeve bearing caster should be selected for racks that must be moved often, or over long distances.
    a. True
    b. False

Trade Terms Introduced in This Module

Backplane: The rear section of a swinging frame rack used to mount the rack to the wall.

British thermal units per hour (Btu/h): The amount of heat required to raise the temperature of one pound of water by 1°F.

Convection: The transfer of heat due to the flow of a liquid or gas caused by a temperature differential.

Ground loops: A condition in which an undesired connection is made between electronic components or enclosures, inducing current flow that can generate noise.

Headroom: The difference between the actual signal level and the maximum level a device is able to handle without distorting the signal.

Rack unit (RU): The height of a single equipment panel; an RU is 1.75".

Separately derived system: An electrical system in which the power is derived from a source that has no direct electrical connection, including connection to a grounded circuit conductor, to supply conductors originating in another system.

Static pressure: The pressure exerted uniformly in all directions within an enclosure, as measured in inches of water column.

Additional Resources

This module presents thorough resources for task training. The following resource material is suggested for further study.

- *Audio Systems Design and Installation*. 1997. Phillip Giddings. Boston, MA: Focal Press.

Middle Atlantic Products, Inc., www.middleatlantic. com

Figure Credits

CONTREN® LEARNING SERIES — USER UPDATE

NCCER makes every effort to keep its textbooks up-to-date and free of technical errors. We appreciate your help in this process. If you find an error, a typographical mistake, or an inaccuracy in NCCER's Contren® materials, please fill out this form (or a photocopy), or complete the online form at www.nccer.org/olf. Be sure to include the exact module number, page number, a detailed description, and your recommended correction. Your input will be brought to the attention of the Authoring Team. Thank you for your assistance.

Instructors – If you have an idea for improving this textbook, or have found that additional materials were necessary to teach this module effectively, please let us know so that we may present your suggestions to the Authoring Team.

NCCER Product Development and Revision
3600 NW 43rd Street, Building G, Gainesville, FL 32606

Fax: 352-334-0932
Email: curriculum@nccer.org
Online: www.nccer.org/olf

❏ Trainee Guide ❏ AIG ❏ Exam ❏ PowerPoints Other _____

Craft / Level: _____ Copyright Date: _____

Module Number / Title: _____

Section Number(s): _____

Description: _____

_____

_____

_____

_____

Recommended Correction: _____

_____

_____

_____

_____

Your Name: _____

Address: _____

_____

Email: _____ Phone: _____

System Commissioning
and User Training

33306-11

Trainees with successful module completions may be eligible for credentialing through NCCER's National Registry. To learn more, go to **www.nccer.org** or contact us at **1.888.622.3720**. Our website has information on the latest product releases and training, as well as online versions of our *Cornerstone* newsletter and Pearson's Contren® product catalog.

Your feedback is welcome. You may email your comments to **curriculum@nccer.org,** send general comments and inquiries to **info@nccer.org**, or use the User Update form at the back of this module.

V.1 4/11

SYSTEM COMMISSIONING AND USER TRAINING

Objectives

When you have completed this module, you will be able to do the following:

1. Describe the phases of the system commissioning process and explain how they apply to the commissioning of specific types of electronic systems.
2. Explain how to develop a user training course.
3. Demonstrate or describe how to prepare for and conduct user training.

Performance Task

Under the supervision of your instructor, you should be able to do the following:

1. Prepare and conduct a user training session.

Trade Terms

Acceptable performance
Authority having
  jurisdiction (AHJ)

Commissioning authority
  (CA)

Commissioning plan

Functional performance
  testing

Contents

Topics to be presented in this module include:

Figures and Tables ────────────

1.0.0 INTRODUCTION

System commissioning is a process by which a formal and organized approach is taken to obtaining, verifying, and documenting the installation and performance of a particular system or systems. The goal of commissioning is to make sure that a system operates as intended and at optimum efficiency. Commissioning is normally required for newly installed or retrofitted commercial and industrial systems. Because each building and its systems are different, the specific elements of the commissioning process must be tailored to fit each situation. However, the objectives for performing system commissioning are the same:

- Verify that the system design meets the functional requirements of the owner.
- Verify that all systems are properly installed in accordance with the design and specifications.
- Verify that all systems and components meet required local, state, and other required codes.
- Verify and document the proper operation of all equipment, systems and software.
- Verify that all documentation for the system is accurate and complete.
- Train building operator and maintenance personnel for efficient operation and maintenance (O&M) on the installed equipment and systems.

This module introduces the elements involved in formal system commissioning. Understanding the process is essential in order to participate in and successfully perform a system commissioning. An important part of the commissioning process involves the training of the user's operator and maintenance personnel. Because selected contractor personnel normally are given the responsibility for performing this training, the second part of this module covers how the designated trainer(s) should prepare for and conduct user training.

2.0.0 COMMISSIONING PROCESS OVERVIEW

A qualified person, company, or agency, commonly referred to as the commissioning authority (CA), has the responsibility for planning and supervising the commissioning of a building and its systems. The CA could be an architect, engineer, or contractor, or their on-site representative, depending on the complexity of the job. Under the CA's guidance and direction, the formal system commissioning process extends through all phases of a project, starting with the design phase and continuing beyond system acceptance (*Figure 1*). Depending on the size and complexity of the system, and the contractual requirements, these activities can include:

- Pre-installation activities
- Commissioning plan development
- Installation activities
- Functional performance testing
- User training
- System acceptance
- Post-acceptance activities

The CA works closely with the authority having jurisdiction (AHJ). The AHJ is the person or department within a municipality or jurisdiction that has the legal authority to approve a system that requires code conformance.

On Site

ASHRAE

The American Society of Heating, Refrigerating, and Air Conditioning Engineers (ASHRAE), founded in 1894, is an international organization that publishes guidelines for building commissioning. These guidelines provide commissioning protocols that cover commissioning requirements for residential and commercial buildings, hospitals, correctional facilities, and even museums.

GOING GREEN

Impact of LEED on System Commissioning

The Leadership in Energy and Environmental Design (LEED) green building rating system has an impact on every phase of building construction, from design through commissioning. Many local energy codes and the guidelines developing through the LEED building certification process are modeling new and varied startup record formats and processes. Under LEED, the building commissioning process has become far more complex. In many cases, a third-party commissioning agent is being employed. These trends and strategies are designed to become part of the long-term building maintenance process, which requires higher levels of training, documentation, system testing, and data reporting than has been traditionally required.

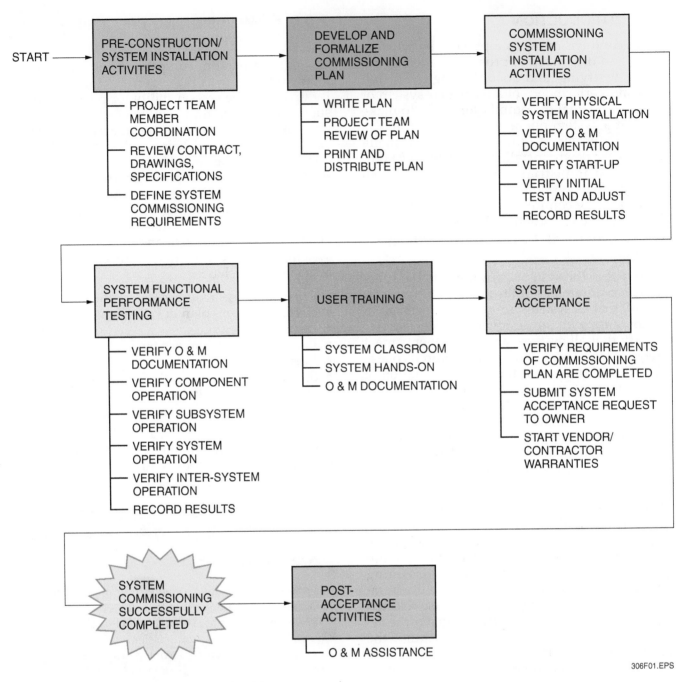

Figure 1 Formal system commissioning activities.

306F01.EPS

2.1.0 Pre-Installation Activities

Ideally, the start of the system commissioning process begins during the system pre-installation phase. This is after the contractors responsible for the installation of the various systems are under contract and the details of the equipment to be installed are firm. During this phase, the CA meets with the project team members, usually consisting of the owner, architect, systems engineers, major system contractors/subcontractors, and equipment vendors. This gives the CA the opportunity to review a full set of design drawings and contract documents and to be briefed by each of the system designers.

The CA can advise the project team on code requirements. The project team can help in determining the commissioning plan requirements. As a result, the team members know what their responsibilities are in support of the commissioning process. Each of the system contractors should prepare preliminary documentation detailing how they intend to comply with the system commissioning requirements. The topics covered by this documentation normally include the following:

- Sequence and schedule of installation and testing activities for their equipment
- Detailed procedures for the tests to be performed on their equipment
- Format of the report forms used to record test data and results
- A list of, and calibration data for, the test equipment used in support of functional performance testing
- Description of and method(s) to be used to accomplish user training

2.2.0 Commissioning Plan Preparation

Using the information and inputs from the pre-installation meetings, the CA formally documents the commissioning requirements for each of the systems covered by the commissioning plan. Draft copies of the commissioning plan are then distributed to all the project team members for their review and comment. At this time, any omissions, conflicts, or ambiguities identified during review of the plan are resolved by the CA and the parties involved. Following this, the commissioning plan is finalized and copies are distributed to all affected parties. At a minimum, the commissioning plan covers the following:

- Responsibilities of each system installer or equipment supplier
- Equipment installation schedules and priorities
- Equipment/systems requiring functional performance testing
- Functional performance test procedures and requirements for each system
- Documentation requirements and format for items such as checklists used to record the results of inspections, startup, and functional performance tests
- Methods and scope of training for operating and maintenance personnel
- Schedule of pre-acceptance test and test schedule

On Site

Communication

Good written and oral communication among all parties involved in commissioning contributes to the success of the commissioning process. Good communication is the key to quickly and efficiently providing clarification about system design and commissioning-related issues.

2.3.0 Installation Activities

During the installation phase, the various systems are installed, tested, and put into operation. It is the CA's responsibility to oversee the on-site installation of the systems while work is in progress. This is done to verify that the physical installation of the system components is being done in accordance with contract requirements. Any approved design changes that occur must be tracked and documented. When system commissioning requires that an installation task or milestone be verified, the CA works with the contractor or vendor to witness and to record its completion on the appropriate form. Throughout the installation phase of the system, the CA participates in job-site meetings to remain informed about the progress of the installation. This permits the CA to provide others with updated information about commissioning related issues.

After the physical installation of a system's hardware and cabling is complete, the CA works with the contractor to witness the startup and initial testing and adjustment of the system and its major equipment. Should some activity of the initial startup not be completed successfully, the CA works with the system contractor to resolve or correct the deficiency. If applicable, retesting is performed to verify correction of the deficiency. Common system activities and commissioning requirements accomplished during the installation phase include the following:

- Checking the physical integrity of system component installations
- Checking system wiring and cable interconnections
- Checking system switch and control settings
- Performing system power-up
- Loading and verifying system software and creating backups
- Performing initial test and adjustment

2.4.0 Functional Performance Testing Activities

Before the start of functional performance testing, the CA should have verified that the physical installation, startup, and all initial testing and adjustment of the system have been completed and the results recorded in accordance with the commissioning plan requirements. The CA also reviews the commissioning documentation requirements for the equipment or system with the system contractor's responsible person. This documentation usually includes O&M data, performance data, and control drawings.

Reporting

During the system installation and testing phases of the project, the CA maintains a log. Entries are made in this log to record the status of documentation and testing for each system and/or piece of equipment, requested schedule changes, and outstanding equipment or documentation deficiencies. Usually, the CA is required to submit a monthly report to the owner. The contents of this monthly report typically include the following:

- Commissioning status
- Noncompliance items
- Current issues of importance
- Recommendations
- Future issues
- Schedule and milestone completion data

The CA witnesses the contractor performing the various functional performance tests on the system and its components. The sequence of testing usually goes from components to subsystems to systems and finally to the interface between systems. The results of the functional performance tests are recorded on the checklists contained in the approved commissioning plan or other forms approved by the CA. It should be pointed out that often several similar pieces of equipment are used in the same system. All must be tested for acceptance, and there should be a separate checklist used to record the test results for each one. Any check or test that fails to meet specifications must be repeated after the necessary corrective action has been taken to resolve the problem encountered with the check or test. This re-testing should be repeated until acceptable performance results are obtained. At the completion of the system functional performance testing process all of the following factors should be verified:

- All equipment and subsystems have been put into operation and are functioning normally
- System alignment/adjustment (where needed) and testing have been completed
- All interfaces with other systems are functioning properly
- All measured values for operating characteristics and parameters are in compliance with those specified in the commissioning plan
- Test results and related data are recorded on appropriate forms included in the commissioning plan or other forms approved by the CA

The specific kinds of installation and functional performance tests done during system commissioning depend on the kind of equipment being commissioned, its complexity, and the prevailing federal and/or local code requirements that may pertain to the equipment. *Table 1* lists some examples of the kinds of tests performed on electronic systems and the test equipment used in support of system commissioning testing. As described earlier, specific information on what is to be tested in a particular system and how the test is to be performed is covered in the commissioning plan and related system operation and maintenance documentation.

2.5.0 User Training and Documentation

Training the user's personnel to operate and maintain the system is an important part of the commissioning process. The CA works with the owner and contractor personnel to prepare a high-quality training program and supporting system documentation. The nature and duration of the training program depends on the size and complexity of the system, and on the availability of the system for hands-on training. For larger systems, it may be scheduled to occur during system installation and continue over a long period. On smaller systems, it can occur immediately after system installation and checkout.

Guidelines for preparing for and conducting a training course are covered later in this module. User training should include a complete overview of all system equipment and components with a focus on the following:

- Contents and scope of the operation and maintenance documentation
- System-operational procedures for all modes of operation
- Acceptable tolerances for system parameters in all operating modes
- Procedures for coping with abnormal and/or emergency situations

Table 1 Examples of Tests and Equipment Used in Support of System Commissioning

System	Commissioning Tests	Tools and Test Equipment
Fire Alarm **Note:** The *National Fire Alarm and Signaling Code, NFPA-72,* is a primary source of testing requirements.	• Power supply input/output voltage • Battery load • Operate notification appliance circuits (NAC) • Voltage drop • Current draw • Pull station operation • Supervision (all circuits) • Ground faults • Dialer/communication monitoring • Detectors • Annunciators • Sequence of operations	• Multimeter (VOM/DMM) • DB meter • Megger • Battery load tester • Smoke pole • Hair dryer • Smoke bomb/fog machine • Pans and fuel • IR lamp • Break rods for pull alarms • Alarm box keys • Beam detection filter (IR) • Cell phones/two-way radios
Audio Sound Systems	• Power supply input/output voltage • Signal processing equipment gain and output level checks • Signal processor and speaker polarity • Average and peak sound pressure levels • Frequency response • Signal-to-noise ratio	• Multimeter (VOM/DMM) • Sine wave generator • Oscilloscope • Phase tester • Sound pressure level (SPL) meter • Real time analyzer (RTA)/spectrum analyzer • Impedance bridge • Loudness (VU) meter • Cell phones/two-way radios
Public Address and Intercommunication Systems	• Power supply input and output voltage • Power amplifier gain and output level • Speaker sound quality and polarity • Speaker impedance	• Multimeter (VOM/DMM) • Sound pressure level (SPL) meter • Impedance bridge • Cell phones/two-way radios
Nurse Call Systems **Note:** The *Standard for Health Care Facilities Code, NFPA-99,* is a primary source of testing requirements.	• Power supply input/output voltage • Leakage current tests of all patient pendant control, pillow speakers, and other patient accessible circuits • Call station load testing • Battery backup testing • Pull cords, dome lamps, master station, auxiliary functions	• Multimeter (VOM/DMM) • Sound pressure level (SPL) meter • Cell phones/two-way radios

The different kinds of documentation used in support of the training program and for operation and maintenance of the equipment are collectively called the operation and maintenance (O&M) manual in the commissioning plan. However, this does not mean that all the documents are contained in a single manual. The commissioning plan describes what documentation is to be provided and how it is to be assembled or packaged for turnover to the owner and/or the system operating personnel. The system and training documentation can include but is not limited to the following:

• Commissioning plan
• Scope of work
• Bill of materials
• Normal system switch and control settings
• Copies of system test and measurement commissioning reports organized by system functional areas
• Electronic or hard copies of system as-built drawings and specifications as modified by approved change orders
• Equipment and system wiring lists and schematics
• Vendor operation and maintenance manuals for major system components (*Figure 2*)
• Vendor equipment data sheets
• Warranty documents

Table 1 Examples of Tests and Equipment Used in Support of System Commissioning (Continued)

System	Commissioning Tests	Tools and Test Equipment
CATV, SMATV, MATV, and CCTV Systems **Note:** FCC Regulations Part 76.605 is a primary source of CATV testing requirements.	• Power supply input/output voltage checks • Audio and video carrier center frequencies • Minimum and maximum visual carrier level • Maximum signal level of adjacent channels • Maximum difference between any video carriers • Minimum and maximum audio/video ratio • Signal-to-noise ratio • Carrier-to-noise ratio • RF signal leakage	• Multimeter (VOM/DMM) • Signal level meter (SLM) • Spectrum analyzer • Cable tone test set • Satellite tester • Portable color TV receiver • Time delay reflectometer
Access Control Systems	• Power supply/ UPS input/output voltage checks • Battery checks • Door release and hold checks • Site access protocols • Door ajar alarms	• Multimeter (VOM/DMM)
Intrusion Detection Systems	• Power supply/ UPS input/output voltage checks • Open and grounds • All sensors • Control panels • Batteries and backup power supplies	• Multimeter (VOM/DMM) • Battery load tester • Cell phones/two-way radios

2.5.1 As-Built Drawings and Documentation

As-built documentation reflects the system as it was actually installed. The as-builts provide information on any changes that were made during the installation. It is not uncommon for system installers to be forced to deviate from the original drawings as a result of unexpected obstacles that interfere with the plan. As-builts also document information, such as wiring details, that could not be completed until the installation was finished.

The important thing is to make sure the documentation that is left behind for the building owner reflects what was actually done. This is usually accomplished by carefully marking up a copy of the original drawings and creating any additional drawings that may be needed. The following is a list of some things that should be documented in the as-built drawings:

• The location of 120VAC sources, including location of breaker boxes and specific breaker number assignments

• The physical layout of low-voltage wiring and cabling, including the locations of conduit, cable tray, and splice boxes
• Detailed point-to-point wiring information, including:
  – Wire color
  – Location and connection information for all system components, junction boxes, and splice points
  – Interconnections at all junction boxes and splice points
  – Conduit routing and distances
• Home run cables and their associated low-voltage riser drawings

Nurse call, fire, and security systems are programmed at the time of installation to meet the needs of the user. A detailed record of this programming must be kept with the system documentation. Manuals and specifications sheets from all equipment manufacturers should be put into a binder or pouch and placed in a location designated by the owner's representative.

TYPICAL CONTENTS

- PRINCIPLES OF OPERATION
- FUNCTIONAL DESCRIPTION OF CIRCUITRY
- CONTROLS AND INDICATORS
- OPERATING PROCEDURES
- MAINTENANCE PROCEDURES
- SAFETY CONSIDERATIONS & PRECAUTIONS
- SCHEMATICS AND WIRING DIAGRAMS
- MECHANICAL LAYOUT AND PARTS LIST

VENDOR O & M MANUAL FOR COMPONENT XYZ

306F02.EPS

Figure 2 Typical contents of vendor O&M manual.

2.6.0 System Acceptance

After all the requirements of the commissioning plan have been completed and correctly recorded, the CA recommends final acceptance of the system. This is done when the CA submits to the owner or designated representative a commissioning report for system acceptance. This report should point out any corrected or uncorrected variations from the design or specifications. In addition, it includes all the related system documents specified in the commissioning plan for inclusion in the system manual. Copies of the warranties are given to the owner, and the appropriate suppliers are notified via a substantial completion form that their equipment has been put into operation.

2.7.0 Post-Acceptance Activities

The system commissioning process ends with acceptance of the system. However, some contracts have a provision for post-acceptance activity to ensure the effective, ongoing operation of the installed system for a predetermined length of time. Post-acceptance activities usually involve providing assistance and training for system preventive maintenance tasks and/or in the correction of malfunctions. It may include the evaluation of any planned modifications to the equipment and the update of as-built documentation to reflect any post-acceptance modifications. It can also involve periodic retesting of the system to measure its actual performance.

3.0.0 USER TRAINING

As described earlier, an important part of the commissioning process involves training of the user's operator and maintenance personnel on how to operate and maintain their new system. One or more of the system contractor's technical personnel are typically given this training responsibility. However, many electronic system technicians have little or no experience in conducting such training. The training course instructor must know what to teach and how to teach. For this reason, the remainder of this section provides information and guidelines about how to prepare for and conduct a successful user training program. The steps involved in preparing for and conducting user training include the following:

- Determine the scope of the training
- Prepare instructor materials
- Establish trainee prerequisites
- Prepare the training site and equipment
- Conduct the training
- Evaluate the trainee and training course
- Course completion

3.1.0 Determining the Scope of the Training

For systems involved in the system commissioning process, the scope of the user's training program was determined as part of the system commissioning plan development. At that time, the CA and owner met with each of the system contractors to first analyze what specific operating and maintenance tasks must be performed on the new system. They then determined what new knowledge and/or skills the system's operating and maintenance personnel must have to perform each of the tasks. This information forms the basis for developing the user training course content.

The analysis to determine the scope of training takes into consideration the extent of user involvement in the operation and maintenance of the system. For example, if the user intends only to operate the system, then training on system maintenance and troubleshooting need not be covered. Similarly, if the user intends to operate and only perform basic preventive maintenance tasks on the system, then troubleshooting need not be covered. Other factors taken into consideration are the current technical skills and capabilities of the user's trainees and their expected roles in operating and/or maintaining the system.

As the scope of the training is explored, there are two approaches to providing facility staff with the knowledge and skills needed to successfully operate and maintain the various systems. The following topics examine how training is conducted in both the classroom and on the job.

3.1.1 Education and Training are Different

Education is a gradual process by which knowledge and facts are transferred, and certain types of skills such as reading, writing, and other cognitive processes are developed. Training, on the other hand, is designed to transfer and develop skills needed immediately.

Training often involves the passing on of expertise that enables the performer to be productive quickly. It is not always necessary for a performer to fully understand the reason for an activity, so long as they can accomplish the task quickly and efficiently. For example, users of an intrusion detection system do not need to understand electronics, actuators, or complex circuitry to turn a howling alarm off. They just need to know which buttons to press, the proper order to press them, and if they get confused, how to clear their actions and start over. On-the-job training is designed to meet these kinds of needs.

There are two types of training requirements. The first kind of training involves informing facility staff about aspects of the various systems. This is often best accomplished in a classroom setting with multiple audiences and is generally more educational in nature. There will also be in-

dividuals or groups who need specific instruction on the operation of the systems. This may be accomplished in a classroom, but, more than likely, you will have to use the installed equipment and perform the instruction at the actual equipment location.

3.1.2 Determine Requirements for Education

One way to think about the difference between education and training is to consider the knowledge requirements. Before an individual can maintain any system, they need certain background knowledge. For example, we can deduce that maintenance personnel must understand electronics, actuators, and complex circuitry to effectively diagnose and troubleshoot an alarm system. These background needs, though obviously skill-based, are probably best thought of as educational in nature.

Ask yourself if the learner can perform a task based only on procedural, step-by-step information. If the answer is no, learners require some form of education to develop the background knowledge and skills.

You can then focus your efforts on identifying what knowledge and information is needed. This doesn't mean you develop or teach the results of this exercise. It simply lays a good foundation for identifying which aspects could or should be covered in your program. You don't want to teach electronics, but if you are providing support for a public address system, you can identify where the background of the learners meets with the requirements of the system. For example, you may need to spend some time discussing balanced lines for connecting microphones, or the use of transformers for constant-voltage systems.

3.1.3 Determine Requirements for Training

If the learner can use only step-by-step instructions to accomplish a given task, then you have a good candidate for on-the-job learning. One of the key elements of training is that learners need to perform certain activities in a step-wise manner. This means that tasks, such as adding new employees to the access control database, must be done in a certain way.

3.2.0 Instructor Preparation

As the instructor, you are seen by the trainees as an expert on the system. Remember, even experts do not know everything there is to know about a system. Regardless of how knowledgeable you are about the system, you still need to

---

On Site

User Training

It is recommended that the user training program make an effort to have the user's operating and maintenance personnel on site during equipment/system startup, testing/adjusting, and functional performance testing so they can observe these activities.

A Rule of Thumb for Learning Requirements

Professional educators and trainers use a three-point model to help determine the learning requirements: knowledge, skills, and attitudes:

- *Knowledge* – What the learner needs to know
- *Skills* – What the learner needs to do with the knowledge
- *Attitudes* – What the learner is willing to do

Education and training requirements are determined by matching the level of the learner's capabilities to the requirements of the job and identifying the gap between them. The gaps in knowledge and skills are what need to be taught. The learner's attitude determines how much they will learn.

thoroughly examine the on-site system and study the information in the system O&M manuals and other documents. This is necessary to become thoroughly familiar with the system and related documentation on which you will be conducting training.

It is recommended that you compare the actual system installation against the as-built drawings. If not already done, make a list of all of the system's components, including their manufacturers and model numbers. This information helps you identify and use the correct vendor documentation when describing the principles of operation and features of each system component. Also, determine what hands-on operation and maintenance training is needed and what materials and test equipment are required to support such training.

Before you start developing a course outline and lesson plans, write down a detailed set of objectives for the course. This involves making a list of the things the trainees need to know about the system, and the skills they need to acquire. When you have finished preparing the course, use the objectives as a checklist to make sure all the required topics have been covered.

As part of your preparation, make notes about key points to cover about the system and/or each component. Plan when and how to introduce these key points relating to system setup, preventive maintenance, troubleshooting, and repair/replace procedures. Think about what training aids to help reinforce your presentation.

You may also want to make notes about any unusual conditions commonly encountered with a particular item of equipment or the system. For example, if you know that a control on a particular unit is especially sensitive and goes out of adjustment easily, this is good information to point out during your instruction. If, through experience, you can offer any tips on how to minimize such problems, plan to do so because the trainees appreciate this kind of input.

If there is a large number of trainees, or if different groups of the trainees have different operation or maintenance responsibilities, you may want to organize the trainees into more than one class and tailor your course content to each audience. For example, if you are doing training on a nurse call system, you may want to have one class consisting of nurses and doctors who will be using the system and another class for maintenance personnel responsible for servicing the system. Another alternative is to have all the trainees attend beginning classes covering system operation and then limit the remainder of the classes to maintenance personnel.

The final part of your preparation is to generate detailed outlines (*Figure 3*) or lesson plans that describe the specific topics you intend to cover during each session of training. As part of this effort, you must also plan and schedule the total time needed to conduct the course. When determining the course duration, be realistic. Avoid being overly optimistic. Remember, you must allow for trainee breaks and classroom discussions, as well as some unplanned activities. While generating the course outline, double-check to make sure that all intended trainee learning objectives and on-equipment performance tasks are included.

Multiple Instructors

Some courses may require more than one instructor. For example, one instructor might teach system operation, servicing, and troubleshooting to customer technicians. A different instructor might cover operation and programming for day-to-day users of the system.

Three rules should be followed when organizing the content of the course:

- Present topics going from the simple to the complex.
- Present topics in an organized step-by-step manner.
- Present topics in the best learning sequence.

3.3.0 Trainee Qualifications

Once the training course content and duration are determined, you should review it with the system owner. Your expectations concerning the basic skills and abilities of the trainees selected to attend the course should also be discussed. If the owner indicates that the abilities of the trainees are less than you have anticipated, you will have to change the course content to provide any needed additional instruction.

Nothing can be more aggravating than having a scheduled training class disrupted by one or more trainees who are absent because their employer considers other work priorities to be more important than their training. For this reason, you should get a firm commitment from the owner/employer that all trainees are available to attend your training sessions, without interruption, for the duration of the course. Failure to get this commitment can disrupt the smooth flow of your instruction and diminish trainee learning.

3.4.0 Equipment/System Preparation

All sessions involving hands-on training on the system should have been previously scheduled with the building owner. This is necessary so that the system is made available without interruption for your training. Failure to pre-schedule the system for training can severely disrupt the

TOPIC	Planned Time
Session III. Control Panels; FACP Primary and Secondary Power	
A. Control Panels	_____
1. User Control Points	_____
a. Keypads	_____
b. Touch Screens	_____
c. Telephone / Computer Control	_____
2. FACP Initiating Circuits	_____
a. Initiating Circuit Zones	_____
b. Alarm Verification	_____
c. FACP Labeling	_____
3. Types of FACP Alarm Outputs	_____
4. FACP Listings	_____
B. FACP Primary and Secondary Power	_____
Session IV. Notification Appliances; Communications and Monitoring; General Installation Guidelines	
A. Notification Appliances	_____
1. Visual Notification Devices	_____
2. Audible Notification Devices	
3. Voice Evacuation System	
4. Signal Consideration	

306F03.EPS

Figure 3 Partial example of a course outline.

Learning Objectives

Learning objectives are statements that reflect the specific learning requirements covered in a course or training session. Objectives have three parts:

- What the learner will accomplish
- What tools, reference materials, and other support will be provided
- How the learner will know when the objective has been accomplished

For example, given an alarm panel, resistors, and a screwdriver, be able to connect the cables for an individual zone within five minutes or less.

- The *goal*: Connect cabling for a single alarm zone
- The *tools or materials*: An alarm panel, resistors, and a screwdriver
- The *measurement*: Five minutes or less

owner's business activities as well as your training program. It is also important that the building occupants and anyone else who may be affected by your training on the system be made aware of the system's operational status.

3.5.0 Conduct the Training

To have a organized and successful training session, it is essential that you arrive early in order to prepare the training location. This means getting all of your training materials organized and any audiovisual equipment set up and operationally checked out before the trainees arrive. The checkout of equipment is important because it can be very disruptive if the equipment you are working with fails during training.

Locate your training materials in a place where they can be retrieved easily without distracting the trainees. This allows you to conduct the class without excessive pauses or disruptions. You need to organize the area from which you will do most of your speaking. In this area, set up a place to keep your notes handy along with any frequently used materials, such as manuals, markers, and a pointer.

At this time, you may also want to distribute trainee materials to each trainee location. These materials can include a notepad, pencil, or pen. They can also include a tent card or name tag for the trainee to print his or her name on. These are useful because they allow you to address each trainee by name throughout your instruction. If applicable, you may also want to distribute copies of handouts, such as wiring diagrams, site drawings, and manuals, at each location. However, it is recommended that you reserve the distribution of these items until the point in your instruction when they are needed. This is because some trainees will be distracted by leafing through and

reading the content of these items rather than focusing on what you are saying while you are giving instruction.

3.5.1 Course Introduction

It is important to set the proper class attitude from the beginning. Make sure that you are well prepared and practiced to handle the opening as smoothly and efficiently as possible. This helps build your credibility with the trainees from the outset. As the class starting time approaches, you may want to stand at the doorway to greet the students as they arrive. If they are unfamiliar with the training site, show them where to put their jackets or other personal items and then invite them to select a seat and sit down.

After all the trainees are seated, introduce yourself to the class. Tell them your name, your company name, and your job title. Also, give them a brief overview of your background as a way of explaining why you are their instructor. Even if you know some or all of the trainees, this step is important to reinforce your credibility and let them know why you are up in front of the classroom.

There are always a few key points of information that students want to know right up front. These can include the training schedule, the training session start and end times, and the frequency and length of any breaks. If the trainees are not familiar with the facility where the classroom training is being conducted, point out the locations of the bathrooms and vending machines.

The key activity in introducing the course is to review the course objectives with the trainees. The objectives typically are shown with a transparency or slide (*Figure 4*). Each objective should consist of a clear, brief statement about what the trainee should be able to do upon completion of the course.

COURSE OBJECTIVES

1. Explain and/or describe the basic principles of fiber optic technology, including:

    - Fundamentals, benefits, and applications of a fiber optic system

    - Operational considerations of a fiber optic system

    - Construction of an optical fiber

    - Various types of fiber optic cable

2. Describe the design, operation, and performance of a fiber optic transmitter and receiver.

3. State the types and construction of fiber optic detectors.

4. Explain the desirable features and connector losses of a fiber optic connector or splice.

5. Demonstrate the installation of fiber optic cabling and support equipment.

6. Describe and/or demonstrate the applications and types of fiber optic splicing/termination.

7. Explain and/or demonstrate the testing procedures for fiber optic systems.

306F04.EPS

Figure 4 Example of training course objectives.

This knowledge makes the training much more effective because it helps the trainees prepare themselves for what they are about to learn. During the presentation of objectives, point out the specific hands-on skills the trainees are expected to acquire during the training.

3.5.2 Demonstration and Hands-On Practice

Demonstration and hands-on practice are essential when teaching people how to operate and maintain equipment. The most effective training course dedicates a large portion of its time to trainee hands-on experience on the equipment.

A very effective approach for teaching procedures or tasks is the know-show-do approach. Using this approach, you first provide the trainees with the information that they need to know in order to perform the task. For example, if you are teaching them how to adjust an audio system mixing board, you want the trainees to know which system component the mixing board is, which controls on the mixing board they must adjust, what is happening when they adjust the controls, and what output they should expect from the mixing board as they adjust it. Next, you show the trainees how to perform the task on the actual mixing board. Finally, you have the trainees do the task several times until they have mastered it.

3.5.3 Safety Considerations

Safety is an important consideration when doing demonstrations and hands-on instruction on the system. When demonstrating procedures, be sure to stress safety. Also, make sure that all trainee hands-on activities are done safely under your direct supervision.

3.5.4 Guiding Learner Performance

Throughout the trainee hands-on exercises, you should provide guidance as appropriate and give the trainees feedback to help them master the task. Ideally, at the completion of the course, each of the trainees will meet a minimum standard of proficiency at performing all the tasks they need to successfully operate and/or maintain the system. The performance of each trainee at completing each of the required tasks should be evaluated by you and recorded on a suitable form. This step is important so that you can gauge the trainee's progress. It also provides a historical record of the trainee's performance for future reference. *Figure 5* shows an example of a typical form that can be used for recording a trainee's performance.

3.5.5 Classroom Instruction

Classroom instruction normally involves the presentation of factual information, such as describing the function and principles of operation for various system components. Factual information is best presented by using the what-why-how approach to instruction. For example, you might describe an audio system mixing board this way:

Step	Example
What	"This is a mixing board."
Why	"It is used to balance and blend the incoming audio signals to create the desired audio output."
How	"Connect each of the input signal devices into its own input channel on the mixing board and then use the channel-specific controllers to adjust each signal to create a pleasant and well-mixed audio output."

On Site

Teaching Methods

There is more to effective training than simply telling a room full of people what you know about the subject or showing and talking about one transparency after another. Studies have shown that what trainees retain depends on several factors. Generally they retain the following:

- 10 percent of what they read
- 20 percent of what they hear
- 30 percent of what they see
- 50 percent of what they see and hear
- 70 percent of what they see, hear, and respond to
- 90 percent of what they see, hear, and do

PERFORMANCE SHEET

TRAINEE NAME: _____

TRAINEE ID NUMBER: _____

CLASS: _____

TRAINING PROGRAM SPONSOR: _____

_____

INSTRUCTOR: _____

Rating Levels: 1. Passed: performed task.

2. Failed: did not perform task.

TASK	RATING
1. Demonstrate the application of a selected connector to fiber optic cable.	
2. Demonstrate the application of a selected splice to fiber optic cable.	
3. Demonstrate a selected testing procedure for fiber optic cable.	

306F05.EPS

Figure 5 Example of a trainee performance sheet.

3.6.0 On-the-Job Learning

On-the-job learning (OJL) is training that is oriented to the performance of specific tasks. It involves both the transfer of factual information and the development of hands-on skills. An emphasis on task performance separates it from classroom activities.

When performing OJL, the trainer is an active participant in almost all activities. Unlike a classroom environment, where you have several learners involved, OJL activities are often performed one-on-one. That is, the trainer and the learner work together. This requires high skill proficiency from the trainer for each of the tasks being presented.

A typical OJL session follows a similar set of methods to those already presented in this section. However, because OJL is performance-based, the model can be simplified. For OJL purposes, restate the process as follows:

- Prepare
- Present
- Try-out performance
- Follow-up

3.6.1 Preparing for OJL

Preparation activities are similar to those outlined previously. You want to practice running through each of the steps in the various tasks to ensure your competency. You also want to pull together any supporting materials, such as bills of materials, job aids such as quick reference guides, blueprints, and any other documentation.

3.6.2 Presenting OJL

Presentation of job skills requires a high degree of patience. You typically present each step and task in its entirety. However, don't expect to simply tell the learner what needs to be done. You actually perform each of the tasks, and demonstrate them several times. The first time you walk through the steps with the learner, you simply focus on the desired outcomes. For example, to set building management parameters for the fire alarm system, you will inform the trainee of the goal, walk through each step, and highlight the end result.

It is a good idea to cover the entire set of tasks in this fashion. You then start over, and perform them all again. On the next several passes, though, you will gradually identify more detail as you progress. Be sure to stress key steps, particularly those that may generate more than one result. Be sure to point out where and when the learner can determine if the step is completed. This may be a specific quality measurement that can be taken or applied to ensure correctness and completeness.

For example, when teaching nurses the procedures for responding to a bed alarm, point out how to turn the alarm off, and how they will know the alarm has actually been terminated at the duty station. Continue with this level of support as you progress through each of the tasks. During these multiple passes through the job elements, you should stop and ask questions of the learner. For example, you may want to ask the learner to tell you the rule-of-thumb to determine if an alarm has been terminated. This feedback provides for two-way communication. You know how well the learner is doing as they integrate the information and form their own mental model of the job. The learner also gains a better sense of the progression of steps as they lead to task mastery.

The key here is obviously not your performance level, though it does serve as model or template for the trainee to learn and apply. At some point you have to let the learner begin performing the various steps and tasks. As the trainer, you must be sensitive to the learning process and recognize that expert performance is the result of continued practice. Don't be surprised if the learner masters one aspect of a given task, but misses another. Your role is to ensure they develop a satisfactory mastery of all the tasks.

3.6.3 Tryout Learner Performance During OJL

The third step in the OJL process is the tryout performance. This is when the learner begins building their individual skills. You begin by encouraging the learner to perform the individual steps that make up the task. You start by turning over the hands-on activities to the learner. You walk them through each step several times by telling them what to do, step-by-step, repeating the key points and quality checks. Then, gradually, you volunteer less and less, as they take over the responsibilities.

Your role at this point is to ensure that the skills are being learned and applied. You need to stop the learner from time to time, and ask questions. For example, you may ask, "Now that you have responded to a bed alarm, how do you ensure that the alarm has been terminated at the duty console?"

If you determine that the learner is about to make a mistake, there are several strategies you can apply. If the mistake is non-threatening, that is, it doesn't create a dangerous condition or result in significant waste, you may let the learner continue. You want to talk through the error, of course, and help the learner identify what they should do differently in the future. The point is, the learner needs to be able to identify possible errors so they can avoid them.

If a mistake will result in a hazardous situation, you must stop the learner immediately. You will have to help the learner recover and restore the task to a pre-hazardous state. Be sure to involve the learner in the process and do not criticize them for the mistake. Rather, talk with them about the conditions, actions, or whatever the root cause of the mistake was. The idea is for them to be able to identify when a given situation is happening so they can either avoid it or recover from it in the future. Remember, you will not be there to guide them after some point in the training process.

Eventually, the learner will be performing all the steps and tasks. Your involvement is then to observe and help the learner fine-tune the application of skills.

3.6.4 Evaluating OJL Performance

The evaluation of the learner's performance is the final step in the OJL process. This involves focusing on the areas where the learner is doing well, as well as the portions of the tasks that need additional support. The most difficult aspect of this is to provide appropriate feedback, both to the learner and to the building owner or agent.

Many tasks require a level of performance that can only improve beyond a certain point after many repetitions. You must be willing to let the learner develop their skills. A good way to evaluate the learner is by developing a checklist that you use to signify that the learner has achieved an acceptable level of performance. Share this with both the CA/owner and the learner. Indicate for each skill that the learner can perform it.

3.7.0 Course Closure

One of the best ways to end a training course is to review with the class the course objectives that were initially listed at the opening session of the course. If in the icebreaker activity you asked the trainees to list what their expectations for the course were, it is a good idea to review if and how their goals were met during the class. This activity helps reinforce the trainees' satisfaction with what they have learned and accomplished by attending the course.

Trainee feedback on your performance as an instructor and about the overall content of the course is another important step in the course closure process. This trainee input is important so that you can judge the quality of your instruction. Should you be given the responsibility to teach another course, this input will help you make improvements. Obtaining trainee feedback about the course can be done easily and informally by passing out an evaluation form to each trainee and asking them to fill it out. *Figure 6* shows a typical evaluation form that can be used for this purpose.

At the end of the course, the instructor usually has a requirement to provide the building owner with a formal verification that the training course has been completed. *Figure 7* shows a typical Training Completion Acknowledgement form. It is also common to give each of the trainees a personalized certificate of course completion to acknowledge their accomplishment. A copy of this certificate should also be sent to their employer's responsible human relations person for inclusion in their employee records.

TRAINING COURSE EVALUATION FORM

Course:

Date:

Instructor:

Name (optional):

Subject Matter	Excellent	Satisfactory	Poor
Useful to my job			
Well organized			
Well presented			

Instructor	Excellent	Satisfactory	Poor
Preparation			
Organization			
Handling questions			
Presentation skills			
Control of group			
Speaking ability			
Eye contact			
Knowledge of the subject			

Documentation	Excellent	Satisfactory	Poor
Covered adequately in the class			
Easy to use			

If you rated anything "poor," please explain.

Did the course move along too slowly, too quickly, or at just the right pace?

What other comments or suggestions do you have regarding this course?

306F06.EPS

Figure 6 Example of a training course evaluation form.

TRAINING COMPLETION ACKNOWLEDGEMENT

Date: _____

Time: _____

Product: _____

Instructor: _____

Customer Name: _____

Attendees:

_____   _____

_____   _____

_____   _____

_____   _____

_____   _____

Owner's Rep: _____

General Contractor: _____

Instructor: _____

306F07.EPS

Figure 7 Example of a training completion form.

SUMMARY

System commissioning is a process in which a formal and organized approach is taken to verifying and documenting the correct installation and performance of a particular system or systems. The goals of the system commissioning process include the following:

- Verifying that the system design meets the functional requirements of the owner
- Verifying that all systems are properly installed in accordance with the design and specifications
- Verifying and documenting the proper operation of all equipment, systems, and software
- Verifying that all documentation for the system is accurate and complete
- Training customer personnel to efficiently operate and maintain the installed equipment and systems

Training the user's personnel to operate and maintain the system is an important part of the commissioning process. The nature and duration of the training program depend on the size and complexity of the system, and on the availability of the system for hands-on training. User training should include a complete overview of all system equipment and components with a focus on:

- Contents and scope of the operation and maintenance documentation
- System operational procedures for all modes of operation
- Acceptable tolerances for system parameters in all operating modes
- Procedures for coping with abnormal and/or emergency situations

1. At what point in the process will the CA get involved?

   a. Before the start of system installation
   b. With the start of system installation
   c. When installation of the equipment and cabling is complete
   d. At system startup

2. The commissioning plan is prepared and distributed by _____.

   a. the building owner
   b. the commissioning authority
   c. the individual system designers
   d. all building project team members

3. All of the following are installation phase commissioning activities *except* _____.

   a. checking integrity of system wiring and cable connections
   b. system startup
   c. initial system testing and adjustments
   d. functional performance testing

4. Successful completion of the functional performance tests is the end of the on-equipment system commissioning activities.

   a. True
   b. False

5. The system acceptance phase is complete when _____.

   a. functional performance testing is complete
   b. user training is complete
   c. both functional performance testing and user training are complete
   d. the CA submits a commissioning report to the owner

6. The first thing that should be done when preparing for a user's course is to _____.

   a. review the system O&M manual
   b. determine the scope of the training
   c. spend time on site learning about the system
   d. review the drawings and specifications

7. During the instructor preparation phase, what should the instructor prepare describing the specific topics to be covered for each training session?

   a. System-related technical documentation
   b. A complete set of overheads
   c. Detailed lesson plans
   d. An end of course test

8. The first thing an instructor should do before the start of any class is _____.

   a. set up and check out the electronic audio-visual equipment
   b. take trainee attendance
   c. introduce himself or herself to the trainees
   d. request that all cell phones and pagers be turned off

9. The key activity in introducing a course is to _____.

   a. establish credibility with the trainees
   b. review course objectives with the trainees
   c. discuss the know-show-do approach
   d. ask the trainees to list their expectations for the course

10. The best way to teach trainees how to perform a new procedural task is to _____.

    a. show them a video
    b. explain what the task is, why it is done, and how to do it
    c. demonstrate the task and then allow the trainees to practice the task
    d. review the task as described in the appropriate O&M manual

Trade Terms Introduced in This Module

Acceptable performance: The condition where a component or system meets specified design parameters under actual operating conditions.

Authority having jurisdiction (AHJ): The person or department within a municipality or jurisdiction having legal authority to approve a system that requires code conformance.

Commissioning authority (CA): A qualified person, company, or agency responsible for planning and carrying out all system commissioning related activities.

Commissioning plan: A document that outlines the organization, scheduling, allocation of resources, method of acceptance testing, and documentation pertaining to the overall commissioning of a system.

Functional performance testing: The collective body of checks and tests performed to verify that all components, subsystems/systems, and interfaces between systems are functioning according to specification requirements.

Additional Resources

This module presents thorough resources for task training. The following resource material is suggested for further study.

Procedural Standards for Whole Building Systems Commissioning of New Construction. Gaithersburg, MD: National Environmental Balancing Bureau.

National Fire Alarm and Signaling Code Handbook, NFPA 72. Quincy, MA: National Fire Protection Association.

Figure Credits

Associated Builders & Contractors, Inc., Module opener

CONTREN® LEARNING SERIES — USER UPDATE

NCCER makes every effort to keep its textbooks up-to-date and free of technical errors. We appreciate your help in this process. If you find an error, a typographical mistake, or an inaccuracy in NCCER's Contren® materials, please fill out this form (or a photocopy), or complete the online form at www.nccer.org/olf. Be sure to include the exact module number, page number, a detailed description, and your recommended correction. Your input will be brought to the attention of the Authoring Team. Thank you for your assistance.

Instructors – If you have an idea for improving this textbook, or have found that additional materials were necessary to teach this module effectively, please let us know so that we may present your suggestions to the Authoring Team.

NCCER Product Development and Revision

3600 NW 43rd Street, Building G, Gainesville, FL 32606

Fax: 352-334-0932
Email: curriculum@nccer.org
Online: www.nccer.org/olf

❏ Trainee Guide   ❏ AIG   ❏ Exam   ❏ PowerPoints   Other _____

Craft / Level: _____   Copyright Date: _____

Module Number / Title: _____

Section Number(s): _____

Description: _____

_____

_____

_____

_____

Recommended Correction: _____

_____

_____

_____

_____

Your Name: _____

Address: _____

_____

Email: _____   Phone: _____

Maintenance and Repair

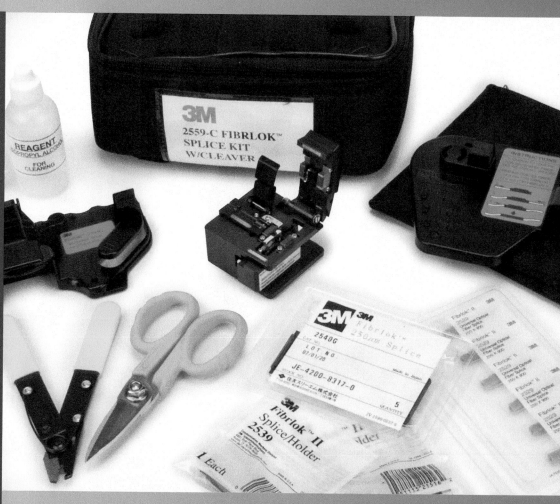

33307-11

Trainees with successful module completions may be eligible for credentialing through NCCER's National Registry. To learn more, go to **www.nccer.org** or contact us at **1.888.622.3720**. Our website has information on the latest product releases and training, as well as online versions of our *Cornerstone* newsletter and Pearson's Contren® product catalog.

Your feedback is welcome. You may email your comments to **curriculum@nccer.org,** send general comments and inquiries to **info@nccer.org,** or use the User Update form at the back of this module.

V.1 4/11

Objectives

When you have completed this module, you will be able to do the following:

1. Explain the difference between maintenance and repair.
2. Identify the common causes of system and equipment failures.
3. Use electrostatic discharge (ESD) control devices and techniques when handling ESD-sensitive equipment and components.
4. Use manufacturers' troubleshooting aids to identify system problem(s).
5. Isolate computer-related problems to hardware or software.
6. Isolate common faults in wiring and equipment.
7. Identify common preventive maintenance measures.
8. Identify and explain preventive maintenance and inspection schedules.

Performance Tasks

Under the supervision of the instructor, you should be able to do the following:

1. Use ESD control devices and techniques when handling and troubleshooting ESD-sensitive equipment or components.
2. Use manufacturers' troubleshooting aids to identify system problem(s).
3. Determine if a power supply is good or bad.
4. Determine if a printed circuit board is good or bad.
5. Isolate the cause of a computer-related problem to the hardware or software.
6. Isolate common faults in copper and fiber optic cable wired networks.

Trade Terms

Fault isolation diagram
Listed

Preventive maintenance
Repair

Troubleshooting

Troubleshooting table

Required Trainee Materials

Latest edition of the *National Electrical Code®*

Note: The designations "*National Electrical Code®*," "*NE Code®*," and "*NEC®*," where used in this document, refer to the *National Electrical Code®*, which is a registered trademark of the National Fire Protection Association, Quincy, MA. All *National Electrical Code®* (*NEC®*) references in this module refer to the 2011 edition of the *NEC®*.

Contents ———————————————

Topics to be presented in this module include:

Figures and Tables

1.0.0 INTRODUCTION

This module introduces background information and the tasks involved in the maintenance and repair of low-voltage systems and equipment. A systematic approach to system and component-level troubleshooting is covered, as are identifying problems requiring common repair techniques. Background information and general guidelines pertaining to the various tasks involved with preventive maintenance are also given.

2.0.0 MAINTENANCE VERSUS REPAIR

There is sometimes confusion in the trade as to the meaning of the terms *maintenance* and *repair*. Maintenance, also known as preventive maintenance, is defined here as any task that is performed on a regular basis in order to ensure the proper operation of the equipment and to prevent equipment failures. Changing the oil in your car in accordance with the manufacturer's schedule is a good example of a preventive maintenance task. Failure to perform this procedure eventually leads to engine failure. Typically, equipment manufacturers recommend that specific equipment maintenance tasks be performed on a daily, monthly, semi-annual, or annual basis. The tasks involved with preventive maintenance are covered in more detail later in this module.

Repair is defined here as those tasks performed to fix a known problem. These tasks can include the replacement or repair of a system or some of its components when they become unreliable or inoperable. Repair is accomplished after troubleshooting procedures are used to locate the source of a problem or malfunction.

3.0.0 CAUSES OF FAILURES

Equipment and cable failures occur naturally over time. However, some failures can occur prematurely as a result of other factors, including the following:

- Environmental conditions
- Improper installation
- Poor power quality
- Electrostatic discharge
- Operator error

3.1.0 Environmental Conditions

Failures in low-voltage systems and equipment can result from exposure to harmful environmental conditions, such as damp or wet locations; exposure to gases, fumes, vapors, or other agents that have a deteriorating effect on the conductors or equipment; or exposure to excessive temperatures. The *NEC*® requires that any equipment or cabling used in harsh environments be listed (suitable) for the intended purpose. When determining the possible cause of an equipment failure, you should be able to recognize when a piece of equipment or its related cabling is not suitable for the environment in which it is installed.

Electrical/electronic equipment and cabling subjected to moisture can fail as a result of short circuits and/or high-resistance connections due to corrosion at equipment terminals, cable terminations, and relay contacts. Definitions of dry, damp, and wet locations are given in *NEC Article 100* (see *Locations*). If the building in which a system is housed has cracks or other unsealed openings in its floors, walls, or ceilings, rain or other weather elements can enter the building and cause moisture problems in the system.

Dirt and other contaminants such as fumes, vapors, abrasives, soot, grease, and oils can cause electric/electronic devices to operate abnormally until they finally break down. The harmful effects of gases, fumes, vapors, and similar agents can cause corrosion at equipment terminals, cable terminations, and relay contacts. In some cases, the wiring insulation can deteriorate, causing a

---

---

short circuit or ground fault. Equipment and wiring that is exposed to the harmful effects of gases, fumes, vapors, and similar agents is subject to the requirements for hazardous locations covered in *NEC Chapter 5*.

Equipment that is subjected to unusual amounts of dust and other airborne particles tends to fail due to overheating caused by clogged air filters or the loss of the free air movement required to dissipate the heat generated in the unit. Papers, boxes, and other materials that block free air movement to equipment are another cause of heat-related failures.

Heat can cause materials to expand, dry out, crack, or blister. Equipment that is exposed to temperature extremes often fails as a result of an open circuit in the printed circuit board wiring, an open connection between an integrated circuit chip pin and the printed board wiring, or a heat-related failure in an integrated circuit. Extreme temperatures can also result in damage to cabling insulation. Other environmental conditions to consider include floor openings above a device or panel assembly. Water or other liquids and debris can fall from above, causing damage. Also be aware of nearby water and steam pipes, which can condense and drip on electrical equipment.

3.2.0 Improper Installation

Improper installation of equipment and cabling by an unqualified or careless installer can result in failures. All systems and equipment must be installed in accordance with the manufacturer's instructions and the current national and local codes. As previously mentioned, installation of the wrong type of equipment for a particular environment can result in failures. Incorrect mounting or support of equipment enclosures can cause vibration

or shock. This can result in increased susceptibility of the equipment to damage and loose connections. Vibrations and physical abuse often cause breakdowns. Failure to properly tighten terminal lugs or to properly terminate a connection can result in a device failing prematurely.

Cable shorts or opens can occur as a result of staples installed too tightly in cable runs, or cables being dragged over sharp edges during installation. Cables installed in high traffic areas can be damaged by being stepped on. The impedance or resistance of some cables installed near electrical equipment can be affected if the cable is nicked, cut, or crimped. Cables installed alongside or over fluorescent lights or too close to electrical junction boxes, hot water pipes, or heating ducts can fail due to overheating. Similarly, cables improperly run in the same raceway with power circuits can cause electrical noise to be induced into voice or data systems. The *NEC®* is very specific regarding the required spacing between power and data/signal cables.

Manufacturing defects are common. For example, you may find a loose circuit board after delivery or installation. Also, circuit boards and components can be damaged during shipment and installation.

3.3.0 Poor Power Quality

Abnormal primary power conditions can cause equipment to operate erratically or become damaged and fail if not protected by the proper devices. The waveform characteristics of the utility power can become distorted by lightning strikes, severe weather, utility power fluctuations, or changing building load conditions, all of which can result in poor power quality. Some of the common types of power abnormalities include the following:

- *Sags* – Sags, also called brownouts, are short-term decreases in voltage levels. Sags are typically caused by the startup power demands of many electrical devices, such as motors, compressors, elevators, and shop tools. Sags can also be deliberately used by the power company as a way of coping with extraordinary power demands. In a procedure called rolling brownouts, the power company systematically lowers voltage levels in certain areas for hours or days at a time. Typically, this occurs in urban areas in the summer, when air conditioning requirements are at their peak. Sags can starve electronic equipment of the power needed to function properly. In computer systems, sags can cause frozen keyboards, unexpected system crashes, and the loss or corruption of data. In other electronic equipment, sags may stress components, causing them to fail prematurely.

- *Blackouts* – A blackout is a total loss of utility power. Blackouts are typically caused by excessive demands on the power grid, lightning storms, and ice on power lines. In computer systems, this can cause the loss of work stored in RAM or cache, and possible loss of the hard drive and its stored data.
- *Spikes or transients* – Spikes or transients are instantaneous, dramatic increases in voltage, typically caused by a nearby lightning strike or utility power coming back online after having been knocked out in a storm. Spikes can cause damage to hardware and loss of data.
- *Surges* – Surges are short-term increases in voltage. They typically occur when large electric motors are switched on or off, resulting in the distribution of excess voltage through the power lines. Surges stress components and may cause premature failures in electronic equipment.
- *Electrical noise* – Electrical noise, also called electromagnetic interference (EMI) and radio frequency interference (RFI), distorts the smooth utility power sine wave. Electrical noise is caused by many factors, including lightning strikes, load switching, generators, radio transmitters, and industrial equipment. Noise introduces glitches and errors into computer software programs and data files.
- *Harmonics* – A harmonic is a waveform with a frequency that is an integral multiple of the fundamental system frequency. For example, the harmonics of a 60Hz sine wave would be 120Hz (second harmonic), 180Hz (third harmonic), 300Hz (fifth harmonic), and so on. When one or more harmonic components are added to the fundamental frequency, a distorted or nonlinear waveform is produced. Harmonics are common where there are several types of equipment (such as switching power supplies, personal computers, adjustable speed drives, or medical test equipment) that draw current in short pulses. Electronic ballasts used in fluorescent and high-intensity discharge (HID) lighting fixtures are also common sources of harmonic generation. The presence of harmonics can result in overheated transformers and neutrals, as well as tripped circuit breakers.

3.4.0 Electrostatic Discharge

Electrostatic discharge (ESD) is the charge produced by the transfer of electrons from one object to another. Static electricity is generated as the result of electron movement between two different materials that come in contact with each other and are then separated. When the two materials are good conductors, the excess electrons in one will return to the other before the separation is complete. However, if one of them is an insulator, both will become charged by the loss or gain of electrons (unless grounded). An example of this is when you get an electric shock from touching a doorknob after walking across a carpet. Static electricity charges can have potentials that exceed several thousand volts. Some examples are as follows:

- Walking across a carpet – 35,000V
- Walking over a vinyl floor – 2,000V
- Vinyl envelopes for work instructions – 7,000V
- Picking up a common plastic bag – 20,000V
- Touching a work chair padded with polyurethane foam – 18,000V

Electrostatic discharges that occur at or near electronic equipment can damage sensitive semiconductor circuits and devices. Most static electricity shocks that humans can feel have a voltage level in excess of 2,000V. Static electricity charges lower than this level are normally below the threshold of human sensation. However, these lower-voltage static electricity charges can damage sensitive semiconductor devices. Some semiconductor devices can be damaged by electrostatic charges as low as 10V. The *Appendix* provides a summary of static electricity charge levels, as well as common semiconductor devices that can be damaged at the various levels.

When subjected to ESD, a sensitive electronic component can experience a catastrophic failure, meaning it is instantly damaged to the point where it is totally inoperative. ESD can also cause a component to fail in a way that allows it to continue to work properly for a while, but results in poor system performance and eventual system failure. This type of failure is called a latent or hidden failure. Another type of failure caused by ESD, called an upstart failure, can cause a current flow in an integrated circuit that is not large enough to cause a catastrophic failure, but may result in gate leakage that causes the intermittent loss of data.

As an electronic systems technician, you must be continually alert to the threat of ESD and understand how to control it. Control of ESD can be accomplished through grounding, isolation, or neutralization.

3.4.1 Grounding

Transients can also be created by static electricity from your body. For this reason, it is important to avoid touching the components, printed

Maintenance and Repair

circuit, and connector pins when handling sensitive printed circuit (PC) boards or components. Always ground yourself before touching the PC board or components. This is normally done by using a grounding device that directly connects you to ground. This eliminates any static charge from being generated by your body.

ESD grounding wrist straps (*Figure 1*) are the most common grounding method used, especially for field work. To provide protection from ESD, the wrist strap must be properly grounded to either the utility ground or to a common ground point on a properly grounded work surface. It must also make good contact with your skin. Other devices used for ESD protection in a shop or lab environment can include conductive footwear and workstations equipped with properly grounded conductive or dissipative work surfaces used in conjunction with a grounded conductive floor or floor mat.

3.4.2 Isolation

Isolation of sensitive components and assemblies during storage and transportation is another way to prevent ESD-related component failures. Static charges cannot penetrate containers that are made of conductive materials. For this reason, sensitive devices are normally contained in metallized shielding bags, conductive bags, and/or conductive tote boxes (*Figure 2*) during storage and/or transportation. It is important to point out that the outside surfaces of these conductive containers can carry static charges that must be removed before the container is opened. This discharge can only be safely accomplished by a properly grounded person opening the container at a static-safe location.

3.4.3 Ionization

Another method of ESD control involves the use of ions to neutralize static charges. Grounding and isolation do not dissipate charges from insulating materials such as synthetic clothing or common plastics. To dissipate these charges, they must be neutralized. The removal or neutralization of the charge on these materials occurs naturally over time by a process called ionization. Ions are charged particles that are always present in the air. They result naturally from sunlight, lightning, and radiation. Ions that get close enough to a charged insulator are attracted to it if they are

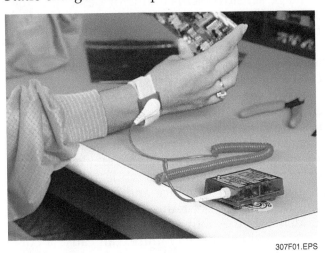

307F01.EPS

Figure 1 ESD grounding wrist strap.

307F02.EPS

Figure 2 ESD prevention tote box.

On Site

ESD Control

Electrostatic discharges can be grounded through a proven building ground. This is done by connecting an ESD grounding wrist strap into a grounding plug. The grounding plug also tests for correct receptacle voltage polarity and ground connections. When plugged into the receptacle, it does this by showing the wiring status of the receptacle via three indicator lamps. A label on the grounding plug relates the lamp status (on or off) to the actual wiring condition of the receptacle. It is important to always test the receptacle wiring integrity first with the grounding plug before you connect the ESD wrist strap to the grounding plug.

oppositely charged. These oppositely charged ions cause the charge on the insulator to be neutralized.

Ions can be artificially produced by a device called an ionizer. It uses high voltage or radiation to generate a balanced mixture of billions of positively charged and negatively charged ions. These ions are blown by fans toward the object or area to be neutralized. The use of an ionizer can cause the charge on an insulator to be neutralized in a matter of seconds, thereby reducing its potential to cause harm.

3.4.4 Guidelines for Preventing Component Damage

Some guidelines that help prevent failures associated with ESD include the following:

- Never enter an ESD-sensitive area without taking the proper precautions.
- When working with ESD-sensitive devices, make sure you ground, isolate, and neutralize each device.
- Keep the work area clean and clear of unnecessary materials, particularly common plastics.
- Test grounding devices such as wrist straps daily and make sure they have not become loose or intermittent.
- Always ship and handle ESD-sensitive components in metallized shielding bags, conductive bags, and/or conductive tote boxes.
- When handling ESD-sensitive printed circuit (PC) boards or components, avoid touching the components, printed circuit, and connector pins.
- When handing an ESD-sensitive PC board to another person, both individuals must be grounded to the same ground point or potential.
- Maintain the room humidity at a minimum of 40 percent, if possible.
- Ensure that all carpets, as well as the clothing of operating personnel, are made of natural materials.
- If carpets are made of synthetic materials, they should have a conductive backing.
- Ionize the air to increase conductivity.
- Ensure that equipment operators wear conductive footwear.
- Use antistatic materials whenever possible to increase conductivity.
- Follow safety guidelines at all times. Use lockout/tagout procedures before working on equipment that must be de-energized before servicing. Be alert to moving parts or machinery in the area where you are working.

3.5.0 Operator Error

Some problems are due to customer or operator error. Such failures can result from inadequate training in the use of the equipment or related software. Failures can occur as a result of the customer or operator not performing the equipment or software maintenance tasks recommended in the operator's/user's manuals. In systems under computer control, the computer operator normally has access to a set of software tools that are run periodically to protect the computer system from crashes and viruses, automatically identify and repair many computer problems, and help maintain the computer to avoid problems in the future.

4.0.0 TEST EQUIPMENT

In order to effectively troubleshoot and test system components and cabling, the technician must have the appropriate test equipment available and be familiar with its operation (*Figure 3*). All items should be operated in accordance with the manufacturer's instructions. Detailed descriptions and uses for various items of test equipment were covered in the Level Two modules, *Electrical Test Equipment* and *Voice and Data Systems*. Depending on the type of troubleshooting or testing involved, the common types of test equipment needed can include the following:

- *Common hand tools* – These include tools such as screwdrivers, pliers, diagonal cutters, and wire strippers.
- *Analog or digital multimeter (VOM/DMM)* – Used for a variety of tests, including voltage, resistance, and current measurements; and for finding opens, shorts, and high-resistance connections in cables and connectors. Make sure the test leads are rated for the voltage being tested.
- *Clamp-on ammeter* – Used to make higher AC current measurements. It can be used to measure the total AC current being drawn by a system, or by the individual loads in a system. The advantage of a clamp-on ammeter is that current measurements can be taken without interrupting the circuit.
- *Oscilloscopes and probes* – Used to display the actual shape of voltages that are changing with time so that waveform measurements can be made, and one waveform can be compared to another. They are also used to measure frequency, duration, or time of occurrence of one or more cycles, phase relationships between voltage waveforms appearing at different points in the circuit, shapes of the waveform, and amplitude of the waveform.

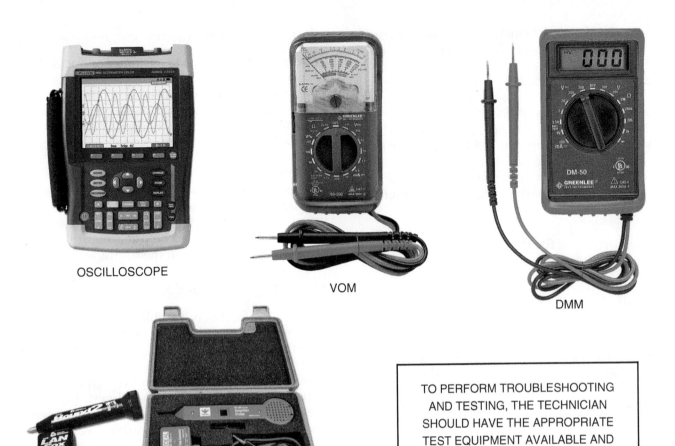

OSCILLOSCOPE

VOM

DMM

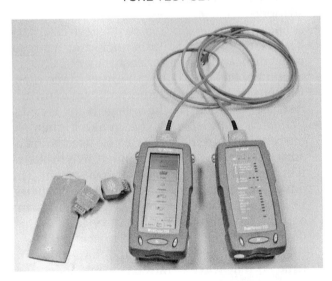

TONE TEST SET

TIME-DOMAIN
REFLECTOMETER

307F03.EPS

CABLE CERTIFICATION TESTER

CLAMP-ON AMMETER

Figure 3 Troubleshooting and testing require the use of the appropriate test equipment.

Digital Multimeter Classifications

Newer digital multimeters are rated according to voltage and current limitations and fault interrupting capacity. Never use a multimeter outside of the limits specified by the manufacturer.

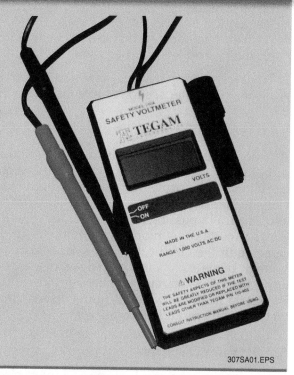

307SA01.EPS

- *Wire map tester* – Used primarily with UTP cabling to test for opens, shorts, crossed pairs, and improper wiring.
- *Handheld cable tester* – Used to test UTP, ScTP, STP-A, and coaxial cabling for opens, shorts, crossed pairs, and improper wiring. This tester also has the ability to measure length, near-end cross talk (NEXT), attenuation, and induced noise.
- *Toner/wand* – Used to identify and trace cables.
- *Telephone test set* – Used to simulate the user's telephone and identify voice circuits.
- *Certification test set* – Used for the same functions as the handheld cable tester, with an additional function that allows verification of whether a cabling system meets the transmission performance requirements of standard TIA/EIA TSB-67.
- *Cable tracer* – Used to generate and receive a signal in order to trace the path of a cable through walls, ceilings, and floors.
- *Optical light source and power meter* – Used with fiber optic cable to perform attenuation measurements.
- *Time-domain reflectometer (TDR)* – Used to locate faults on almost all types of metallic cables, including power cables, data cables, and CATV/CCTV cables. It can also be used to measure the length of cable in a cable run. Similarly, an optical time-domain reflectometer (OTDR) is used to test fiber optic cables.

- *Optical fiber flashlight* – Used to generate a safe light that can be viewed with the naked eye in order to test the continuity of fiber optic cable strands.
- *Impedance meter* – Used to check the impedance of constant-voltage speakers.
- *Low-intensity laser* – Used to generate a red laser beam in order to identify individual optical fibers within a cable, or to locate an open in an optical fiber.
- *Megohmmeter* – Used to test the high resistances of conductor insulation and transformer windings.
- *Frequency meter/counters* – Used to measure the frequency of carrier signals and other signals.
- *Appropriate test adapters, leads, and cables* – Used to make test connections.
- *Sound pressure level (SPL) meter* – A dB meter used for checking NFPA limits for fire alarm ambient noise levels. It can also be used on audio systems.

5.0.0 COMMON CAUSES OF ELECTRICAL EQUIPMENT FAULTS

As described earlier, equipment and cable failures can result from factors such as environmental conditions, improper installation, and poor power quality. The most common types of electrical equipment faults that result from such factors include the following:

Maintenance and Repair

- Short circuits
- Open circuits
- Ground faults
- Mechanical failures
- Environmental problems, such as temperature, humidity, and loose airborne debris

Troubleshooting of shorts, opens, and grounds is covered in a later section of this module.

5.1.0 Short Circuits

A short circuit results when the current takes a direct path across a source. For example, a short circuit in a transformer is caused by a defect in the transformer where two wires of the circuit touch and cause a bypassing of the normal current flow. Short circuits draw more current because the resistance in the circuit decreases. As a result, the voltage decreases. Common signs of short circuits are tripped circuit breakers or blown fuses, increased heat, low voltage, high current, smoke, charred embers, and/or burning smells.

5.2.0 Open Circuits

Open circuits are a result of an incomplete circuit. An open circuit will have infinite resistance and zero current because its path has been broken. A device that is completely or partially inoperable is a common sign of an open circuit.

5.3.0 Ground Faults

A ground fault results when a defect in the insulation or placement of a wire or component causes the current to take an abnormal route in the circuit. In a control circuit, for example, an isolation power transformer with a grounded circuit can result when part of the isolated secondary winding makes electrical contact with the iron core or frame of the transformer. A ground fault is similar to a short, but has different characteristics. Typically, a short circuit causes circuit devices to stop operating and trips a circuit breaker. With a grounded circuit, the devices often continues to operate, but may function poorly or draw abnormal currents and voltages. Depending on the voltages involved, ground faults of isolated circuits can be dangerous because the devices keep working, possibly subjecting an operator to shocks.

Ground faults typically result from wires with poor or damaged insulation, pinched wires, or improperly placed components. Common signs of ground faults are abnormal current and voltage readings, abnormal resistance readings from the devices or the circuit wiring to ground, shocks, abnormal circuit performance, and the intermittent blowing of fuses and/or tripping of circuit breakers.

5.4.0 Mechanical Failures

Mechanical failures occur when excess friction, wear, abuse, or vibration causes the physical breakdown of an electrical or electronic device. Broken or loose belts, worn bearings, worn contacts, a damaged chassis, and broken controls are typical examples of mechanical problems. Common signs of a mechanical failure include noisy or abnormal operation.

6.0.0 USING A SYSTEMATIC APPROACH TO TROUBLESHOOTING

While the reliability of electrical/electronic systems, equipment, and cabling is quite high, failures do occur. These failures can be the result of defective parts, improper installation, manufacturers' defects, or poor maintenance. To locate and correct a problem, you must be able to do the following:

- Understand the purpose and function of each assembly or component in a particular system
- Observe and identify the symptoms that point to the improper operation of any part of the system
- Use the proper procedures and test equipment to correct the malfunction quickly and efficiently

Troubleshooting can be defined as a process by which the technician locates the source of an equipment, cable, or software problem, then performs the necessary repairs or adjustments to correct the problem. A systematic approach to troubleshooting can be divided into the following five basic elements:

- Customer interface
- Physical examination of the system
- Basic system analysis
- Manufacturer's troubleshooting aids
- Fault isolation in system/unit problem area

NOTE

The vast majority of problems can usually be found quickly. Only the most stubborn and difficult problems require the use of all five elements.

7.0.0 CUSTOMER INTERFACE

The process of troubleshooting should always begin with obtaining all the information possible regarding the equipment problem or malfunction. Asking questions of a customer with first-hand knowledge of the problem is always recommended. This can provide valuable information on equipment operation, which can aid in the troubleshooting process. The customer interview may sometimes determine that the source of a problem is not related to the low-voltage equipment, thereby eliminating unnecessary equipment troubleshooting or maintenance. When interviewing the customer about a specific fault or problem, try to determine the following:

• The scope of the problem and when and how it began
• How often the problem occurs
• If the problem occurred after some specific event, such as the addition of new equipment, relocation of equipment or new construction work
• Whether the problem affects all activity or only specific applications
• If the system, equipment, or channel with the problem has been tested recently
• Whether environmental conditions could have caused this problem

Assuming that the customer interview points to the equipment as being the source of a problem, always explain to the customer what course of action you intend to take and what system impact might be encountered. This could include service interruptions or total system downtime. If required, notify off-site monitoring agencies or authorities (such as the police or fire department) that you will be working on the system.

8.0.0 PHYSICAL EXAMINATION OF THE SYSTEM

Problems can sometimes be identified by checking for anything unusual around the equipment, cabling, and adjacent area. When performing a physical examination of the system:

• Look for loose or disconnected cables or power cords, or for power or other switches set to the wrong position.
• Check the work area for signs of user intervention, such as the addition or deletion of cables, terminators, and computers.
• Examine telecommunications closets and similar areas for incorrect wiring or signs of recent changes, such as construction or electrical work.
• Look for loose or defective terminations and improperly installed connectors.
• Check to make sure the appropriate types of cable, patch cable, and plugs have been installed for the application.
• For operating equipment, check for odors of burning insulation, sounds of arcing, and abnormally warm units.

9.0.0 BASIC SYSTEM ANALYSIS

A proper diagnosis of a problem requires that you know what the system or unit should be doing when it is operating properly.

If you are not familiar with how a particular system or unit should operate, you must first study the manufacturer's service literature to learn the equipment modes and sequence of operation.

The second part of the diagnosis is to find out what symptoms are exhibited by the system or unit. This can be done by carefully listening to the customer's complaints and analyzing the operation of the system or unit yourself. This usually means making electrical measurements at key points in the system or at the unit. Measured values can be compared with a set of previously recorded values or with values given in the manufacturer's service literature.

When troubleshooting, avoid making assumptions about the cause of problems. Always secure as much information as possible before arriving at your diagnosis.

10.0.0 USE OF MANUFACTURERS' TROUBLESHOOTING AIDS

A good troubleshooter knows how and where to locate information, including vendor technical support lines, online resources, and the use of vendor-supplied manuals and other product documentation.

Technical/service manuals for specific types of equipment are invaluable sources of information (*Figure 4*). These manuals typically contain the following information:

• Principles of operation
• Data sheets
• Functional description of the circuitry
• Illustrations showing the location of each control or indicator and its function
• Step-by-step procedures for the proper use and maintenance of the equipment

- Safety considerations and precautions that should be observed when operating the equipment
- Schematic diagrams, wiring diagrams, mechanical layouts, and parts lists for the specific unit

To aid in the isolation of faults in a specific system or equipment, manufacturers typically provide troubleshooting aids for their equipment. Depending on the complexity of the equipment, these aids can include wiring diagrams, troubleshooting tables, and diagnostic tests.

10.1.0 Wiring Diagrams

Wiring diagrams attached to the equipment or contained in technical/service manuals are the primary troubleshooting aid when isolating electrical problems.

Wiring diagrams (*Figure 5*) typically show the actual external and internal connections between circuit boards and other components within equipment, including the wire color and physical connection points. In the case of telecommunications and other systems, schematic diagrams, building cabling diagrams, and floor plans can also provide additional help when troubleshooting.

- PRINCIPLES OF OPERATION
- FUNCTIONAL DESCRIPTION OF CIRCUITRY
- CONTROLS AND INDICATORS
- OPERATING PROCEDURES
- MAINTENANCE PROCEDURES
- SAFETY CONSIDERATIONS AND PRECAUTIONS
- SCHEMATICS AND WIRING DIAGRAMS
- MECHANICAL LAYOUT AND PARTS LIST

307F04.EPS

Figure 4 Technical/service manuals contain maintenance and repair information.

10.2.0 Troubleshooting Tables and Fault Isolation Diagrams

Troubleshooting tables or fault isolation diagrams are usually contained in the manufacturer's installation and service instructions for a particular unit. Troubleshooting tables (*Figure 6*) are intended to guide you to a corrective action based on your observations of system operation. Fault isolation diagrams (*Figure 7*), also called troubleshooting flowcharts, normally start with a failure symptom observation and take you through a logical decision-action process to isolate the failure.

10.3.0 Diagnostic Equipment and Tests

For more complex systems or equipment, manufacturers may incorporate special testing features to help isolate malfunctions. Depending on the complexity of the equipment, these built-in diagnostic devices can be simple or complex. Some microprocessor-controlled units can run a complete checkout of all system or unit functions. Normally, these built-in test functions isolate a fault to a functional problem area. Following this, the technician must perform additional troubleshooting within the problem area to pinpoint the exact location of the fault.

11.0.0 FAULT ISOLATION IN THE SYSTEM/UNIT PROBLEM AREA

Once the use of troubleshooting aids has helped isolate a problem to a functional system or equipment area, it may be necessary to make additional measurements and use a step-by-step process of elimination to isolate the specific cause of the problem. This is usually the case for complex systems.

Troubleshooting problems in electrical/electronic equipment may appear difficult, but it can be simplified if the components in a system or assembly are divided into smaller functional circuits based on the operation they perform. Most systems can be divided into four broad areas for the purpose of troubleshooting (*Figure 8*).

- Input power and/or power supply circuits
- Control/sensor circuit inputs and outputs
- Central processing circuits
- System cabling

Guidelines for troubleshooting the electrical components used in each of these functional areas are provided in the following sections of this module.

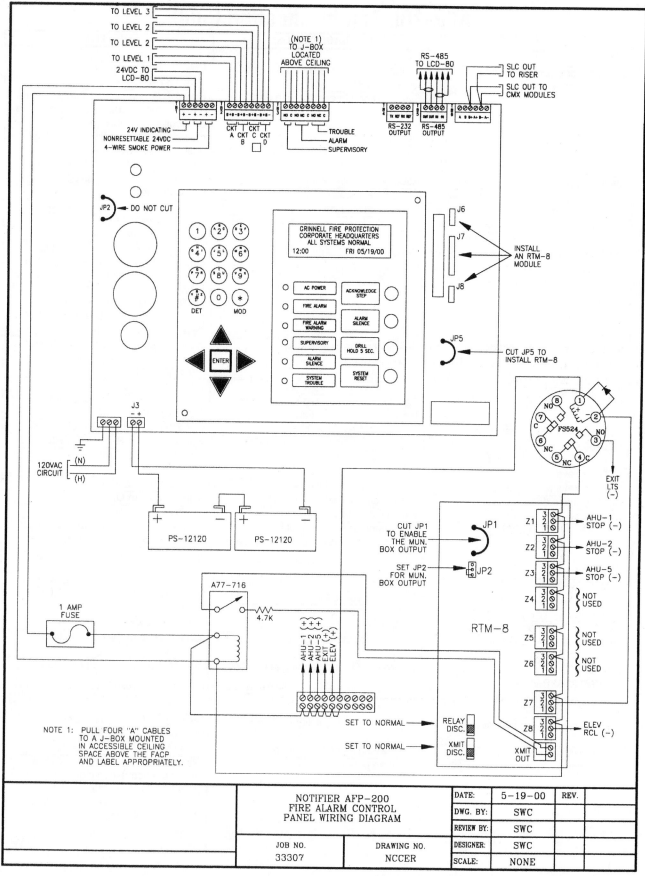

Figure 5 Example of a wiring diagram.

307F05.EPS

AFP-200 Troubleshooting Guide

SYMPTOM	PROBABLE CAUSE	SOLUTION
Annunciator goes into trouble mode upon system reset	Annunciator is powered from resettable power on TB1	Rewire annunciator power source from TB1-3(+) & 4(-).
Ground fault	System wiring in contact with earth ground	Remove terminals TB1 thru TB6 until ground fault clears. Remove portions of circuit wiring until ground fault clears. Re-attach portions of circuit until condition re-appears. Repeat until ground fault can be pinpointed.
Invalid reply from a single field device	Address used twice on field devices	Identify devices and reprogram one.
Invalid reply from a single field device	Device not installed	Install the device.
Invalid reply from a single field device	Device unaddressed	Dial-in correct address at the device.
Invalid reply from a single field device	Wrong device type code programmed	Program correct type code.
Invalid reply from all field devices	Reverse SLC polarity at AFP-200	Correct polarity at TB6.
Invalid reply from all field devices	Reverse SLC polarity at field device	Starting in the middle of the SLC circuit and working back toward the panel, remove portions of field SLC wiring until condition clears. Re-attach portions of SLC until condition re-appears. Repeat until reverse polarity device can be pinpointed.
Invalid reply from all field devices	Short circuit on SLC wiring	Same as above. Voltmeter may be used instead of relying on panel to indicate problem.
Invalid reply from all field devices	SLC landed on TB6 A+ & A-	Land SLC on TB6 B+ & B-.
LCD-80 does not operate properly	Panel not programmed for annunciator terminal mode operation	Reprogram AFP-200 for annunciator type TERM(LCD-80).
LCD-80 does not operate properly	SW1-7 on the AFP-200 is OFF	Set SW1-7 on the AFP-200 is ON.
LCD-80 does not operate properly	SW2 on AFP-200 not set to TERM	Set SW2 on AFP-200 to TERM.
LCD-80 keys do not function	AKS jumper is installed	Remove AKS jumper.
LCD-80 keys do not function	AKS-1 switch in disable position	Move keyswitch to enable position.
LCD-80 keys do not function	SW1-4 on LCD-80 is ON	Turn SW1-4 to OFF.
LCD-80 piezo does not sound	SW1-3 on LCD-80 is ON	Turn SW1-3 to OFF.
LCD-80 screen is blank	Field wiring error or damage	Repair field wiring problems.
MAINTENANCE ALERT	Dirty detector head, panel has adjusted drift compensation to the acceptable limits	Clean or replace detector. Delete detector from programming, power cycle panel, then reprogram detector.
MAINTENANCE ALERT	Faulty detector	Replace the detector.
MAINTENANCE ALERT	Unsatisfactory ambient conditions	Adjust sensitivity or relocate detector.
Open circuit on CMX-2 circuit	Power not applied to A77-716B EOLR	Apply power to end of line relay.
Open circuit on panel bell / CMX-2 ckt	Field wiring error or damage	Repair field wiring problems.
Open circuit on panel bell circuit	Incorrect end of line resistor value	Replace 47K with 4.7K.
Verifier-200 or PK-200 does not connect to AFP-200	Error in Eia-232 wiring	Check DB-9 to TB4 connections.
Verifier-200 or PK-200 does not connect to AFP-200	Incorrect COM port setting in PK200	Select opposite COM port address.
Verifier-200 or PK-200 does not connect to AFP-200	MODEM connection selected in PK200	Select DIRECT_CONNECT in PK200.
Verifier-200 does not connect to AFP-200	Software incompatibility	Assure software is Revision 2 or greater. If not, use PK-200.
Short circuit on panel bell / CMX-2 ckt	Field wiring error or damage	Repair field wiring problems.
Short circuit on panel bell / CMX-2 ckt	Reverse polarity strobe	Remove portions of strobe circuit wiring until condition clears. Re-attach portions of circuit until condition re-appears. Repeat until reverse polarity device can be pinpointed.
		SWC

307F06.EPS

Figure 6 Example of a troubleshooting table.

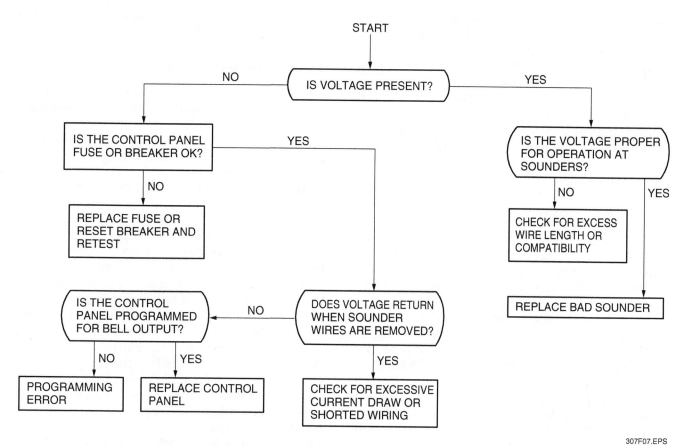

Figure 7 Troubleshooting flowchart for a bad sounder.

307F07.EPS

11.1.0 Troubleshooting Input Power and Power Supply Circuits

The utility power source for a low-voltage system or unit typically consists of a single-phase branch circuit running from the main electrical service panel either directly to the equipment or to an outlet into which the equipment power cord is plugged. In some cases, the utility power source may be connected to the primary windings of a step-down transformer. The branch circuit feeding the equipment or transformer consists of the branch circuit wiring and a protective device (normally a circuit breaker).

11.1.1 Input Voltage Check

All electrical/electronic equipment is designed to operate within a specific range of system voltages including a safety factor (typically ±10 percent). For example, equipment made to operate at 120VAC is designed to operate adequately with voltages ranging from 108VAC to 132VAC. This safety factor is added to compensate for temporary supply fluctuations that might occur. Contin-

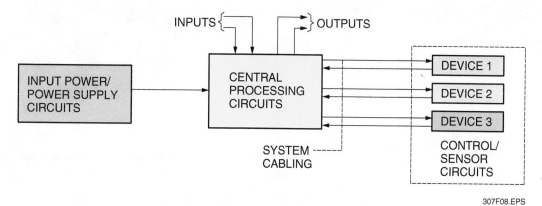

Figure 8 Most electrical/electronic systems can be divided into four functional areas.

307F08.EPS

uous operation of equipment outside the specified range can cause improper operation and premature failure of the equipment.

In the event of a system failure or recurring problems, the utility voltage applied to the equipment should be measured (*Figure 9*). The result should be checked against the specified supply voltage indicated on the equipment nameplate or in the service literature for the equipment. If the voltage is too low or too high, contact a qualified person to arrange repair.

11.1.2 Low-Voltage Transformer Checks

Transformers are usually checked by measuring the voltage across the primary and secondary windings (*Figure 10*). Typically, the secondary winding is measured first. The VOM/DMM should be set to a range that is higher than expected for the secondary voltage. If the voltage measured across the secondary winding is within ±10 percent of the required voltage, the transformer is good. If no voltage is measured at the secondary winding, the voltage across the primary winding must be measured. If there is a fuse in the secondary, check it to see if it is blown.

If the voltage measured at the primary winding is within ±10 percent of the required voltage, the transformer is most likely bad. This can be confirmed by performing a continuity check of the primary and secondary windings. If no voltage or low voltage is measured across the primary

winding, the AC voltage input to the transformer should be checked as previously described.

11.1.3 Troubleshooting Power Supplies

DC power supplies are commonly used in telecommunications and other types of equipment. They can be stand-alone units, or they can be an integral part of an item of equipment. For the purpose of troubleshooting, the testing of power supplies is reviewed here.

The first thing you should do when checking an inoperative power supply is to make sure the input power cord is plugged in (or input power is being applied at the unit's input terminals) and the circuit breaker has not tripped. Follow this by looking for obvious defects, such as burned resistors, broken wires, or poor solder connections. For an intermittent type of failure in which the power supply works properly for a while and then fails, look and feel for overheated components.

If a power supply is operating, it should be tested to see if it is providing the required noise-free DC output. A complete checkout of a power supply normally involves performing voltage, current, regulation, and ripple checks in a bench-testing environment. However, this is not always practical. An online check of a power supply can be done to determine whether or not it is operating properly. This test should be done with the power supply connected in the system and under normal load conditions. Basically, the test involves locat-

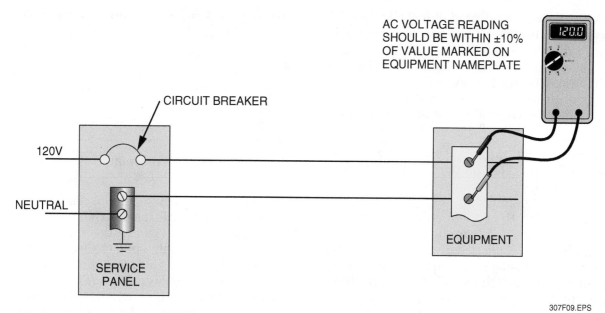

AC VOLTAGE READING SHOULD BE WITHIN ±10% OF VALUE MARKED ON EQUIPMENT NAMEPLATE

Figure 9 Single-phase input voltage check.

ing a suitable measurement point (terminal board or connector pin) where the power supply output voltage can be accessed for testing. Then the output voltage should be measured with a VOM/DMM and the ripple checked with an oscilloscope.

To measure the amount of ripple voltage (sometimes called output noise) in the power supply output, the scope AC/DC mode controls should be set to the AC mode. This blocks the DC output of the supply. Adjust the scope controls to produce two or three stationary cycles on the screen, then measure the peak amplitude of the ripple on the scope's voltage-calibrated vertical scale.

Once the value for the ripple is known, the percentage of ripple relative to the full output voltage can be calculated. For example, if 0.03V of ripple is measured with a 5V output, the ratio is 0.03/5 or 0.006, which can be converted to a percentage (0.006 × 100 = 0.6 percent).

The percent of ripple voltage should be small compared to the supply voltage. The specifications for the power supply being tested should be consulted to determine the maximum ripple. Keep in mind that all switching-type power supplies will have ripple no matter how good the regulation and filtering. Any power supply with excessive ripple should be replaced or repaired.

11.2.0 Troubleshooting Control/Sensor Circuits

Switches and relay contacts controlled by sensors are commonly used in low-voltage equipment to provide switched control signals to a load such as an alarm or siren. Once the source of a problem

has been isolated to an inoperative load and its control circuit, the next step is to make a series of voltage measurements across the various switch or relay contacts in the circuit to find the faulty component. As shown in *Figure 11*, the measurements can start from the line or control voltage side of the circuit and move toward the load de-

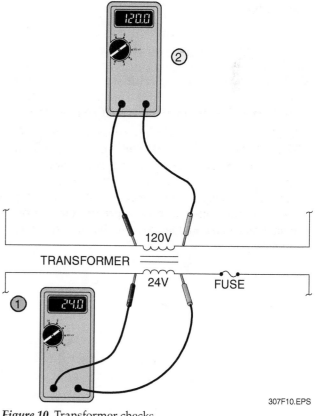

Figure 10 Transformer checks.

vice. Measurements are made until no voltage is observed, or until the voltage has been measured across all the components in the circuit. The technique shown in *Figure 11* is often called the hopscotch method of troubleshooting.

As a result of taking these voltage measurements, one of the two following situations should exist:

- At some point in the circuit under test, no voltage will be indicated on the VOM/DMM. This pinpoints the open set of switch or relay contacts. *Figure 11* shows an example of this situation in which the switch contacts are open, preventing the load from being energized. If the problem is a switch, both the on and off positions of the contacts can usually be checked by setting the switch to the on or off position, respectively.

> **NOTE**
>
> If the open occurs at a set of relay contacts, don't assume the relay has failed. You must first determine if the open contacts are a result of the related relay coil or its controlling sensor being faulty.

- If voltage is measured at the input to the load device, this means that the switch or relay contacts are closed and that the load device is most likely bad.

Once a set of relay contacts has been identified as the probable cause of an electrical problem, the contacts can be tested to confirm their position. With the power to the circuit turned off, the relay contacts can be tested by making a continuity measurement to determine whether the contacts are open or closed (*Figure 12*). If the contacts are open, the VOM/DMM indicates an infinite re-sistance reading. If the contacts are closed, the VOM/DMM indicates a short (0).

> **WARNING**
>
> Never perform continunity checks with power applied.

When working with relay contacts, remember the following:

- Contacts that open when the relay is energized are called normally closed (NC) contacts.
- Contacts that close when the relay is energized are called normally open (NO) contacts.
- The contacts of a relay may not be marked NO or NC on a schematic or wiring diagram. However, a schematic or wiring diagram should show the contacts as they are positioned in the de-energized condition of the relay. Therefore, a contact shown closed is considered a normally closed contact, and a contact shown open is considered a normally open contact.

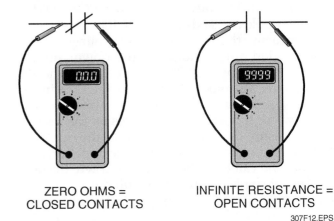

ZERO OHMS = CLOSED CONTACTS

INFINITE RESISTANCE = OPEN CONTACTS

307F12.EPS

Figure 12 Checking the continuity of relay contacts.

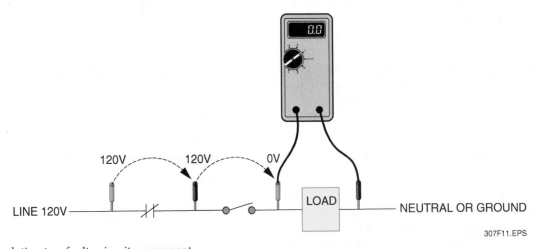

Figure 11 Isolating to a faulty circuit component.

307F11.EPS

NCCER – *Contren® Learning Series* 33307-11

11.3.0 Troubleshooting Central Processing Circuits

The central processing circuits used in low-voltage systems and units are normally contained on printed circuit boards. Once a board is installed and operating, it is unlikely to fail unless the failure is caused by some outside influence. Because of their complexity, many of the components on boards are difficult to troubleshoot and repair in the field. It often takes special automated test equipment at the manufacturer's plant to accomplish this task. For this reason, if a board fails it is normally replaced with a new one.

Many boards are expensive, so you want to be absolutely sure that the board is bad before replacing it. A board should only be replaced when all the input signals to the board are known to be good, but the board fails to generate the proper output signal(s). Many boards used in complex units contain a microprocessor control that controls all sequences of operation for the system or unit. To determine the condition (good or bad) of the board, it is necessary to check for the presence of the proper input signals. It is also necessary to verify that the board generates the proper output signal(s) in response to the input signal(s). Some boards have test points where measurements of key input and output signals can be made. The schematic will identify the signal that should be available at each point.

To isolate a problem to a board correctly, you must understand the sequence of operation. If you are not familiar with the system or board being serviced, study the schematic and sequence of operation described in the manufacturer's service literature. Once you understand what the system or board should be doing, you must find out what the board is doing or what symptoms are present in the unit. To aid you in this analysis, many microprocessor-controlled boards have built-in diagnostic circuits that can run a check of all system functions and report their status. When troubleshooting, always use the built-in diagnostic features (if any) and follow the manufacturer's related troubleshooting instructions.

Before replacing a board that seems to have failed, try to determine if an external cause might account for the problem. There are a variety of such sources:

- Electrical noise (EMI/RFI) from communications equipment or microwave ovens can cause microprocessors to fail or operate erratically. The source of interference should be removed, or the board shielded.

- Microprocessors and other integrated circuit chips can be fatally damaged by static electricity from sources such as lightning or people. It is important when handling circuit boards to avoid touching components, the printed circuit, and the connector pins. Also, you should ground yourself before touching a board. Disposable grounding wrist straps are sometimes supplied with boards containing electrostatic-sensitive components. If a wrist strap is not supplied with the equipment, use a wrist strap grounding system. It allows you to handle static-sensitive components without fear of electrostatic discharge (ESD). Unused boards should always be kept inside their special conductive plastic storage bags. Other precautions for preventing ESD-related problems were given earlier in this module.

- Voltage surges and excessive voltage can also cause board failures. Make sure that the applied voltage is within allowable limits and that the unit, and therefore the board, is properly grounded.
- Excessive heat can damage the board. The unit in which the board is installed should not be altered in any way that will restrict airflow to the board.

Many boards have dipswitch-selectable options or operating modes used to select site-specific parameters. When replacing a board with dipswitch-selectable options, make sure to set the dipswitch(es) on the new board to the correct position for your site situation. Determine the dipswitch position on the failed board, then set the dipswitch on the replacement board to the same position. Other boards may have jumpers that need to be added or wires that may need to be cut. Some devices must have the power reset. Check the manufacturer's instructions.

11.4.0 Troubleshooting Computer-Related Problems

Because of the wide variety of computer systems and software available, there is no single method for troubleshooting computer problems. A brief overview, with an emphasis on software, is given here.

Once a computer-related problem is detected, the first determination that must be made is whether the problem is in the hardware or software. A record of the software programs resident in the computer and their version numbers should be available to facilitate troubleshooting. Software problems usually leave the rest of the computer running, although you may have to reboot the system. Software problems are usually identified by error messages telling you that the computer cannot execute a part of the program, or a cryptic message such as Invalid Function Call, Missing Library File, or any other message that does not specifically identify a piece of hardware that is malfunctioning.

Some software errors on a computer occur when the system has been running too long, or when several different programs have been run either at the same time or in succession. Every program uses some of the computer's resources, such as memory or file handles. A file handle is a specific internal memory location that is used to read or write information from a file. A properly written program will reserve the resources that it needs when it starts, and releases those resources for other programs to use when it is terminated. Unfortunately, there are many improperly written programs that do not release all of their resources. If many programs like this are run, or even if the same program is run many times, the system's available resources run low, causing other, seemingly unrelated programs to fail. Problems of this type can often be solved by rebooting the computer.

Computer and/or system problems often occur as a result of a computer operator entering incorrect system operating parameters or configuration commands into an application program. The values or settings for all such optional parameters should be checked.

A problem with a hard drive can be the most difficult to locate. It may seem like a software problem, by stating that a part of the program cannot be found or that a data file cannot be read. This could be due to a problem with the software or with the disk on which the software is stored. Most computers have some sort of disk testing

On Site

Printed Board Repair

Today it is standard practice to replace a faulty printed circuit board with another board known to be good. If the board is still under warranty, the manufacturer will generally replace the board at no cost. Do not try troubleshooting components on a board unless instructed to do so. Once a board has been worked on, the manufacturer has no way of knowing if the board was damaged through a manufacturing error or because you worked on it. In these cases, many manufacturers will not replace the board without payment.

307SA03.EPS

tool, often called a disk check program or disk scanner, that can be used to diagnose these problems.

Computer hardware problems typically leave the computer or some peripheral device, such as a printer, inoperable. Troubleshooting a computer hardware problem is done by following the same systematic approach and using the same techniques as described earlier for troubleshooting other types of electrical/electronic hardware. Many computer hardware and software problems can be difficult to diagnose, so do not hesitate to seek technical help. If necessary, contact the vendor's tech support to get help diagnosing the problem. Their expertise in a particular type of equipment or software can save you a great deal of time when troubleshooting.

11.5.0 Troubleshooting and Testing Copper Cable

Troubleshooting cabling requires a systematic approach and the appropriate test equipment, cabling diagrams, and other relevant documentation. Troubleshooting and testing of cables is accomplished by using the appropriate cable tester, operated according to the manufacturer's instructions, then interpreting the results.

Figure 13 shows a systematic approach for diagnosing copper cable problems. As shown, the troubleshooting process begins by locating the cable run or segment associated with the problem. *Table 1* shows the types of test instruments typically used to test the various types of cables for each troubleshooting or testing task identified in *Figure 13*.

As shown in *Figure 13*, problems encountered with copper cable can result from the following:

- Incorrect connections
- Incorrect cable length
- Incorrect impedance
- Excessive cable attenuation
- Excessive near-end crosstalk (NEXT)
- Noise

11.5.1 Incorrect Connections, Opens, Shorts, and Grounds

Incorrect connections include unwanted opens, shorts, crossed pairs, split pairs, and grounds. Opens and shorts can be caused by damaged cable terminations. Crossed and split pairs can be caused by incorrect installation. Incorrect connections can also result from the use of wrong patch cords, mismatched terminations, or T568A and T568B wiring differences. All area outlets, patch panels, and termination blocks must be compatible with the transmission performance requirements for the desired category of cable.

Opens, short circuits, or crossed conductors are easy to detect using a cable tester. The cable tester is placed on both ends of the cable and turned on. The tester indicates the type of fault, which in turn determines if the cable needs to be replaced. When testing for an open, be sure to move the cable around to eliminate the possibility of an intermittent problem.

A TDR can also be used to detect and locate cable opens, shorts, and grounds. The fault will

Table 1 Test Instruments and Troubleshooting Tasks

Troubleshooting Task	Cable (UTP/ScTP/ STP-A/Coaxial)
Identify problem cable run or segment	Toner/wand or cable tracer
Determine if cable run or segment is active	Telephone test set
Determine if cable is correctly connected	Wire map test set
Determine if cable length is correct	VOM/DMM or TDR
Determine if cable impedance is correct	Certification test set
Determine if there is excessive cable attenuation	Certification test set
Determine if there is excessive near end crosstalk (NEXT)	Certification test set
Determine if noise is being induced into the cable	Certification test set or field-strength meter

307T01.EPS

cause a reflection to be displayed on a TDR connected to one end of the cable. The position of the reflection on the display is directly related to the location of the fault in the cable.

Grounds or short circuits are also easily detected using a multimeter set up to measure continuity. To test for a short or ground, disconnect both ends of the cable. To check for shorts, place the test leads of the multimeter across a pair of the conductors. If continuity exists, the conductors are shorted together. To check for grounds, place one test lead of the multimeter on a conductor and the other on a common ground. If continuity exists, the conductor is grounded. Finding the ground can sometimes be time consuming because the entire cable length may have to be

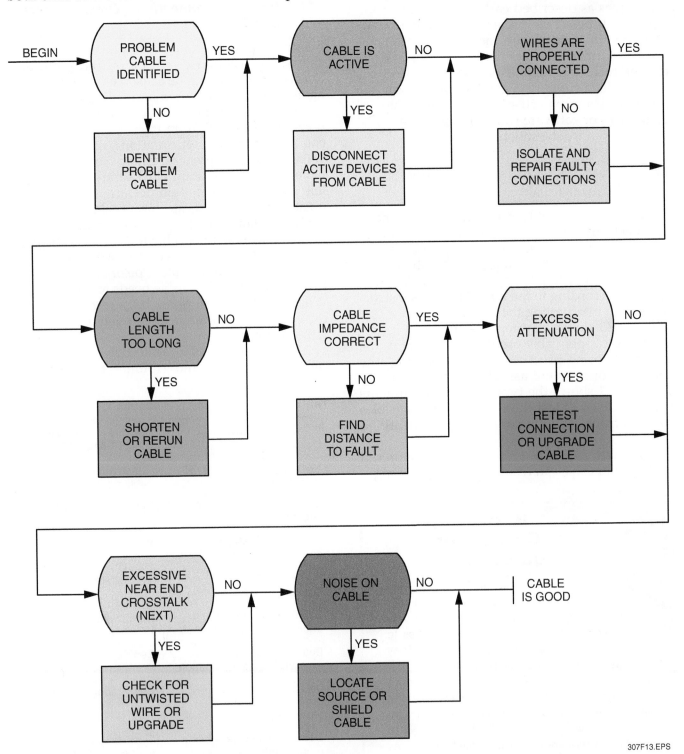

Figure 13 Systematic approach for diagnosing copper cable problems.

307F13.EPS

NCCER – *Contren® Learning Series* 33307-11

physically examined to find where the contact with ground is taking place. However, a TDR can be used to locate the exact position of a ground in any transmission line.

11.5.2 Signal Loss

The required physical length of cable runs can be determined from drawings and cabling diagrams. When a cable test shows an open or a short, the problem can actually be caused by a poor termination at an intermediate connection point, such as a patch panel or wall outlet. A cable that is too long when tested can be the result of having excess cable coiled in a ceiling or other location, or the use of several excessively long patch cords. It should be noted that because of the twisting of pairs, the actual physical length of a twisted-pair cable appears slightly shorter than the electrically measured length.

Normally, the length of a balanced pair of conductors (twisted pair) in a cable or a coaxial cable can be accurately and easily found using a TDR. However, in the event that a TDR is not readily available, the DC loop resistance of a pair of conductors or coaxial cable can be measured using a multimeter set up to measure resistance, and the cable length can be approximated as described in the following procedures.

The following are guidelines for performing a DC loop resistance test for a balanced pair of conductors:

Step 1 If testing in order to verify the total length of a balanced pair of conductors, short out one end of the pair of conductors.

> **NOTE**
> Some technicians prefer to connect a terminating resistor instead of a short at one end of the balanced conductor pair. If this is done, you must make sure to subtract the terminating resistor value from the measured loop resistance value before attempting to determine the approximate length of the cable in accordance with Step 6 of this procedure.

Step 2 Set the multimeter to the lowest ohms scale and connect the leads of the meter together. Zero out any meter reading.

Step 3 Connect the multimeter leads to the non-shorted end of the pair of conductors.

Step 4 Read and record the DC loop resistance of the pair of conductors.

Step 5 Determine the DC resistance per foot of the single AWG wire size being checked (*Table 2*). Multiply the value by two to obtain the nominal DC loop resistance of a pair of the conductors for one foot.

> **NOTE**
> The nominal DC loop resistance of a pair of solid 24 AWG wires at 68°F (20°C) is 0.0514Ω/ft. The nominal DC loop resistance of a pair of 22 AWG wires is 0.0324Ω/ft.

Step 6 Divide the DC loop resistance reading recorded from the meter in Step 4 by the nominal DC loop resistance per foot of the pair of conductors determined in Step 5 to obtain the approximate length of the pair of conductors or the distance to a short.

The following are guidelines for performing a DC loop resistance test for coaxial cable:

Step 1 If testing in order to verify the total length of a coaxial cable, fabricate a shorting jack by connecting the center conductor of an appropriate jack to the outer body of the jack with a very short length of wire. Connect the shorting jack to one end of the cable.

Step 2 Set the multimeter to the lowest ohms scale and connect the leads of the meter together. Zero out any meter reading.

Step 3 Connect the multimeter leads between the center conductor and the shield of the non-shorted end of the coaxial cable.

Step 4 Read and record the DC loop resistance of the center conductor plus the shield.

Step 5 From the cable manufacturer's data (*Table 3*), determine the nominal DC loop resistance per foot of cable.

Step 6 Divide the DC loop resistance reading recorded from the meter in Step 4 by the nominal DC loop resistance per foot of the coaxial cable determined in Step 5 to obtain the approximate length of the coaxial cable or the distance to a short.

11.5.3 Incorrect Impedance

Impedance is a measurement of the opposition to current flow for alternating current signals. Impedance mismatches can occur as a result of poor installation techniques, the use of incorrect or defective cable, devices with mismatched impedances, or the use of the wrong termination.

 Maintenance and Repair

Table 2 Typical Conductor DC Resistance

Resistance in Ohms per 1000 feet per conductor at 20°C and 25°C of solid wire and class B concentric strands copper and aluminum conductor

Conductor Size, AWG or kcmil	ANNEALED UNCOATED COPPER ANNEALED ALUMINUM								ANNEALED COATED COPPER			
	Solid				Stranded Class B				Solid		Stranded Class B	
	20°C		25°C*		20°C		25°C*		20°C	25°C*	20°C	25°C*
	CU	AL	CU	AL	CU	AL	CU	AL	CU	CU	CU	CU
24	25.7000	—...	26.2000	—...	—...	—...	—....	—...	26.8000	27.3000	—...	—...
22	16.2000	—...	16.5000	—...	—...	—...	—...	—...	16.9000	17.2000	—...	—...
20	10.1000	—...	10.3000	—...	10.30000	—...	10.50000	—...	10.5000	10.7000	11.00000	11.20000
19	8.0500	—...	8.2100	—...	—	—...	—...	—...	8.3700	8.5300	—...	—....
18	6.3900	—...	6.5100	—...	6.51000	—...	6.64000	—...	6.6400	6.7700	6.92000	7.05000
16	4.0200	—...	4.1000	—...	4.10000	—...	4.18000	—...	4.1800	4.2600	4.35000	4.44000
14	2.5200	4.1400	2.5700	4.220	2.57000	—...	2.62000	—...	2.6200	2.6800	2.68000	2.73000
12	1.5900	2.6000	1.6200	2.660	1.62000	2.65000	1.65000	2.70000	1.6200	1.6800	1.68000	1.72000
10	0.9990	1.6400	1.0200	1.670	1.02000	1.67000	1.04000	1.70000	1.0400	1.0600	1.06000	1.08000
9	0.7920	1.3000	0.8080	1.320	0.80800	1.33000	0.82400	1.35000	0.8160	0.8310	0.84000	0.85700
8	0.6280	1.0300	0.6410	1.050	0.64100	1.05000	0.65400	1.07000	0.6460	0.6590	0.66600	0.67900
7	0.4980	.8170	0.5080	.833	0.51800	.83300	0.51800	0.85000	0.5130	0.5230	0.52800	0.53900
6	0.3950	.6480	0.4030	.661	0.40300	.66100	0.41000	0.67400	0.4070	0.4150	0.41900	0.42700
5	0.3130	.5140	0.3190	.524	0.32000	.52400	0.32600	0.53500	0.3230	0.3290	0.33300	0.33900
4	0.2480	.4070	0.2530	.415	0.25300	.41600	0.25900	0.42400	0.2560	0.2610	0.26400	0.26900
3	0.1970	.3230	0.2010	.330	0.20500	.33000	0.20500	0.33600	0.2030	0.2070	0.20900	0.21300
2	0.1560	.2560	0.1590	.261	0.15900	.26200	0.16200	0.26700	0.1610	0.1640	0.16600	0.16900
1	0.1240	.2030	0.1260	.207	0.12600	.20600	0.12900	0.21100	0.1280	0.1300	0.13100	0.13400
1/0	0.0982	.1610	0.1000	.164	0.10000	.16500	0.10200	0.16800	0.1010	0.1030	0.10400	0.10600
2/0	0.0779	.1280	0.0795	.130	0.07950	.13100	0.08110	0.13300	0.0798	0.0814	0.08270	0.08430
3/0	0.0618	.1010	0.0630	.103	0.06300	.10300	0.06420	0.10500	0.0633	0.0645	0.06560	0.06680
4/0	0.0490	.0803	0.0500	.082	0.05000	.08210	0.05090	0.08360	0.0502	0.0512	0.05150	0.05250
250	—...	—...	—...	—...	0.04230	.06950	0.04310	0.07080	—...	—...	0.04400	0.04490
300	—...	—...	—...	—...	0.03530	.05790	0.03600	0.05900	—...	—...	0.03670	0.03740
350	—...	—...	—...	—...	0.03020	.04960	0.03080	0.05050	—...	—...	0.03140	0.03200
400	—...	—...	—...	—...	0.02640	.04340	0.02700	0.04420	—...	—...	0.02720	0.02780
500	—...	—...	—...	—...	0.02120	.03480	0.02160	0.03540	—...	—...	0.02180	0.02220
600	—...	—...	—...	—...	0.01760	.02900	0.01800	0.02950	—...	—...	0.01840	0.01870
750	—...	—...	—...	—...	0.01410	.02320	0.01440	0.02360	—...	—...	0.01450	0.01480
1000	—...	—...	—...	—...	0.01060	.01740	0.01080	0.01770	—...	—...	0.01090	0.01110
1250	—...	—...	—...	—...	0.00846	.01390	0.00863	0.01420	—...	—...	0.00871	0.00888
1500	—...	—...	—...	—...	0.00705	.01160	0.00719	0.01180	—...	—...	0.00726	0.00740
1750	—...	—...	—...	—...	0.00604	.00992	0.00616	0.01010	—...	—...	0.00622	0.00634
2000	—...	—...	—...	—...	0.00529	.00869	0.00539	0.00885	—...	—...	0.00544	0.00555
2500	—...	—...	—...	—....	0.00427	.00702	0.00436	0.00715	—...	—...	0.00440	0.00448

307T02.EPS

11.5.4 Excessive Cable Attenuation

Attenuation is the loss in power of a signal as it travels along a cable. Higher-than-normal cable attenuation can be a result of cable exposure to high temperatures, excessive cable length, or transmission of higher-frequency signals than expected. It can also be the result of using the wrong grade or category of cable for a particular application, use of the wrong terminations, or use of improper patch cables for the installation.

To measure attenuation in a cable, a signal of known frequency and power is applied from a signal generator into one end of the cable. At the other end, a measurement is taken using a power meter or spectrum analyzer. The difference in power levels in dB represents the amount of attenuation for the cable. The acceptable attenuation loss for a fixed length (typically 100 feet or 100 meters) of a specific cable type can be obtained from the cable manufacturer's product data.

11.5.5 Excessive Near-End Crosstalk

Excessive near-end crosstalk (NEXT) is the undesirable signal transfer between pairs at the end of a cable nearest the point of transmission. NEXT

Table 3 Typical Coaxial Cable Attenuation and Loop Resistance

Frequency (MHz)	Drop Cable				Semiflex Cable									
	RG59	RG6	RG7	RG11	412	500	625	750	875	1,000	565	700	840	1,160
5	0.77	0.57	0.56	0.36	0.20	0.16	0.13	0.11	0.09	0.08	0.14	0.11	0.09	0.07
55	1.88	1.50	1.22	0.95	0.68	0.55	0.45	0.37	0.32	0.29	0.47	0.37	0.32	0.24
211	3.59	2.87	2.29	1.81	1.35	1.08	0.89	0.73	0.64	0.58	0.93	0.74	0.64	0.48
250	3.89	3.12	2.49	1.98	1.49	1.19	0.98	0.81	0.70	0.64	1.03	0.82	0.70	0.53
300	4.27	3.43	2.74	2.17	1.64	1.31	1.08	0.89	0.78	0.72	1.13	0.90	0.77	0.59
350	4.64	3.72	2.98	2.36	1.78	1.43	1.18	0.97	0.84	0.78	1.23	0.98	0.84	0.65
400	4.88	4.00	3.20	2.53	1.91	1.53	1.27	1.05	0.91	0.84	1.32	1.05	0.91	0.70
500	5.50	4.51	3.61	2.85	2.15	1.73	1.43	1.18	1.03	0.96	1.49	1.19	1.03	0.80
600	6.18	4.98	3.99	3.16	2.37	1.91	1.58	1.31	1.14	1.06	1.64	1.31	1.14	0.89
750	6.96	5.62	4.50	3.58	2.68	2.16	1.79	1.48	1.29	1.21	1.85	1.49	1.30	1.01
870	7.54	6.09	4.87	3.90	2.90	2.35	1.95	1.61	1.41	1.33	2.01	1.62	1.41	1.11
950	7.90	6.39	5.11	4.10	3.03	2.49	2.04	1.72	1.50	1.35	2.15	1.75	1.51	1.15
1,000	8.09	6.54	5.25	4.23	3.13	2.53	2.11	1.74	1.53	1.44	2.17	1.75	1.53	1.20
1,200	8.91	7.18	5.77	4.71	3.44	2.83	2.32	1.96	1.72	1.55	2.45	2.00	1.72	1.33
1,450	9.82	7.89	6.34	5.29	3.81	3.12	2.61	2.16	1.90	1.81	2.66	2.13	1.90	1.52
1,750	10.92	8.74	6.93	5.95	4.23	3.47	2.92	2.41	2.13	2.03	2.96	2.36	2.13	1.71
1,850	11.23	8.99	7.13	6.12	4.36	3.60	2.97	2.52	2.22	2.07	3.13	2.57	2.23	1.74
2,000	11.67	9.34	7.41	6.36	4.55	3.76	3.12	2.64	2.32	2.11	3.27	2.69	2.33	1.82
Loop Resistance	59.9	39.6	26.8	19.5	2.5	1.7	1.1	0.8	0.4	1.3	0.9	0.9	0.6	0.3

Typical Cable Attenuation [dB/100' @ 68°F (20°C)]

Note: Loop resistance is shown in Ω/1,000'.

307T03.EPS

can be caused by the use of the wrong grade of cable for the application, improper termination practices, and/or split pairs. It can also be caused by the use of incorrect or substandard components.

To measure NEXT, a signal is injected into a wire pair. A measurement is then taken of the signal that is passed into an adjacent wire pair. This passed signal must be below a certain level; for example, in Category 5 cable at 10MHz, the passed signal must be a minimum of 47dB below the original signal level.

11.5.6 Noise

Noise can be induced into a cable directly by a connected signal source, such as a computer, or it can be radiated from some external source, such as an electrical power cable installed too near a data cable. Low-voltage continuous noise (less than 3V) can affect data transmission on a cable. The frequency of continuous noise gives a clue as to its source. A spectrum analyzer can be used to observe the nature and frequency of the noise and to help identify its source. Noise under 150kHz is usually from sources such as AC power lines, fluorescent lighting, and machinery. Noise in the

On Site

Coaxial Cable DC Loop Resistance

The center conductor and the shield of a coaxial cable have two different values of resistance per unit length. Therefore, the DC loop resistance of a coaxial cable must be determined by adding the nominal DC resistance of the center conductor per foot to the nominal DC resistance of the shield per foot. Fortunately, coaxial cable manufacturers have already done this when giving loop resistance data in their reference tables. In *Table 3*, for example, the loop resistance value for each type of coaxial cables is given as a single (combined) value at the bottom of the table.

150kHz to 20MHz range is typically caused by light dimmers, medical equipment, computers, copiers, laser printers, and similar equipment. Noise above 20MHz is typically from radios, cellular phones, wireless phones, television sets, microwave ovens, and broadcast equipment.

High-voltage impulse noise can be caused by building loads, such as air conditioners and motors, which produce a large power disruption on startup or shutdown. This type of noise can damage equipment. It also frequently causes computer system lockups and corrupts data. Guidelines for separating data cables from common sources of noise are contained in *ANSI/TIA/EIA-569, Commercial Building Standard for Telecommunications Pathways and Spaces.*

11.6.0 Troubleshooting and Testing Fiber Optic Cable

> **WARNING**
>
> When working with fiber optic cable, never look at the end of a fiber or connector without first making sure that there is no dangerous light source on the other end. Laser sources are potentially dangerous because the output level can be very high and the wavelength is out of the visible range of a person's eyes.

Figure 14 shows a general troubleshooting diagram used for isolating a fault in fiber optic cable. As shown, troubleshooting begins by performing a visual inspection that may locate cable problems caused by incorrect wiring or obvious physical damage, such as the following:

- Disconnected or incorrectly connected jumpers

- Jumper to a hub or router that has been unplugged
- Hub or router that has been turned off
- Damaged cables inside a telecommunications closet, equipment room, or entrance facility
- Broken fiber inside a fiber distribution unit

When performing a visual inspection of a fiber optic cable, it is important that each connector be cleaned before inspection. This is because any dust particle on the end of a connector acts as a very efficient attenuator. Connectors can be cleaned using isopropyl alcohol (99 percent pure) applied to a clean, lint-free cloth. Using moderate pressure, wipe the end of the connector with the alcohol-soaked area of the cloth in one direction only for about five seconds. Following this, wipe the face of the connector in one direction only with a dry area of the lint-free cloth.

A light source and power meter (*Figure 15*) are used to test fibers. A multimode light source uses a light-emitting diode that provides an output of either 850nm (nanometers) or 1,300nm. A single-mode light source uses a laser diode to provide an output of either 1,310nm or 1,550nm. Both types of light sources produce a very stable output signal, allowing precise measurements to be made with a power meter.

Power meters contain a photo detector sensitive to a wide range of wavelengths. A switch is used to select the operating wavelength. When using a light source and power meter to test the condition of a fiber from its origin to its extreme, then from its extreme back to its origin, the results of the two readings should be averaged. If the result is within the standard (typically 2dB for the horizontal or 11dB for the backbone), the fiber is probably good and should be placed back in service. However, it should be pointed out that the maximum loss for horizontal (2dB) and backbone (11dB) is based on the worst case maximum length. These values may not be acceptable for a particular system design. In this case, you must use the appropriate system value to determine whether the system is within an acceptable range.

A visible laser light source is a laser diode device that produces a low-intensity visible light that can be easily seen when identifying fibers or when

---

On Site

Damaged Fiber Optic Cables

A damaged area in a fiber optic cable can often be felt with your fingers. Any compression severe enough to damage the fiber will also distort the plastic buffer.

On Site

Troubleshooting Fiber Optic Cables

A fault finder can also be used in place of an OTDR to troubleshoot a fiber for a no-light or high-attenuation condition. It is similar to an OTDR except it does not have an X-Y screen. The test results are displayed one at a time as alphanumeric characters.

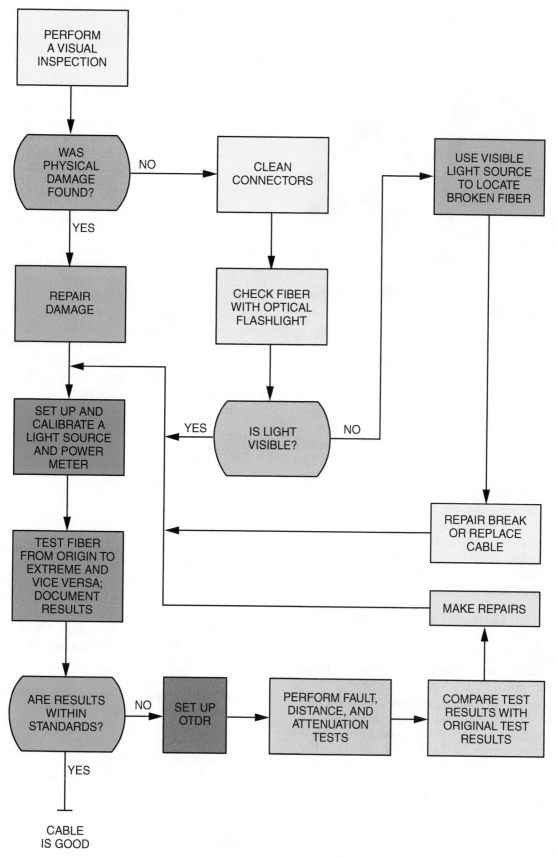

Figure 14 Systematic approach for diagnosing fiber optic cable problems.

307F14.EPS

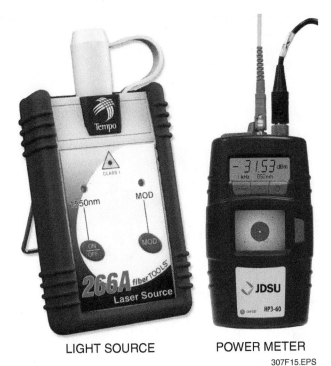

LIGHT SOURCE   POWER METER

307F15.EPS

Figure 15 Light source and power meter.

trying to locate a break in a fiber. When used, a red glow is evident at any point where light is escaping from the fiber. All such locations should be marked on the sheath. If a broken cable is detected in the horizontal, replace the cable; if it is in the backbone, slack should be pulled from the service loop and the cable spliced.

An optical time-domain reflectometer (OTDR) is a test instrument that sends out a light pulse on a fiber and then measures the time and amplitude of the reflected signal. The results are displayed as an X-Y plot on a screen. When using an OTDR to test a fiber for a no-light or high-attenuation condition, it is recommended that a copy of the original OTDR test results (traces) for the fiber under test be obtained, if possible. This information is invaluable. If the original traces are available, take special care to set the OTDR refractive index, pulse width, pulse duration, and sampling rate to that of the original trace. Compare the new test results with the original results and determine any point of difference between the two tests. These differences represent faults (broken, partially broken, or high-impedance locations, for example).

11.7.0 Testing After Repair

After the repair of any fault, it is a good practice to test the system to verify that it is operating correctly. For safety-related systems, such as fire alarm systems, testing after a repair is mandatory. For example, the *National Fire Alarm and Signaling Code (NFPA 72)* requires that reacceptance testing

be performed after system components are added or deleted; after any modification, repair, or adjustment to system hardware or wiring; or after any change to software. It states that all components, circuits, system operations, or site-specific software functions known to be affected by the change be 100 percent tested. In addition, 10 percent of initiating devices that are not directly affected by the change (up to a maximum of 50 devices) must also be tested and proper system operation verified. Testing is described in more detail later in this module.

12.0.0 PREVENTIVE MAINTENANCE

As defined earlier, preventive maintenance is any task that is performed on a regular basis in order to ensure the proper operation of the equipment and prevent anticipated equipment failures. Typically, equipment manufacturers recommend that specific maintenance tasks be performed on a daily, monthly, semi-annual, or annual basis. Many safety and building codes also mandate periodic inspection, maintenance, and testing of equipment. The following tasks relate to preventive maintenance:

- Inspection
- Cleaning
- Lubrication of mechanical assemblies
- Testing and adjustment

12.1.0 Inspection

A thorough visual inspection of equipment on a scheduled basis can detect conditions that can cause an equipment failure if not corrected. Visual inspection guides specific to particular equipment are commonly contained in the equipment manufacturer's service literature. Guides for mandatory inspections of safety equipment are also contained in related codes, such as the *National Fire Alarm and Signaling Code*. *Figure 16* shows an example of a typical visual inspection frequency table for fire alarm equipment.

Performing a visual inspection basically involves using common sense and your senses of sight, hearing, and smell. These are some common conditions to look for during a visual inspection of electronic equipment:

- Dust and dirt inside panels and enclosures and on circuit boards and components
- Dirty or missing air filters
- Dirty, gummy, or noisy fan motors
- Disconnected or damaged system cables
- Overheated cabinets, circuit boards, or components

Component	Init./Reacpt.	Monthly	Quarterly	Semiann.	Ann.
1. Control Equipment: Fire Alarm Systems Monitored for Alarm, Supervisory, Trouble Signals					
a. Fuses	X				X
b. Interfaced Equipment	X				X
c. Lamps and LEDs	X				X
d. Primary (Main) Power Supply	X				X
2. Control Equipment: Fire Alarm Systems Unmonitored for Alarm, Supervisory, Trouble Signals					
a. Fuses	X (weekly)				
b. Interfaced Equipment	X (weekly)				
c. Lamps and LEDs	X (weekly)				
d. Primary (Main) Power Supply	X (weekly)				
3. Batteries					
a. Lead-Acid	X	X			
b. Nickel-Cadmium	X			X	
c. Primary (Dry Cell)	X	X			
d. Sealed Lead-Acid	X			X	
4. Transient Suppressors	X			X	
5. Control Panel Trouble Signals	X			X	
6. Fiber-Optic Cable Connections	X				X
7. Emergency Voice/Alarm Communications Equipment	X			X	
8. Remote Annunciators	X			X	
9. Initiating Devices					
a. Air Sampling	X			X	
b. Duct Detectors	X			X	
c. Electromechanical Releasing Devices	X			X	
d. Fire Extinguishing System(s) or Suppression System(s) Switches	X			X	
e. Fire Alarm Boxes	X			X	
f. Heat Detectors				X	
g. Radiant Energy Fire Detector			X		
h. Smoke Detectors					
i. Supervisory Signal D...					
j. Waterflow D...					

Figure 16 Example of a visual inspection table.

307F16.EPS

- Loose or corroded terminals or connectors
- Evidence of heat damage around a terminal (usually a sign of a loose connection)
- Cracked or broken standoffs on terminal boards
- Equipment wiring harnesses and wires that are frayed, cut, discolored by heat, or have crushed insulation
- Damaged circuit boards
- Inoperative lamps or broken lenses
- Smells or sounds of arcing
- Damaged lightning/surge suppressors
- Chattering, noisy controls
- Rusty, corroded, or improperly secured antennas

Heat-related stress is probably the most common cause of failures in electronic circuitry. The components that are the most vulnerable to heat-related failures are those that must dissipate a great deal of power. Surrounding devices are also likely to be at risk. Such equipment must have adequate ventilation. This is typically provided by convection through louvered or perforated cabinets or by small fans. Equipment should be inspected to ensure that the means of dissipating heat is not being defeated by any of the following conditions:

- Papers or other items stacked on top of a vented chassis
- Papers or other items blocking side vents
- An inoperative vent fan
- A dirty air filter

Many safety and standby systems use batteries and battery banks. Visual inspections should be performed in accordance with the battery manufacturer's recommendations and/or the user facil-

Maintenance and Repair

ity schedule. Such inspections typically include checking for:

- Corroded or leaking batteries
- Loose battery connections
- Level of electrolyte in lead-acid batteries

- General appearance of the battery/rack/area
- Dirt or electrolyte on jars or covers
- Charger voltage and current output
- Cracks or leaks in the jar or cover
- Clogged flame arrester
- Insulating covers on racks
- Seismic rack parts and spacers
- Detailed cell inspection
- Cracks in plates
- Abnormal sediment accumulation
- Jar and post seals
- Excessive jar or cover distortion
- Excessive gassing
- Signs of vibration

12.2.0 Cleaning

Cleaning is important, particularly for parts such as equipment air filters and screens, circuit boards, keyboards, blowers, and fan motors.

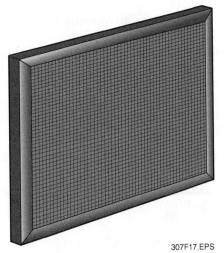

307F17.EPS

Figure 17 Clean air filter at recommended intervals to prevent equipment overheating.

Most electrical/electronic equipment uses a permanent type of air filter or steel mesh screen (*Figure 17*). This air filter or screen should be cleaned at the maintenance interval recommended by the manufacturer. Clean air filters or screens are important in order to allow proper air flow in equipment to prevent overheating.

Air filters and screens should be cleaned with the equipment turned off. After the filter or screen has been removed from the unit, excess dust and lint can be removed by lightly rapping the filter, dirty side down, into a suitable container. A vacuum cleaner can also be used. The cleanliness of a filter or screen can be checked by placing a strong light on one side and looking through it from the other side to determine how much light can be seen and how uniform the pattern is. When necessary, steel mesh screens can be washed with water. A mild detergent can be used to wash extremely dirty screens. After the screen has been sufficiently cleaned, it should be rinsed with clean water and allowed to thoroughly dry.

Unless directed by the manufacturer's instructions, do not oil or coat a filter or screen with adhesive. If a filter/screen frame is bent or warped, or the material is ripped or punctured, it should be replaced.

Circuit boards in personal computers and other electronic units should be thoroughly cleaned. If too much dust is allowed to accumulate on circuit boards it acts as an insulating layer, trapping heat inside the board components. For this reason, the case or housing of the unit should be opened periodically and low-pressure compressed air should be used to blow the dust out. Vent holes in the unit's case should be cleaned of any dust or other obstructions. Computer keyboards should be vacuumed to remove dirt and debris that may have fallen between the keys.

Blower and fan motors and their blades should be cleaned to remove dirt, grease, and other contaminants. If a motor has vent holes, make sure they are open and clean. Small fans installed in a chassis often develop bearing problems and start to slip. This means that they are no longer rotating as fast as they should, and are moving less air. They should be replaced.

Periodic vacuuming and cleaning of equipment rooms, cable trays, and cable troughs should also be performed in order to reduce the chance of dirt, dust, and other contaminants getting into the equipment.

12.3.0 Lubrication

Lubrication of electronic equipment is limited primarily to the lubrication of blower and fan motor bearings in order to reduce friction and prevent wear. Many motors used in electronic equipment have permanently lubricated bearings and therefore require no lubrication. The manufacturer's service literature for the particular unit should specify if motor bearing lubrication is necessary.

For those that require lubrication, the bearings must be lubricated using the type of lubricant specified in the manufacturer's service literature. The most common problem with lubricating bearings is over-lubrication, so be careful to use the amount specified by the manufacturer. In some larger equipment, drawer slides and other sliding surfaces might also require lubrication. Again, use only the lubricant type and amount specified by the manufacturer.

12.4.0 Testing and Adjustment

Preventive maintenance can also involve the testing and calibration of equipment. This is especially true in safety-related systems such as fire alarm and detection systems. Testing and calibration of nonsafety-related types of equipment should be performed in accordance with the manufacturer's instructions and at the intervals specified by the manufacturer or the authority having jurisdiction (AHJ). Making periodic checks of standby units, detectors, sensors, and alarms is necessary to make sure that they are operating properly. Make sure to follow the manufacturer's instructions for shutting down and turning on equipment with batteries.

In the case of fire alarm systems, mandatory testing of systems and components must be performed in accordance with the schedules given in the *National Fire Alarm and Signaling Code (NFPA 72)* or more frequently if required by the AHJ. Approved methods of testing fire alarm equipment and the frequency of tests are specified in *NFPA 72*.

Figures 18 and *19* show examples of test methods and frequency tables for fire alarm systems. Similarly, testing of engine-driven generators dedicated to a fire alarm system or hospital backup power system are mandated by *NFPA 110, Standard for Emergency and Standby Power Systems*. *NFPA 99, Standard for Health Care Facilities* requires that both the physical and operating conditions of the various laboratory and patient-care related electrical equipment used in hospitals, nursing homes, and other health care facilities also be tested and/or calibrated on a regular basis.

Test Methods

Device	Method
1. Control Equipment a. Functions	At a minimum, control equipment shall be tested to verify proper receipt of alarm, supervisory, and trouble signals (inputs), operation of evacuation signals and auxiliary functions (outputs), circuit supervision including detection of open circuits and ground faults, and power supply supervision for detection of loss of AC power and disconnection of secondary batteries.
b. Fuses	Remove fuse and verify rating and supervision.
c. Interfaced Equipment	Integrity of single or multiple circuits providing interface between two or more control panels shall be verified. Interfaced equipment connections shall be tested by operating or simulating operation of the equipment being supervised. Signals required to be transmitted shall be verified at the control panel.
d. Lamps and LEDs	Lamps and LEDs shall be illuminated.
e. Primary (Main) Power Supply	All secondary (standby) power shall be disconnected and tested under maximum load, including all alarm appliances requiring simultaneous operation. All secondary (standby) power shall be reconnected at end of test. For redundant power supplies, each shall be tested separately.
2. Engine-Driven Generator	Where an engine-driven generator dedicated to the fire alarm system is used as a required power source, operation of the generator shall be verified in accordance with NFPA 110, *Standard for Emergency and Standby Power Systems,* by the building owner.
3. Secondary (Standby) Power Supply	Disconnect all primary (main) power supplies and verify that required trouble indication for loss of primary power occurs. Measure or verify system's standby and alarm current demand and, using manufacturer's data, verify whether batteries are adequate to meet standby and alarm requirements. Operate general alarm systems for a minimum of 5 minutes and emergency voice communications systems for a minimum of 15 minutes. Reconnect primary (main) power supply at end of test.
4. Uninterrupted Power Supply (UPS)	Where a UPS system dedicated to the fire alarm system is used as a required power source, operation of the UPS system shall be verified by the building owner in ... *al Energy Emerg...*

307F18.EPS

Figure 18 Example of a test methods table.

13.0.0 INSPECTING AND TESTING FORMS

All activity relating to inspection and testing should be documented on the appropriate forms. While performing inspections and operational checks, the equipment condition and values for measured parameters should be recorded on a log. This log should remain attached to the equipment or placed in a designated location. It serves as a history of the equipment's operating conditions. It is also proof that the appropriate inspections and testing have been performed. Data that was previously recorded in the log can be compared against current data to help identify any degradation in system or equipment operation. This enables early detection of potential problems, allowing corrective action to be taken before a system failure results. *Figure 20* shows a partial example of a typical inspection and testing form.

Testing Frequencies

Component	Init./Reaccpt.	Monthly	Quarterly	Semiann.	Ann.
1. Control Equipment: Fire Alarm Systems Monitored for Alarm, Supervisory, Trouble Signals					
a. Functions	X				X
b. Fuses	X				X
c. Interfaced Equipment	X				X
d. Lamps and LEDs	X				X
e. Primary (Main) Power Supply	X				X
f. Transponders	X				X
2. Control Equipment: Fire Alarm Systems Unmonitored for Alarm, Supervisory, Trouble Signals					
a. Functions	X		X		
b. Fuses	X		X		
c. Interfaced Equipment	X		X		
d. Lamps and LEDs	X		X		
e. Primary (Main) Power Supply	X		X		
f. Transponders	X		X		
3. Engine-Driven Generator	X (weekly)				
4. Batteries — Central Station Facilities					
a. Lead-Acid Type					
1. Charger Test (Replace battery as needed)	X				X
2. Discharge Test (30 min.)	X	X			
3. Load Voltage Test	X	X			
4. Specific Gravity	X			X	
b. Nickel-Cadmium Type					
1. Charger Test (Replace battery as needed)	X		X		
2. Discharge Test (30 min.)	X				X
3. Load Voltage Test	X				X
c. Sealed Lead-Acid Type	X	X			
1. Charger Test (Replace battery as needed)		X	X		
2. Discharge Test (30 min.)	X	X			
3. Load Voltage Test	X	X			
5. Batteries — Fire Alarm Systems					
a. Lead-Acid Type					
1. Charger Test (Replace battery as needed)	X				X
2. Discharge Test (30 min.)	X			X	
3. Load Voltage Test	X			X	
4. Specific Gravity					
b. Nickel-Cadmium Type					

307F19.EPS

Figure 19 Example of a testing frequency table.

INSPECTION AND TESTING FORM

SERVICE ORGANIZATION	PROPERTY NAME (User)
Name:	Name:
Addr:	Addr:
City, State:	City, State:
Representative:	Owner Contact:
License #:	Telephone:
Telephone:	

MONITORING ENTITY	APPROVING AGENCY
Contact:	Contact:
Telephone:	Telephone:
Monitoring Account Ref. No.:	

TYPE TRANSMISSION		SERVICE	
McCulloh		Weekly	
Multiplex		Monthly	
Digital		Quarterly	
Reverse Priority		Semi-Annually	
Radio Frequency		Annually	
Other (specify)		Other (specify)	

PANEL MANUFACTURER:_____ Model #: _____
Circuit Styles: _____ Number of Circuits: _____
Software Rev.: _____ Last system service performed: _____
Last software or configuration revision: _____

ALARM INITIATING DEVICES AND CIRCUIT INFORMATION

QTY	CIRCUIT STYLE		SPECIAL TESTING INFORMATION
		MANUAL STATIONS	
		ION DETECTORS	
		PHOTO DETECTORS	
		DUCT DETECTORS	
		HEAT DETECTORS	
		WATERFLOW SWITCHES	
		SUPERVISORY SWITCHES	
		OTHER:	

ALARM INDICATING DEVICES AND CIRCUIT INFORMATION

QTY	CIRCUIT STYLE		SPECIAL TESTING INFOMATION
		BELLS	
		HORNS	
		CHIMES	
		STROBES	
		SPEAKERS	
		OTHER:	
	TOTAL NUMBER OF ALARM INDICATING CIRCUITS		YES NO ARE CIRCUITS SUPERVISED?

307F20A.EPS

Figure 20 Example of an inspection and testing form (1 of 2).

SUPERVISORY SIGNAL INITIATING DEVICES AND CIRCUIT INFORMATION

QTY	CIRCUIT STYLE		COMMENTS
		Building Temp.	
		Site Water Temp.	
		Site Water Level	
		Fire Pump Power	
		Fire Pump Running	
		Fire Pump Auto Position	
		Fire Pump or Pump Controller Trouble	
		Generator Running	
		Generator in Auto Position	
		Generator or Generator Controller TBL	
		Switch Transfer	
		Other:	

SIGNALING LINE CIRCUITS See NFPA 72 table 6.6.1 QTY: _____ Style(s): _____

SYSTEM POWER SUPPLIES

    a) **Primary** (Main): Nominal Voltage _____, Amps _____
       Overcurrent Protection: Type _____, Amps _____
       Location (Panel Number) _____ Circuit Number _____
       Disconnecting (Means) Location: _____

    b) **Secondary** (Standby):

       _____Storage Battery: Amp-hr rating _____
       Calculated capacity to operate system, in hours: _____24 _____60
       _____ Engine-driven generator dedicated to fire alarm system:
       Location of fuel storage: _____
       Type Battery ___Dry Cell ___Nickel Cadmium ___ Sealed Lead-Acid ___Lead-Acid
          ___Other (specify): _____

    c) **Emergency or Standby system used as a backup to primary power supply, instead of using a secondary power supply:**
    ____ Emergency system described in NFPA 70, Article 700
    ____ Legally required standby described in NFPA 70, Article 701
    ____ Optional standby system described in NFPA 70, Article 702, which also meets the
        performance requirements of Article 700 or 701.

PRIOR TO ANY AND ALL TESTING

NOTIFICATION ARE MADE:	YES	NO	WHO	TIME
Monitoring Entity				
Building Occupants				
Building Management				
Other (specify)				
AHJ (Notified) of any impairments				

307F20B.EPS

Figure 20 Example of an inspection and testing form (2 of 2).

SUMMARY

One of the more important duties of an EST is to properly perform preventive maintenance tasks so that the customer can enjoy safe and efficient equipment operation.

Preventive maintenance is any task performed on a regular basis in order to ensure the proper operation of the equipment and prevent anticipated equipment failures. Repair includes those tasks performed to fix a known problem. This includes the replacement of equipment, components, or cabling when for any reason they become undependable or inoperable.

Effective troubleshooting is a process by which the technician listens to a customer's complaint, performs an independent analysis of a problem, and then initiates and performs a systematic approach to troubleshooting and repair. You must understand the purpose and principles of operation of each component in the equipment to be serviced. You must also be able to tell whether a given device is functioning properly, and recognize the symptoms arising from the improper operation of any part of the equipment.

1. Which of these tasks is considered to be a repair?

   a. Inspecting a patch panel
   b. Replacing a faulty relay
   c. Cleaning an air filter
   d. Completing scheduled equipment calibration

2. The root cause of an electrical/electronic failure resulting from loose connections or vibration is most likely _____.

   a. environmental conditions
   b. poor power quality
   c. operator error
   d. improper installation

3. The root cause of an electrical/electronic failure resulting from harmonics is most likely _____.

   a. environmental conditions
   b. poor power quality
   c. static electricity
   d. improper installation

4. The device used primarily for testing UTP cable for shorts, crossed pairs, and improper wiring is the _____.

   a. wire map tester
   b. handheld cable tester
   c. toner/wand
   d. cable tracer

5. A hot, smoky unit or device is often a sign of a(n) _____.

   a. short circuit
   b. ground fault
   c. open circuit
   d. mechanical failure

6. A circuit that measures infinite resistance and zero current is _____.

   a. a short circuit
   b. a ground fault circuit
   c. an open circuit
   d. either a short or ground fault circuit

7. When using a systematic approach to troubleshooting, the first step should be to _____.

   a. consult the manufacturer's troubleshooting aids
   b. perform a basic system analysis
   c. physically inspect the system
   d. interview the customer about the problem

8. To achieve proper operation when using an RTM-8 module in conjunction with an AFP-200 fire alarm control panel, it is required that _____. (Hint: Refer to the AFP-200 troubleshooting aids in the module.)

   a. jumper JP1 be in place
   b. jumper JP1 be cut
   c. jumper JP5 be in place
   d. jumper JP5 be cut

9. When troubleshooting an AFP-200 fire alarm control panel to find a problem involving an invalid reply from a single field device, which of the following causes can be eliminated? (Hint: Refer to the AFP-200 troubleshooting aids in the module.)

   a. The address is used twice on field devices
   b. Reverse SLC polarity at the AFP-200
   c. A wrong device type code is programmed
   d. The device is not addressed

10. The service manual for a piece of equipment shows that it is designed to operate on 120VAC. What is the minimum voltage that can be applied to this equipment and still have it operate properly?

    a. 105V
    b. 108V
    c. 110V
    d. 115V

11. When measuring the ripple in the output of an 18VDC power supply, you measure 0.08V. What percentage of ripple is this?

    a. 0.2 percent
    b. 0.4 percent
    c. 2 percent
    d. 4 percent

12. When using an oscilloscope to check power supply ripple, the oscilloscope AC/DC mode control should be set to _____.
    a. AC
    b. DC
    c. ripple test
    d. output noise test

13. If the contacts of a relay are shown closed on a schematic diagram, when the relay is energized, the contacts remain closed.
    a. true
    b. false

14. A circuit board should be replaced only when _____.
    a. the board fails to generate the expected output signal(s)
    b. one or more input signal(s) are missing
    c. one or more input and/or output signal(s) are missing
    d. the board fails to generate the proper output signal(s) when the input signal(s) are known to be good

15. Noise in the 150kHz to 20MHz range is usually caused by fluorescent lighting.
    a. True
    b. False

Trade Terms Introduced in This Module

Fault isolation diagram: A troubleshooting aid usually contained in the manufacturer's installation and/or service instructions for a particular unit. Normally, fault isolation diagrams begin with a failure symptom, then guide the technician through a logical decision-action process to isolate the cause of a failure.

Listed: Equipment or material that is identified by a recognized agency as meeting certain standards, or equipment and material that has been tested and found suitable for a particular purpose.

Preventive maintenance: Any task that is performed on a regular basis in order to ensure the proper operation of the equipment and/or prevent anticipated equipment failures.

Repair: The replacement of a system or some of its components when they become undependable or inoperable.

Troubleshooting: A procedure by which the technician locates the source of an equipment, cable, or software problem, then performs the necessary repairs and/or adjustments to correct the problem.

Troubleshooting table: A troubleshooting aid usually contained in the manufacturer's installation and/or service instructions for a particular unit. Troubleshooting tables are intended to guide the technician to a corrective action based on observations of system or unit operation.

DEVICE SENSITIVITY TO ELECTROSTATIC DISCHARGE

DEVICE SENSITIVITY TO ELECTRO-STATIC DISCHARGE

Class 1: 0 to 1,999 Volts

The following devices or microcircuits were identified by test data as Class 1:

Microwave and high-frequency devices
(Schottky barrier diodes, probe contact diodes, other detector diodes)
Discrete MOSFET devices
SAW
JFETS
CCDs
Precision voltage regulator diodes
OP AMP
Thin film resistors
Integrated circuits
Hybrids utilizing Class 1 parts
VHSIC
SCRs

Class 2: 2,000 to 3,999 Volts

Devices or microcircuits when identified by Appendix A test data as Class 2:

Discrete MOSFET devices
JFETs
OP AMPS
ICs
VHSIC
Precision resistor networks (Type RZ)
Hybrids utilizing Class 2 parts
Low power bipolar transistors

Class 3: 4,000 to 15,999 Volts

Devices or microcircuits when identified by Appendix A test data as Class 3:

Discrete MOSFET devices
JFETs
OP AMPS
ICs
VHSIC
All other microcircuits not included in Class 1 or Class 2
Small signal diodes
General purpose silicon rectifiers
Opto-electronic devices (LEDs, phototransformers, opto couplers)
Resistor chips
Piezo electric crystals
Hybrids using Class 3 parts

307A01.EPS

Maintenance and Repair

Additional Resources

This module presents thorough resources for task training. The following resource material is suggested for further study.

Mike's Basic Guide to Cabling. Englewood, CO: Global Engineering Documents.

Cabling: The Complete Guide to Network Wiring. San Francisco, CA: Sybex.

Figure Credits

CONTREN® LEARNING SERIES — USER UPDATE

NCCER makes every effort to keep its textbooks up-to-date and free of technical errors. We appreciate your help in this process. If you find an error, a typographical mistake, or an inaccuracy in NCCER's Contren® materials, please fill out this form (or a photocopy), or complete the online form at www.nccer.org/olf. Be sure to include the exact module number, page number, a detailed description, and your recommended correction. Your input will be brought to the attention of the Authoring Team. Thank you for your assistance.

Instructors – If you have an idea for improving this textbook, or have found that additional materials were necessary to teach this module effectively, please let us know so that we may present your suggestions to the Authoring Team.

NCCER Product Development and Revision
3600 NW 43rd Street, Building G, Gainesville, FL 32606

Fax: 352-334-0932
Email: curriculum@nccer.org
Online: www.nccer.org/olf

❏ Trainee Guide   ❏ AIG   ❏ Exam   ❏ PowerPoints   Other _____

Craft / Level: _____ Copyright Date: _____

Module Number / Title: _____

Section Number(s): _____

Description: _____

_____

_____

_____

_____

Recommended Correction: _____

_____

_____

_____

_____

Your Name: _____

Address: _____

_____

Email: _____ Phone: _____

Glossary

Acceptable performance: The condition where a component or system meets specified design parameters under actual operating conditions.

Ad hoc: A Latin term meaning formed for a special purpose.

Address: A unique identifier for a node or device on a network. On the Internet, the address is composed of four numbers, which define the location of the node in a hierarchical numbering system.

Addressable: A device with a discrete identification that identifies the type of device and its location.

Attenuation: The degradation of a signal as it moves through a medium.

Authority having jurisdiction (AHJ): The person or department within a municipality or jurisdiction having legal authority to approve a system that requires code conformance.

Backbone: A high-capacity network link between two major sections of a network.

Backbone provider: One of a very few select network service providers that own and operate the major Internet backbone networks.

Backplane: The rear section of a swinging frame rack used to mount the rack to the wall.

Baseband: A transmission technique in which all of the available bandwidth is dedicated to a single communications channel. Only a single message transfer can occur at a given time.

Basic input/output system (BIOS): A core set of instructions defining how a computer system performs read-and-write operations on its components.

Baud rate: The rate at which information moves on the network.

Binary: Having one of only two possible values, such as on or off, or 1 or 0.

Bit: A single binary value within a computer. Computer equipment, communications paths, and software are described according to the number of bits they handle.

Bridge: A device for transferring information from one network to another.

British thermal units per hour (Btu/h): The amount of heat required to raise the temperature of one pound of water by 1°F.

Broadband: A transmission technique in which the bandwidth is divided into multiple channels. This allows multiple message transfers to take place simultaneously.

Broadcast: Transmission from a single source to all locations on a network.

Bus: A communication channel made up of multiple parallel paths.

Byte: A logical grouping of bits within a computer, representing a value.

Carrier sense multiple access with collision detection (CSMA/CD): A common method used for controlling access to the network medium.

Centralized control: Network access control provided by a single node or entity on the network.

Change order: The formal document that modifies the original agreement covering work to be performed. This includes, but is not limited to, work added or deleted, changes to location of services, different materials, and the time allotted for the project. It can also include damages to your work caused by others or by natural forces.

Channel: A path for information transfer.

Checksum: A value calculated from the content of a message. It is used by the receiving device to verify that the data has not been altered during its transfer from source to receiver.

Circuit switching: A method for routing information on the network by establishing a virtual direct connection between two nodes.

Cladding: A material that encloses the fiber in fiber optic cable, allowing the signal to be transmitted without escaping the cable.

Collision: A condition where two or more messages attempt to use the network at the same time.

Commissioning authority (CA): A qualified person, company, or agency responsible for planning and carrying out all system commissioning related activities.

...hat
..., alloca-
...nce testing,
... to the overall

...iod used for compress-
...ding a wireless transmis-
...crophones compress the signal
...itter and expand it at the receiver.

...ct: The legal framework of an agree-
...t between at least two parties in which one agrees to perform some task for the other in return for a specified payment.

Convection: The transfer of heat due to the flow of a liquid or gas caused by a temperature differential.

Deterministic: Based upon a predictable set of rules.

Direct cost: The cost for labor wages, fringe benefits, and wage-related taxes and insurance for all workers and supervisors directly engaged in the work; also, the cost for all materials, engineering and design, permits and fees, and tools and equipment used in an installation to accomplish the finished job.

Distributed control: Network access control split up among the individual nodes on the network, in which each node is responsible for determining when it may place a message on the medium.

Duplex: A term used to indicate two-way transmission.

Emitter: A semiconductor device that emits an infrared signal. It is used as a transmitting device in IR systems.

Ethernet: A common networking protocol that led to the growth and acceptance of local area networks. It uses CSMA/CD access control and was originally based on bus topology.

Exclusions: A designated section or statements in contractual documents that identify work, items, and other entities exempted from or not covered by the requirements of the contract or scope of work.

Fault isolation diagram: A troubleshooting aid usually contained in the manufacturer's installation and/or service instructions for a particular unit. Normally, fault isolation diagrams begin with a failure symptom, then guide the technician through a logical decision-action process to isolate the cause of a failure.

File transfer protocol (FTP): A message format for packaging information and transferring files over the Internet.

FireWire®: Trade name for the IEEE 1394 standard for high-speed serial communications.

Functional performance testing: The collective body of checks and tests performed to verify that all components, subsystems/systems, and interfaces between systems are functioning according to specification requirements.

Gateway: A hardware device that can pass information packets from one network environment to another (such as from an Ethernet network to a token ring).

Ground loops: A condition in which an undesired connection is made between electronic components or enclosures, inducing current flow that can generate noise.

Headroom: The difference between the actual signal level and the maximum level a device is able to handle without distorting the signal.

Hub: A device for networking two or more similar devices.

Hybrid topology: A network that uses elements of two or more topologies together to configure the nodes.

Hypertext transfer protocol (HTTP): A method of requesting and sending text files that include page layout information and hypertext links to other documents.

Hypertext: Text containing active links to other documents.

Insertion loss: The amount of optical power lost through a connection or splice.

Internet mail access protocol (IMAP): A protocol for handling email messages while they remain on a remote server.

Internet Relay Chat (IRC): An Internet-based system for communicating with others in real time through text-based chat screens.

Laser: A source of high-power coherent light (derived from light amplification by stimulated emission of radiation).

Listed: Equipment or material that is identified by a recognized agency as meeting certain standards, or equipment and material that has been tested and found suitable for a particular purpose.

Media Access Control (MAC): The means of controlling access to the actual medium or wire of the network.

Micron: A unit of length equal to one-millionth of a meter.

Microwave: A radio frequency in the range of 0.3 to 300GHz. In alarm technology, an RF system using motion detectors that operate in the microwave frequency range.

Mode: A path that can be taken by a ray of light through a material.

Multi-mode: The term used when there is more than one way that light can travel through a material.

Multicast: Transmission from a single source to multiple, finite destinations on a network.

Multiplexing: The broadcast of several signals over a line simultaneously.

Network basic input/output system (NetBIOS): A core set of instructions required by an operating system for communicating with the network hardware.

Network news transfer protocol (NNTP): A protocol for sending, retrieving, and transferring bulletin-board messages across the Internet.

Network operating system (NOS): A program consisting of the commands and instructions that allow computers to function as a network.

Network software: Software that runs on a computer to provide additional network services that are not provided by either the operating system or a network operating system.

Network switch: A device used to manage traffic between devices on a network.

Node: An element on the network that functions as an autonomous part of the network.

Numerical aperture: The range of angles over which a system can accept or emit light.

Open Systems Interconnection (OSI) Reference Model: A seven-layer model developed by the International Standards Organization to describe how to connect any combination of devices for the purposes of communication.

Operating system (OS): The program that runs constantly on a computer to provide the basic services and control needed to execute other programs.

Oscillator: An electronic device that produces a pure sine wave at a specified frequency or range of frequencies.

Packet: A data unit created at the network layer of the OSI model. It contains the data and control information necessary to transfer a message from one network to another.

Packet sniffer: Also called packet analyzer. It is a hardware or software tool used to intercept network traffic.

Packet switching: A method for sharing a data transmission medium by breaking each message into smaller pieces.

Parallel: Two or more paths running alongside each other without meeting.

Point of presence (POP): A computer system equipped with multiple modems that allows users to dial in and connect to the Internet.

Polling: A method for controlling access to the network by individually checking each node to see if it needs to send a message.

Port: Either a physical connection to a system, or, in the case of an Internet server, a logical address where data for a specific application or protocol is routed.

Post office protocol (POP): A protocol for storing email messages on a remote server and retrieving them with a client program.

Preventive maintenance: Any task that is performed on a regular basis in order to ensure the proper operation of the equipment and/or prevent anticipated equipment failures.

Propagation: The method of traveling through space or a specific material.

Protocol: A common language or set of rules allowing controllers connected in a network to share information.

Punch list: A list of repairs or other work that must be done to satisfy the requirements of the contract and obtain customer acceptance of the job.

Rack unit (RU): The height of a single equipment panel; an RU is 1.75″.

Receiver sensitivity: The term used to describe the minimum signal that a receiver can detect.

Record drawings: Drawings that show major changes and modifications made from the original job plan to the finished state.

Register: A special, high-speed area of memory that is typically used as a staging area for either processing by the CPU or communication with another device.

Repair: The replacement of a system or some of its components when they become undependable or inoperable.

Return loss: Light reflected back through a fiber toward the source. Return losses are kept to less than −55dB to ensure speed and clarity of transmission.

Ring: A network topology in which all of the nodes are wired in a continuous sequence.

Router: A hardware device that provides a communication path from one node of a network to another, with both nodes using the same communication protocol.

RS-232: An interface standard for connecting serial devices. The standard supports a 25-pin or 9-pin D-type connector.

Ruggedized: Refers to equipment built using materials and methods that allow it to be used in harsh environments.

Scope of work: A written document that defines the quantity and quality of work to be done and the materials to be used in the construction of a building or other structure. Its content and format are generally of a less formal nature than those used in the specifications.

Separately derived system: An electrical system in which the power is derived from a source that has no direct electrical connection, including connection to a grounded circuit conductor, to supply conductors originating in another system.

Serial: A communication path with a single channel or stream, where all information must flow sequentially, one bit at a time.

Server: A computer used to manage a network. It generally contains high-reliability components and larger, faster storage devices.

Simple mail transfer protocol (SMTP): An Internet protocol for passing messages to an email server.

Simplex: A term used to indicate one-way transmission.

Specifications: A formal written document that defines the quantity and quality of work to be done and the materials to be used in the construction of a building or other structure. It also provides additional information not contained on a related set of working drawings.

Splitter: A device that connects three or more fiber ends to create one or more outputs.

Static pressure: The pressure exerted uniformly in all directions within an enclosure, as measured in inches of water column.

Stochastic: Based upon random time values.

Stream: A continuous flow of data through a channel or path.

Takeoff: The process of surveying, measuring, itemizing, and counting all materials and equipment needed for a construction job as indicated by the drawings or job site survey.

Telnet: A program that allows a user to log in to a remote computer as if they were physically present at a terminal on the computer.

Token: A data packet passed from node to node in a specific sequence. The node that has the token is allowed to transmit data on the network.

Token ring: A network topology in which a token must be passed to a terminal or workstation by the network controller before it can transmit.

Topology: The physical layout of a network, especially how the nodes are connected to one another.

Transducer: A device that converts one form of energy to another. A microphone and a speaker are transducers used for opposite purposes.

Transfer medium: The physical medium, usually some type of wire or cable, that connects the nodes on a network.

Transmission Control Protocol/Internet Protocol (TCP/IP): The set of protocols that forms the foundation for the Internet and defines how messages are passed between nodes on the Internet.

Troubleshooting: A procedure by which the technician locates the source of an equipment, cable, or software problem, then performs the necessary repairs and/or adjustments to correct the problem.

Troubleshooting table: A troubleshooting aid usually contained in the manufacturer's installation and/or service instructions for a particular unit. Troubleshooting tables are intended to guide the technician to a corrective action based on observations of system or unit operation.

Universal Serial Bus (USB): A communication standard for connecting peripherals to a computer.

Usenet: A worldwide bulletin board system that can be accessed through the Internet.

Voltage standing wave ratio (VSWR): A measure of the impedance match between two RF components.

Web browser: A program used for viewing documents on the Internet using the hypertext transfer protocol.

Wide area network (WAN): A network whose elements are separated by distances great enough to require the use of telephone lines.

Word: In computer terminology, a grouping of bits to form a value. A word may consist of one or more bytes.

Index

Water-to-air cooling (33305): 24
Waveguide (33303): 6
Wavelength-division multiplexing (WDM) (33302): 23
Wavelength-division multiplexing (WDM) splitters (33302): 23
Web browser (33301): 24, 40
Wheeler, Henry (33302): 1
Wide area network (WAN) (33301): 21, 29, 40
Wi-fi hotspots (33303): 16
Wild Blue™ (33303): 18
Windows® (33301): 1, 19, 28
Windows® 7 (33301): 1
Windows® NT (33301): 28
Windows® XP (33301): 2
Wireless access point (WAP) (33303): 12
Wireless communication networks
    local area (LAN)
        applications (33303): 11–12
        collision avoidance (33303): 13
        equipment (33303): 14–15
        historically (33303): 12–14
        IR connections (33303): 8, 9
        IR data transmission (33303): 11
        range (33303): 14
        security (33303): 14–15
        standards (33303): 19
    planning (33301): 5
    wide area (WAN) (33301): 21, 29, 40
Wireless communications
    analog signals (33303): 2–3
    antennas (33303): 22–24
    applications (33303): 11–12
    defined (33303): 1
    digital signals (33303): 2–3
    electromagnetic interference (33303): 24–25
    historically (33303): 1, 4
    modulation (33303): 2
    multiplexing (33303): 3
    noise in (33303): 24–25
    principles (33303): 1
    test equipment
        RF analyzer/standing wave meter (33303): 21
        RF field strength analyzer (33303): 20
        RF power meter (33303): 21
        satellite signal meter (33303): 22
Wireless communication systems
    components (33303): 1
    frequency bands (33303): 1

infrared (IR)
    applications (33303): 8
    beam-break alarm systems (33303): 11
    components (33303): 8, 9–10
    range (33303): 9
    remote control circuits (33303): 10
    remote control distribution systems (33303): 10–11
    RS-232 data transmission interface systems (33303): 11
    types of (33303): 8–9
radio frequency (RF)
    antennas (33303): 6–8
    receivers (33303): 4–6
    repeaters (33303): 6
    transceivers (33303): 6
    transmitters (33303): 3–4
    VSWR (33303): 8
    waveguide (33303): 6
satellite
    background (33303): 16
    orbits (33303): 19
    overview (33303): 17–19
signal loss, causes of (33303): 8
standards (33303): 19
Wire map tester (33307): 7
Wiring certification diagrams and lists (33304): 25
Wiring diagrams (33307): 10, 16
WLANS. See Local area network (LAN): wireless
Women
    communication styles (46101): 4
    sexual harassment of (46101): 5
Words (33301): 3, 40
Workers compensation
    denial of benefits (46101): 31, 38
    statistics (46101): 29
Workforce diversity (46101): 3–6, 17
Workforce shortages (46101): 1
Work-related injury statistics (46101): 29
World wide web port address (33301): 19
Worms (33301): 20
Written communication (46101): 16–17

X

X.25 (33301): 2
Xerox Corporation (33301): 2

Y

Younger workers, training (46101): 2